HIGH-ENERGY PHYSICS IN THE EINSTEIN CENTENNIAL YEAR

Studies in the Natural Sciences

A Series from the Center for Theoretical Studies
University of Miami, Coral Gables, Florida

Recent Volumes in this Series

Volume 7 — TOPICS IN ENERGY AND RESOURCES
Edited by Behram Kursunoglu, Stephan L. Mintz, Susan M. Widmayer,
Chui-Shuen Hui, Joseph Hubbard, Joseph Malerba, and George Soukup

Volume 8 — PROGRESS IN LASERS AND LASER FUSION
Edited by Behram Kursunoglu, Arnold Perlmutter, Susan M. Widmayer,
Uri Bernstein, Joseph Hubbard, Christian Le Monnier de Gouville,
Laurence Mittag, Donald Pettengill, George Soukup, and M. Y. Wang

Volume 9 — THEORIES AND EXPERIMENTS IN HIGH-ENERGY PHYSICS
Edited by Behram Kursunoglu, Arnold Perlmutter, Susan M. Widmayer,
Uri Bernstein, Joseph Hubbard, Christian Le Monnier de Gouville,
Laurence Mittag, Donald Pettengill, George Soukup, and M. Y. Wang

Volume 10 — NEW PATHWAYS IN HIGH-ENERGY PHYSICS I
Magnetic Charge and Other Fundamental Approaches
Edited by Arnold Perlmutter

Volume 11 — NEW PATHWAYS IN HIGH-ENERGY PHYSICS II
New Particles—Theories and Experiments
Edited by Arnold Perlmutter

Volume 12 — DEEPER PATHWAYS IN HIGH-ENERGY PHYSICS
Edited by Behram Kursunoglu, Arnold Perlmutter, Linda F. Scott,
Mou-Shan Chen, Joseph Hubbard, Michel Mille, and Mario Rasetti

Volume 13 — THE SIGNIFICANCE OF NONLINEARITY IN THE NATURAL SCIENCES
Edited by Behram Kursunoglu, Arnold Perlmutter, Linda F. Scott,
Mou-Shan Chen, Joseph Hubbard, Michel Mille, and Mario Rasetti

Volume 14 — NEW FRONTIERS IN HIGH-ENERGY PHYSICS
Edited by Behram Kursunoglu, Arnold Perlmutter, Linda F. Scott,
Osman Kadiroglu, Jerzy Nowakowski, and Frank Krausz

Volume 15 — ON THE PATH OF ALBERT EINSTEIN
Edited by Behram Kursunoglu, Arnold Perlmutter, and Linda F. Scott

Volume 16 — HIGH-ENERGY PHYSICS IN THE EINSTEIN CENTENNIAL YEAR
Edited by Behram Kursunoglu, Arnold Perlmutter, Frank Krausz, and
Linda F. Scott

A Continuation Order Plan is available for this series. A continuation order will bring delivery of each new volume immediately upon publication. Volumes are billed only upon actual shipment. For further information please contact the publisher.

ORBIS SCIENTIAE

HIGH-ENERGY PHYSICS IN THE EINSTEIN CENTENNIAL YEAR

Chairman
Behram Kursunoglu

Editors
**Arnold Perlmutter
Frank Krausz
Linda F. Scott**

*Center for Theoretical Studies
University of Miami
Coral Gables, Florida*

PLENUM PRESS • NEW YORK AND LONDON

Library of Congress Cataloging in Publication Data

Orbis Scientiae, University of Miami, 1979
High-energy physics in the Einstein centennial year.

(Studies in the natural sciences; v. 16)
"A part of the proceedings of Orbis Scientiae 1979 held by the Center for Theoretical Studies, University of Miami, Coral Gables, Florida, January 15–18, 1979."
Includes index.
1. Particles (Nuclear physics) — Congresses. I. Kurşunoğlu, Behram, 1922-   II. Perlmutter, Arnold, 1928-   III. Krausz, Frank. IV. Scott, Linda F. V. Miami, University of, Coral Gables, Fla. Center for Theoretical Studies. VI. Title. VII. Series.
QC793.07 1979   539.7'21   79-18441
ISBN 0-306-40297-1

A part of the proceedings of Orbis Scientiae 1979,
held by the Center for Theoretical Studies,
University of Miami, Coral Gables, Florida, January 15–18, 1979

© 1979 Plenum Press, New York
A Division of Plenum Publishing Corporation
227 West 17th Street, New York, N.Y. 10011

All rights reserved

No part of this book may be reproduced, stored in a retrieval system, or transmitted, in any form or by any means, electronic, mechanical, photocopying, microfilming, recording, or otherwise, without written permission from the Publisher

Printed in the United States of America

PREFACE

The editors are pleased to submit to the readers the state of the art in high energy physics as it appears at the beginning of 1979.

Great appreciation is extended to Mrs. Helga S. Billings and Mrs. Connie Wardy for their assistance with the conference and skillful typing of the proceedings which was done with great enthusiasm and dedication.

Orbis Scientiae 1979 received some support from the Department of Energy.

The Editors

CONTENTS

Evidence for Quarks from Neutrino-Nucleon Scattering...... 1
   F. Sciulli

Direct Experimental Evidence for Constituents in the
     Nucleon from Electromagnetic Scattering
     Experiments... 31
   Karl Berkelman

Physics After τ and T 79
   A. Pais

Protons Are Not Forever...................................... 91
   D.V. Nanopoulos

Gauge Hierarchies in Unified Theories....................... 115
   Itzhak Bars

Anomalies, Unitarity and Renormalization.................... 133
   Paul H. Frampton

Charm Particle Production by Neutrinos...................... 139
   N.P. Samios

Techniques to Search for Proton Instability to 10^{34}
     Years... 157
   David B. Cline

Charged and Neutral-Current Interference: The Next
     Hurdle for Weinberg-Salam............................... 175
   S.P. Rosen

The Quark Model Pion and the PCAC Pion...................... 191
   K. Johnson

Quark Model Eigenstates and Low Energy Scattering........... 207
   F.E. Low

On the Equations of State in Many Body Theory............ 221
   R.E. Norton

Dyson Equations, Ward Identities, and the Infrared
      Behavior of Yang Mills Theories.................... 247
   M. Baker

Instantons and Chiral Symmetry........................... 267
   Robert D. Carlitz

QCD and Hadronic Structure............................... 285
   Laurence Yaffe

High Energy Predictions from Perturbative Q.C.D.......... 303
   A. Mueller

Perturbative QCD: An Overview............................ 313
   Stephen D. Ellis

Topics in the QCD Phenomenology of Deep-Inelastic
      Scattering... 327
   L.F. Abbott

Quantum Chromodynamics and Large Momentum Transfer
      Processes.. 347
   J.F. Owens

Testing Quantum Chromodynamics in Electron-Positron
      Annihilation at High Energies...................... 373
   Lowell S. Brown

Spin Effects in Electromagnetic Interactions............. 395
   P.A. Souder and V.W. Hughes

Very Recent Spin Results at Large $P_\perp^2$ 441
   A.D. Krisch

Hadroproduction of Massive Lepton Pairs and QCD.......... 455
   Edmond L. Berger

Orbis Scientiae 1979, Program............................ 515

Participants... 519

Index.. 523

EVIDENCE FOR QUARKS FROM NEUTRINO-NUCLEON SCATTERING*

F. Sciulli

California Institute of Technology

Pasadena, California 91125

I. POINT-LIKE STRUCTURE - QUARKS

There is near unanimity in the particle physics community that hadrons (neutrons, protons, pions, etc.) are made up, at least in part, by quarks - point-like objects with some well-defined properties (spin - 1/2 $\hbar$, charge - 1/3, + 2/3, etc.). As with all scientific concepts, there is no way to <u>prove</u> it. We can only ask whether the concept of quarks provides an acceptable explanation of the known facts, whether the ideas have led to new predictions which were verified, and whether there are any other experimental facts which conflict with that idea. If there are alternative concepts that provide all three of these, none in my knowledge brings the simplicity to the problem that the quark concept brings.

When Gell-Mann and Zweig invented[1] the concept more than 15 years ago, the idea was revolutionary: quarks were considerably less real to us then than they are today. Fractional charge indeed! And yet, the tables of hadron masses could be grouped into predictable sets, relations between certain masses could be found, other properties of the hadrons could be predicted. The prediction and discovery of the

---

*Work supported by the U.S. Department of Energy under Contract NO. EY76-C03-0068 for the San Francisco Operations Office.

Ω^- was one of the most exciting scientific enterprises of this century.[2] We cannot ignore the brilliant successes of the quark-concept ... they are too definitive and too numerous.

The single most intellectually difficult problem with quarks is that we have not been able to study them in isolation from hadrons - we cannot examine their properties with the same techniques used for the other particles that we study - therefore, they assume less reality in our minds. There is no question that the experiment which succeeds in separating a quark from all other quarks will completely solve this conceptual difficulty.[3] But we must face the possibility that nature will not allow such a thing to happen: that quarks might never be completely removed from the environment of other quarks.

To some extent, we have already faced this problem. We are studying the spins and charges of quarks; we don't use the traditional tried-and-true techniques of passing beams of them through electric and magnetic fields, however. We can't make such beams. We use other techniques, some of which will be described today. Those techniques have already established the quark charge and spin and are presently being used to unravel the quark-quark force. Table I shows the properties of those quarks for which there presently exists evidence.

Table I

Quark	Spin	Charge	Nucleon Number
u	$1/2\,\hbar$	$-1/3\,e$	$+1/3$
d	$1/2\,\hbar$	$+2/3\,e$	$+1/3$
s	$1/2\,\hbar$	$-1/3\,e$	$+1/3$
c	$1/2\,\hbar$	$+2/3\,e$	$+1/3$
b	$1/2\,\hbar$	$-1/3\,e$	$+1/3$

The quantum properties of the ordinary nucleons, neutron or proton, are presumed to be a consequence of their composition by the first two entries, u and d. That is the proton has p - (uud), and the neutron has n - (ddu). There may, of course, be other quarks in the nucleon if they are accompanied by their anti-quark partner, and there may be other kinds of constituents (e.g. gluons) which do not carry the quantum numbers in the table.

II. THE NEUTRINO PROBE

Neutrinos are particles with very low mass, no charge, spin of $1/2\ \hbar$, and with very small interaction strength. Because of their spin, they have two helicity states: right-handed and left-handed. They are of importance in many areas of physics and cosmology, so I will assume that the audience is familiar with them. Suffice it to say that we believe that we have a very good theory for neutrino interactions with "point-like" (i.e. lepton-like) particles, V-A theory, that should work up to neutrino energies over 500 GeV, and may work well beyond that with simple known modifications. Figure 1 shows how we picture a neutrino interaction with a spin 1/2 point-like target, T.

These are the reactions:

$$\nu_\mu + T \to \mu^- + T' \qquad (1a)$$

$$\bar{\nu}_\mu + T' \to \mu^+ + T \ . \qquad (1b)$$

For our purposes, using neutrinos as a tool, let me state the three relevant properties of such collisions that come from the V-A theory:

1. As the velocities of all the particles in the center-of-mass system approach the speed of light ($\beta \to 1$),
   (a) only left-handed T-particles interact
   (b) only right-handed T-antiparticles interact.
   (It is a consequence of this fact that neutrinos only exist in the left-handed state, and anti-neutrinos only exist in the right-handed state).

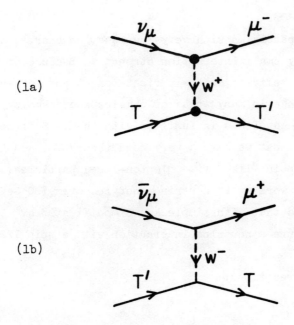

Figure 1a: The charged-current weak collision of a neutrino with a point-like target, T. The weak force or current is transmitted by means of the virtual boson, W^+ or W^-. These have rest masses much larger than present center-of-mass energies, so that the weak coupling of the Fermi-constant, G_F, has the dimensions of energy^{-2}.

Figure 1b: The antineutrino collision.

2. The cross-sections for the reactions (1) must depend on the square of the weak interaction coupling constant, G_F, which has the dimensions of inverse energy squared (see figure 1). The cross-section itself has the dimension of length squared, or by using $\hbar=c=1$, the dimensions of inverse energy squared. So $\sigma \propto G^2 S$ implies, in dimensions, that $E^{-2} \propto E^{-4} \dim(S)$ so that cross-sections must depend on a parameter S, which has the dimensions of either energy-squared or inverse length squared. The V-A theory tells us that this parameter is just the center-of-mass energy squared for point-like targets, T, whose mass is small compared to $\sqrt{S}$. (For extended, non-point-like objects, S could as example be proportional to the inverse area of the scattered target.) Since a neutrino of high energy E impinging on a target, of mass M, has $S = 2ME$, cross-sections should rise linearly with E.

3. Because the neutrino is changed to a negative muon, the target must raise its charge by one unit. Conversely, antineutrino scattering <u>lowers</u> the charge by one unit.

III. NEUTRINO-NUCLEON SCATTERING AT LOW ENERGIES

Let's look at a familiar example of point-like scattering. If the neutrino has enough energy ($\gtrsim$ 110 MeV), processes (2a) and (2b) will occur from nucleon targets:

$$\nu_\mu + n \rightarrow \mu^- + p \qquad (2a)$$

$$\bar{\nu}_\mu + p \rightarrow \mu^+ + n \qquad (2b)$$

Indeed, these processes have been observed, and their behavior is quite consistent with what is expected.

Figure 2 shows the experimental behavior[4] for the process (2a). After an initial rise from threshold, the total cross-section becomes constant.

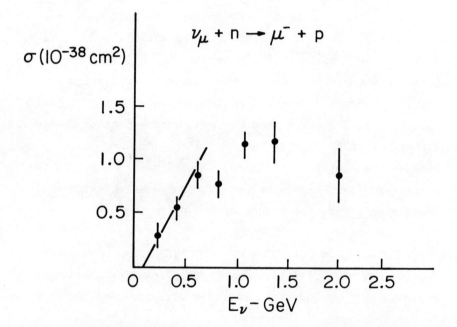

Figure 2: ANL data[4] on the exclusive process $\nu_\mu + n \to \mu^- + p$. Cross-section versus neutrino energy.

IV. NEUTRINO-NUCLEON TOTAL CROSS-SECTION

The flat behavior in fig. 2 has set in by ~1 GeV, where the wavelength of the probe is becoming smaller than the radius of the nucleon. But an interesting thing happens. Even though the exclusive reaction (2a) becomes independent of energy, the <u>total</u> cross-section for neutrino-nucleon scattering keeps increasing. This can happen because new final states become possible, e.g.

$$\nu_\mu + n \to \mu^- + p + \pi^0$$
$$\to \mu^- + n + \pi^+$$
$$\vdots$$
$$\to \mu^- + n + \pi^+ + \pi^- + \pi^0 + \pi^+ + K^+ + K^- + \ldots \text{etc.}$$

Figure 3 shows a sketch of the sum of various exclusive channels, as the energy passes their thresholds; a monotonic rise is clearly visible so long as data exists that covers all of these channels.

There is clearly a rich array of final states that can be studied. But is there some <u>simple</u> behavior for the total neutrino-nucleon cross-section that can be predicted from a simpler picture? We call this generalized process as shown in (3) $\nu_\mu + n \to \mu^- + X$ where X is the sum of all possible final states. If we picture the nucleon as being made from smaller, point-like objects, there is a very simple picture, indeed. Let's call one of these constituents q; then the fundamental process is

$$\nu_\mu + q \to \mu^- + q'$$

where this scattering is point-like. Then the <u>total</u> neutrino cross-section is just a sum of many such cross-sections that increase linearly with neutrino energy and will itself be linear with energy.

Of course, there is a lot neglected here. To mention just a couple of things: the finite masses of the constituents, the effects of the forces binding the constituents inside, the radiative (elec-

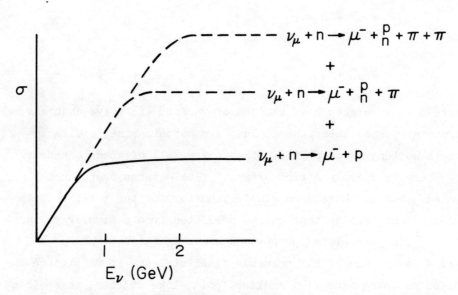

Figure 3: Sketch of how various final states enter the total neutrino cross-section plotted versus neutrino energy.

tromagnetic) forces. These latter effects[5] are known to change the linear behavior by roughly 5%, so we should not expect our simplistic picture without strong forces to work better than, say, 20%. Then a crucial question to answer is: "Does σ rise linearly with E, as expected from point-like target constituents?"

The answer is yes. Figure 4 shows the data on ν_μ and $\bar{\nu}_\mu$ scattering from roughly equal mixtures of neutrons and protons as of 1974; the rise is to a good approximation linear with E. Figure 5 shows the most recent data[7] where deviations from precise linearity are made more apparent by plotting σ/E versus E. Some deviation is apparent, but linearity is certainly true approximately (~20%). The striking difference between the ν_μ and $\bar{\nu}_\mu$ slopes (upper and lower data sets, respectively) should be noted. We will return to it later. This behavior is certainly consistent with the idea of point-like structure inside the nucleon.

V. COMPARISON OF NEUTRON AND PROTON TARGETS

With more optimism, we might even call it good evidence. There is another piece of evidence form total cross-sections which, together with the linear rise, gives qualitative evidence that the point-like objects have charges consistent with quarks.

This is the comparison of the anti-neutrino cross-sections from neutron and proton targets

$$\bar{\nu}_\mu + n \to \mu^+ + X \tag{4a}$$

$$\bar{\nu}_\mu + p \to \mu^+ + X \tag{4b}$$

We picture the fundamental process as

$$\bar{\nu}_\mu + u \to \mu^+ + d \tag{5a}$$

The corresponding reaction ($\bar{\nu}_\mu$ + d) cannot occur, since there is no quark with - 4/3 charge in the picture. In this model, reaction (4a) will occur half as often as reaction (4b) since there are half as

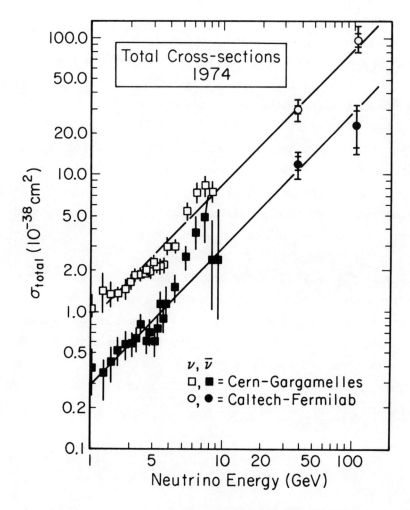

Figure 4: Behavior of the total neutrino and antineutrino cross-sections versus energy. The lines indicate linear dependence on incident neutrino energy.

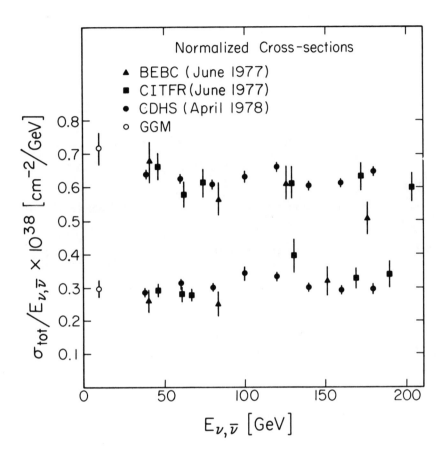

Figure 5: Most recent data on σ/E versus E. Precise point-like behavior would have the data, as plotted, be independent of E. The data reflect the simple predicted behavior at the 20% level. The upper data are for a ν_μ beam and the lower data are for a $\bar{\nu}_\mu$ beam.

many u quarks in the nucleon. Again, this might be only approximate; pairs of quark-antiquark pairs would modify the result. Figure 6 shows data[8] on the reaction. The ratio seems to approach a value close to 1/2 at high energy, very consistent with the picture.

VI. COMPARISON OF WEAK AND ELECTROMAGNETIC SCATTERING

Another qualitative check on the charge of these constituents is the comparison of weak (figure 1) scattering of neutrinos from nucleons with the analogous electromagnetic scattering of electrons or muons from nucleons.

Whereas all constituents scatter with the same probability in fig. 1, the probability associated with electromagnetic scattering is proportional to the square of the electric charge of T. After taking into consideration the known differences between weak and electromagnetic propagators, one can form a ratio from results of the two experiments which is proportional to the mean square of the constituents:

$$<q^2> = \frac{\sigma^{eN}}{\sigma^{\nu N}} \times \text{known corrections} = \frac{1}{N} \sum_{i=1}^{N} q_i^2$$

For simple quark pictures, this is 5/18. Figure 7 shows this comparison as of 1974, with the available eN data (SLAC-MIT) and νN data. Again, agreement with the simplest quark picture is apparent. Karl Berkelman will elaborate on this point today with the most recent electron and muon data.

VII. NEUTRINO-NUCLEON DIFFERENTIAL CROSS-SECTION

If the total neutrino cross-section did not have the predicted linear rise with energy, people would probably not have pursued this whole line of thought. But, given that linearity, one can try to carry the idea one step further. We can ask how the cross-section behaves differentially. Richard Feynman looked at this question[9] a number of years ago, calling the hypothetical point-like constituents "partons". The behavior of the high energy differential cross-

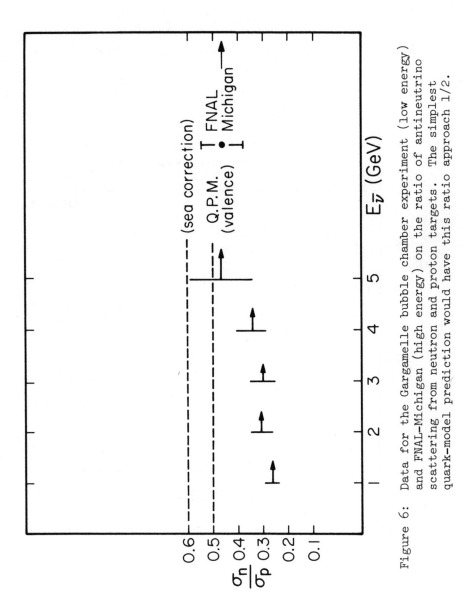

Figure 6: Data for the Gargamelle bubble chamber experiment (low energy) and FNAL-Michigan (high energy) on the ratio of antineutrino scattering from neutron and proton targets. The simplest quark-model prediction would have this ratio approach 1/2.

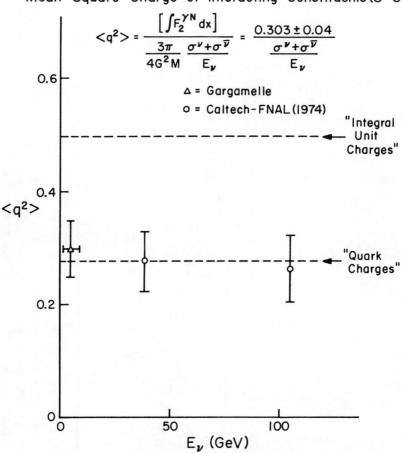

Figure 7: Comparison of νN and eN data to provide mean square charge as of 1974. More recent data on this question will be provided in K. Berkelman's talk.

section, he showed, should reflect the properties of these partons. Previously, of course, the periodicity of nucleon and meson states had been explained by the quark concept (i.e. point-like spin 1/2, fractionally charged). It was to be an interesting exercise to determine if the partons had the properties of quarks.

First, let's look at the way such experiments are done, and try to determine what it is we can measure. Figure 8 shows a computer reconstruction of an actual event induced by a 200 GeV neutrino. The target is made of steel, but imbedded in the steel are numerous scintillation counters for energy detection of the final shower hadrons, as well as spark chambers for tracking the muon, which penetrates through steel. The momentum, or energy, of the muon can also be measured by measuring its deflection in a magnetic field. These laboratory quantities E(neutrino energy), E_μ (muon energy), E_h (shower energy), and θ_μ (muon angle) are quite enough to tell us a lot about the fundamental process (reaction 1), i.e.

$$\nu_\mu + T \to \mu^- + T' \tag{1a}$$

where T is the point-like constituent (parton...quark?) inside the nucleon. This information is usually related in terms of so-called scaling variables: x and y. Let's look at the second of these

$$y \equiv E_h/E = 1 - E_\mu/E \tag{6}$$

This variable actually tells us the angle of scattering in the center-of-mass (θ^*) of reaction (1a). (See figure 9). But the angular distribution is a straightforward consequence of the V-A neutrino theory and the spin of the target T. (See Table II).

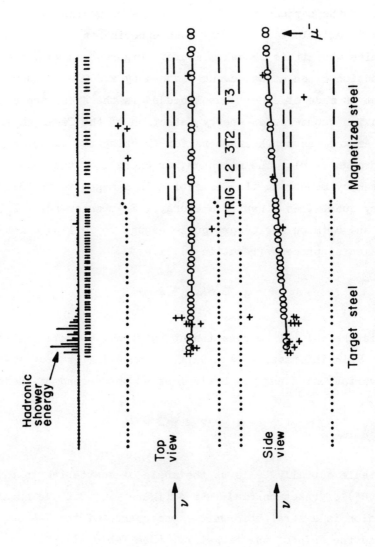

Figure 8: Computer reconstruction of a neutrino interaction in steel.

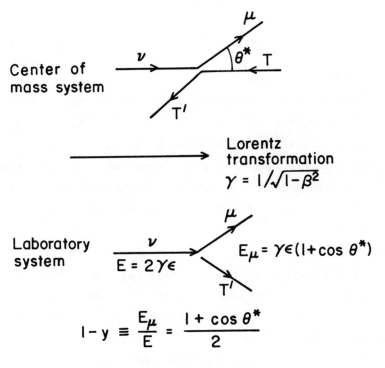

Figure 9: Neutrino collision with low mass point-like constituent, T. The y-variable gives directly the center-of-mass scattering angle, θ^*.

Table II

Type		V-A Cross Section	Scaling Form	y-function $f(y)$
(1)	νT or $\bar{\nu}\bar{T}$	$\dfrac{d\sigma}{d\Omega^*} = \dfrac{G^2 S_T}{4\pi^2} T$	$\dfrac{d\sigma}{dy} = \dfrac{G^2 S_T}{\pi} T$	$f_1(y) = 1$
(2)	$\nu \bar{T}$ or $\bar{\nu} T$	$\dfrac{d\sigma}{d\Omega^*} = \dfrac{G^2 S_T}{4\pi^2} T \left(\dfrac{1+\cos\theta^*}{2}\right)^2$	$\dfrac{d\sigma}{dy} = \dfrac{G^2 S_T}{\pi} T (1-y)^2$	$f_2(y) = (1-y)^2$
(3)	$\nu V, \bar{\nu}\bar{V}$ $\bar{\nu}V, \nu\bar{V}$	$\dfrac{d\sigma}{d\Omega^*} = \dfrac{G^2 S_T}{4\pi^2} T \left(\cos\dfrac{\theta^*}{2}\right)^2$	$\dfrac{d\sigma}{dy} = \dfrac{G^2 S_T}{\pi} T (1-y)$	$f_3(y) = 1-y$

S_T is the square of the center-of-mass energy in the ν_μ - T system. Here we take T to be a spin 1/2 target constituent, and V to be a spin 0 target constituent. The three y-functions shown turn out to be the only possible ones. This is a consequence of the fact that the spin 1 boson exchanged in the V-A interaction has only three possible states of polarization. <u>Any</u> target spin will show up as a linear combination of the three functions shown.

The other so-called scaling variable used to describe these processes is $x = q^2/2ME_h$, described in fig. 10. Another way of writing this which has the virtue of being Lorentz invariant is $x = q^2/2p \cdot q$ where q and p are the 4-vectors carried by the propagator and proton, respectively. For these physical processes, q^2 is always a negative number, meaning that the momentum transfer is spacelike (momentum always larger than the energy in any frame of reference). If we transform to another frame where $q = \vec{q}$, i.e. the transfer is <u>all</u> momentum-like, we have the picture shown in fig. 10. We see here that $\vec{q} - 2\xi\vec{p}$, or

$$-q^2 = 2\xi\vec{p} \cdot \vec{q} = 2\xi p \cdot q \qquad (7)$$

Hence, the fraction of the momentum carried by the struck parton is

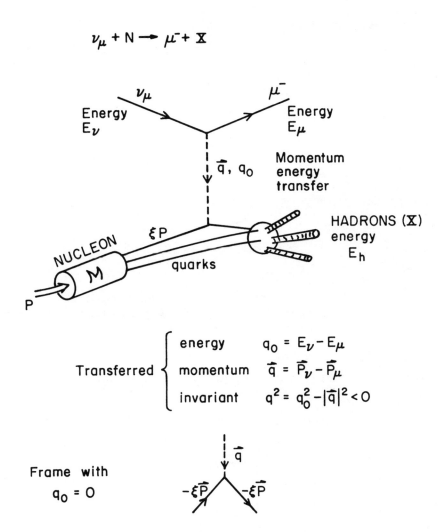

Figure 10: A view of the neutrino-nucleon collision as a collision with quarks. The fraction of the nucleon momentum carried by the struck quark, ξ, is given by the experimental parameter, $x = -q^2/2ME_h$ (see text). Here $-q^2 = E_\nu E_\mu \theta^2$, where θ is the laboratory angle of the final state muon.

$$\xi = \frac{-q^2}{2p \cdot q} = \frac{-q^2}{2ME_h} \equiv x \qquad (8)$$

and is directly measured in each event.

In this picture, then, the cross-section for neutrino scattering will have the form

$$\frac{d\sigma^{\nu N}}{dy} = \frac{G^2 S}{2\pi} \; [Q + \bar{Q}(1-y)^2 + K(1-y)] \qquad (9)$$

$$\frac{d\sigma^{\bar{\nu} N}}{dy} = \frac{G^2 S}{2\pi} \; [Q(1-y)^2 + \bar{Q} + K(1-y)] \qquad (10)$$

where, for example,

$$Q = \int_0^1 x \, \rho_1(x) \, dx \qquad (11)$$

is the nucleon momentum fraction carried by point-like particle constituents of spin 1/2.

The forms (9) and (10) are especially simple in the case of the simplest quark model, where the spin 0 component should be small ($K \approx 0$), while the quark component should be dominant ($Q \gg \bar{Q}$). In that case, the νN distribution should be dominantly flat in y, and the $\bar{\nu} N$ distribution should be dominantly $(1-y)^2$. Figure 11 shows that this form is generally observed.[10]

VIII. CONSEQUENCES OF THE SIMPLE QUARK MODEL

This form has some additional consequences. For the simplest quark model, we expect $K = 0$, $\bar{Q} \ll Q$. Then the total cross-sections and mean y-values will be

$$\sigma^{\nu N} = \frac{G^2 S}{2\pi} (Q + \frac{1}{3} \bar{Q}) \qquad \qquad <y>^\nu = \frac{\frac{1}{2} Q + \frac{1}{12} \bar{Q}}{Q + \frac{1}{3} \bar{Q}}$$

EVIDENCE FOR QUARKS FROM NEUTRINO-NUCLEON SCATTERING

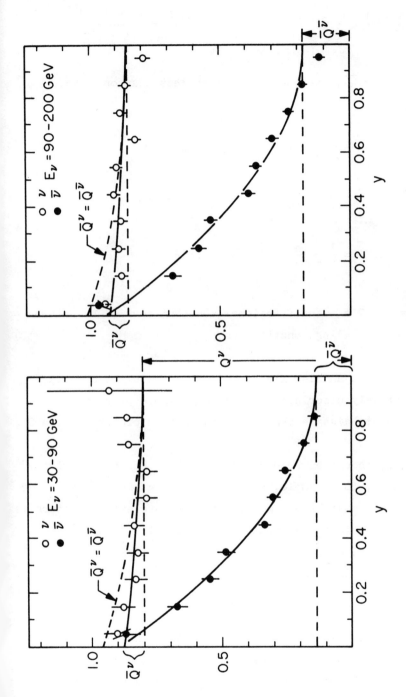

Figure 11: CDHS data on νN, $\bar{\nu} N$ scattering in two different neutrino energy regimes. The general form of dominantly flat for ν and dominantly $(1-y)^2$ for $\bar{\nu}$ is seen. In detail, the data do not fit the forms (9) and (10), although the comparison is quite compatible with the 20% agreement expected.

$$\sigma^{\bar{\nu}N} = \frac{G^2 S}{2\pi} \left(\frac{1}{3} Q + \bar{Q}\right) \qquad \langle y \rangle^{\bar{\nu}} = \frac{\frac{1}{12} Q + \frac{1}{2} \bar{Q}}{\frac{1}{3} Q + \bar{Q}} ,$$

And for $\alpha = \bar{Q}/(Q + \bar{Q})$, expected to be much less than one,

$$\frac{\sigma^{\nu N}}{\sigma^{\bar{\nu}N}} = \frac{1 + 2\alpha}{3 - 2\alpha} \qquad \text{see figure 14}$$

$$\langle y \rangle^{\nu} = \frac{6 - 5\alpha}{12 - 8\alpha} \qquad \text{see figure 12}$$

$$\langle y \rangle^{\bar{\nu}} = \frac{1 - 5\alpha}{4 + 8\alpha} \qquad \text{see figure 13}$$

$$\frac{3}{4} \frac{\sigma^{\nu N} + \sigma^{\bar{\nu}N}}{G^2 S/2\pi} = Q + \bar{Q} = \text{total fractional momentum carried by interacting constituents.} \tag{12}$$

Figures 12, 13, and 14 illustrate that quite a bit of data[11] can be qualitatively described by this very simple picture. But there are certainly disagreements from the simple picture at the 20% level.

However, it should be remembered that a lot has been neglected here. There must, for example, be forces binding the quarks inside the nucleon. They must have some effect. An interesting feature is that the total fractional momentum (eq. 12) carried by the quarks, $Q + \bar{Q}$, is about 0.5, so that there is a lot more inside the nucleon which is not seen. This is not explained without more than the simple quark model.

IX. EVIDENCE FOR q^2-DEPENDENCE = QUANTUM CHROMODYNAMICS?

The most striking experimental evidence for the deviations from the simplest model is in the behavior of Q and $\bar{Q}$ as a function of x

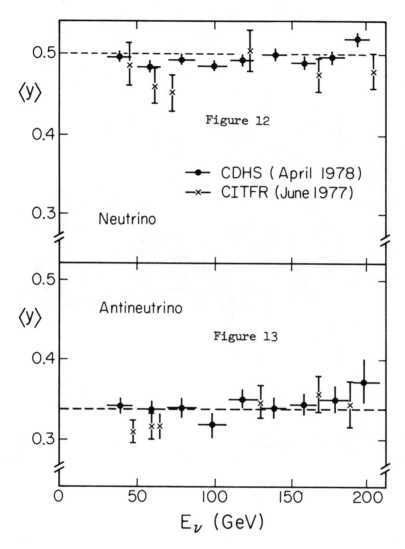

Figures 12 and 13: Lines correspond to an antiquark fraction α = 0.16, (see text). The data are for mean y for neutrino events (12) mean y for antineutrino events (13).

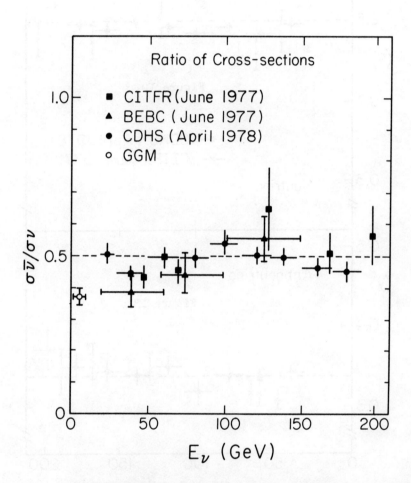

Figure 14: Lines correspond to an antiquark fraction α = 0.16, (see text). The ratio of antineutrino to neutrino cross-section.

and q^2. The simplest forms (eqs. 9 and 10) would have a neutrino cross-section linearly dependent on energy, and mean x and y values independent of energy. In figure 15, we show the mean value, <xy> vs. neutrino energy. There seems to be a gentle, probably logarithmic effect with neutrino energy, E.

It is hoped that this behavior will find a natural explanation in the field theory for strong interactions (QCD). In that theory, the forces between quarks are presumed to be carried by vector propagators, called gluons. In that case, new diagrams must be invoked which give the qualitative feature of the data shown in figure 15.

X. THE PROBLEM OF R

One qualitative feature of this whole picture which remains a potential problem is the function which we referred to in equations (9) and (10) as K. Its differential form would appear most simply in the sum of the ν_μ and $\bar{\nu}_\mu$ cross-sections

$$\frac{d\sigma^{\nu+\bar{\nu}}}{dxdy} = \frac{G^2 S}{2\pi} \left[(Q + \bar{Q}) [1 + (1-y)^2] + 2K(1-y) \right]$$

The parameter R is defined as

$$R = K/(Q + \bar{Q}) \qquad (13)$$

Table III gives the presently measured values of R, as measured in eN, μN, and νN scattering up until recently.

Table III: Measurements of R with Statistical and Systematic Errors

Technique	x-range	q^2-range	R
eN scattering[12]	0.1 - 0.9	2 - 20 GeV^2	0.2 + 0.1
μN scattering[13]	0 - 0.1	1 - 12.5 GeV^2	0.44 ± 0.25 ± 0.19
νN scattering[14]	0 - 1.	0.1 - 50 GeV^2	0.15 ± 0.10 ± 0.04

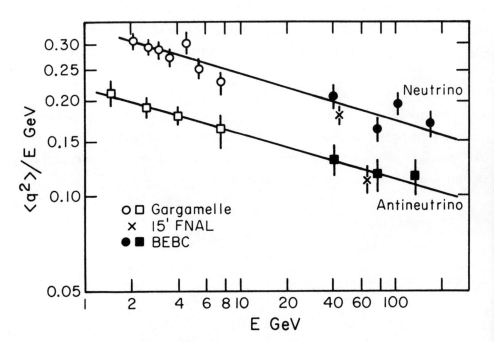

Figure 15: The mean value of xy is seen to charge logarithmically with neutrino energy, E. Such behavior does not occur in the simplest quark model.

Within our 20% criterion, there would seem to be no real problem. There are mechanisms (e.g. QCD) that are expected to produce values of R that are finite, even from spin 1/2 constituents (see later this section). However, the first entry of the table has a rather curious behavior in x. This is shown in figure 16. The behavior is as large at large x as at small x. This is not expected from any anticipated mechanism.

There has been a more recent measurement reported that colors this concern differently. That value[15] is reported as

$$R^{\nu N} = -.03 \pm .04 \tag{14}$$

averaged over x for typical q^2 of about 10 GeV^2. If this value holds up, it would appear that the very simplest quark model predictions are still trustworthy, but it would also mean that some expectations of QCD are probably wrong.[16]

XI. CONCLUSIONS

We have seen that the simplest quark model predictions for the behavior of neutrino and antineutrino cross-sections prevail at the 20% level. This means that the constituents act
 (1) point-like
 (2) fractionally charged
 (3) spin 1/2

There are some uncertain experimental aspects to the situation, such as the need for more precise measurements of the structure functions, particularly R. But there are some clear ~ 20% effects which the simplest model does not easily incorporate:
 (1) Where the other 50% of the nucleon momentum is.
 (2) How to explain the gentle (logarithmic) dependence on energy.
 (3) How to incorporate finite R values if they turn out to be experimentally present.

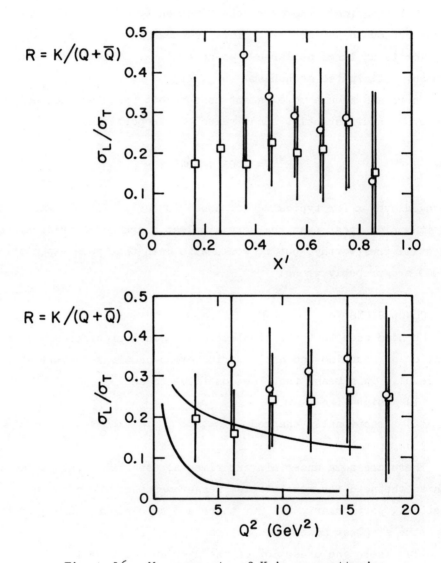

Figure 16: Measurements of K is ep scattering.

We hope that these questions will be answered when we understand the nature of the forces between the quark constituents. There is a candidate theory for that force (QCD) which has had some striking confirmations, as well as some problems. However, there is little question that the qualitative aspects of neutrino scattering are quark-like. The problem before us is to understand the force.

REFERENCES

1. M. Gell-Mann, Phys. Lett. $\underline{8}$, 214 (1964); G. Zweig, CERN Rep. TH-401 (1964); CERN Rep. TH-412 (1964).
2. V. E. Barnes et al., Phys. Rev. Lett. $\underline{12}$, 204 (1964).
3. Recently, results have been reported that could be interpreted as free quarks, see G. S. LaRue et al., Phys. Rev. Lett. $\underline{42}$, 142 (1979).
4. W. A. Mann, et al., Phys. Rev. Lett. $\underline{31}$, 844 (1973).
5. R. Barlow and S. Wolfram, Oxford Preprint 24/78.
6. B. C. Barish, et al., Phys. Rev. Lett. $\underline{35}$, 1316 (1975).
   T. Eichen, et al., Phys. Lett. $\underline{46B}$, 274 (1973).
7. B. C. Barish, et al., Phys. Rev. Lett. $\underline{39}$, 1595 (1977).
   P. C. Bossetti, et al., Phys. Lett. $\underline{70B}$, 273 (1977).
   CERN-Dortmund-Heidelberg-Saclay Collaboration, Proceedings of the Topical Conference on Neutrino Physics at Accelerators, Oxford, July 4-7, 1978 (p. 11), presented by F. Eisele.
8. Gargamelle Antineutrino Collaboration, Proceedings of the Topical Conference on Neutrino Physics at Accelerators, Oxford, July 4-7, 1978 (p. 68), presented by M. Rollier; private communication, B. Roe.
9. See, e.g. R. P. Feynman, Neutrinos-1974, ed. by C. Baltay (AIP 1974), p. 299.
10. CDHS collaboration results presented by F. Eisele, Proceedings of the Topical Conference on Neutrino Physics at Accelerators, Oxford, July 4-7, 1978 (p. 11). Also, see ref. 11.

11. Summary talk by F. Sciulli, Proceedings of Topical Conference on Neutrino Physics at Accelerators, Oxford, July 4-7, 1978 (p. 405)
12. M. Mestayer, SLAC Report No. 214.
13. B. A. Gorden, et. al., Phys. Rev. Lett. $\underline{41}$, 615 (1978).
14. P. C. Bossetti, et al., Nuclear Physics $\underline{B142}$, 1 (1978).
15. P. Block, et al., Inclusive Interactions of High Energy Neutrinos and Antineutrinos in Iron, submitted to Zeitschrift fur Physik.
16. Contributions to R may occur even with all of the scattering occurring from spin 1/2 constituents, if the scattered quark has finite transverse momentum. QCD gives calculable contributions from known diagrams that is approximately 0.1 in this q^2-region and falls roughly as $\ln^{-1} q^2$. Other expected contributions, which fall like q^{-2}, would add to that.

DIRECT EXPERIMENTAL EVIDENCE FOR CONSTITUENTS IN THE NUCLEON

FROM ELECTROMAGNETIC SCATTERING EXPERIMENTS

        Karl Berkelman

        Laboratory of Nuclear Studies

        Cornell University

        Ithaca, New York 14853

Since the earliest high-energy electron scattering experiments carried out at Stanford by Hofstadter and others,[1] we have seen nearly three decades of experimental and theoretical work on the structure of the nucleon. I am sure you have all heard the subject discussed many times, so it's going to be hard for me to find something new to say about it. Let's try to look at the data from a new point of view, though. The usual approach goes like this. You take your favorite theory - involving quarks, gluons, asymptotic freedom, QCD if you're in the mainstream nowadays - and adjust it to see if you can fit the latest data. I'm going to ask you to forget all that for now, and start out fresh by looking at just the experimental results and trying to see, with a minimum of theoretical prejudice, what the data directly imply about the constituents of the nucleon. And because of the constraints imposed on this talk, we're going to consider just the electron and muon scattering data.

ELASTIC SCATTERING

    To lowest order in the electromagnetic coupling the differential cross section for electron scattering from an extended charge distribution is given by

$$\frac{d\sigma}{dQ^2} = \left[\frac{4\pi\alpha^2}{Q^4} \frac{E'}{E} \cos^2\frac{\theta}{2}\right] F^2(Q^2), \qquad (1)$$

the product of the point-charge cross section (in square brackets) and the square of a momentum transfer dependent form factor $F(Q^2)$, which contains the information on the unknown structure of the target. In the nonrelativistic domain, $Q^2 \ll M^2$, the form factor is a kind of amplitude for the final state of the target to resemble the initial state, i.e., hold together,

$$F(Q^2) = \int <f|e^{i\vec{q}\cdot\vec{r}}|i> d^3\vec{r}, \qquad (2)$$

or equivalently, the Fourier transform of the static charge distribution of the target,

$$F(Q^2) = \int \rho(r) e^{i\vec{q}\cdot\vec{r}} d^3\vec{r}. \qquad (3)$$

Before discussing the electron-nucleon scattering data I have to mention some complications. The fact that the nucleon has a magnetic moment as well as a charge implies that the scattering must be described by two form factors. In one way of defining them, the cross section becomes

$$\frac{d\sigma}{dQ^2} = \left[\frac{4\pi\alpha^2}{Q^4} \frac{E'}{E} \cos^2\frac{\theta}{2}\right]\left\{\frac{G_E^2(Q^2) + \tau G_M^2(Q^2)}{1+\tau} + 2\tau G_M^2(Q^2)\tan^2\frac{\theta}{2}\right\}, (4)$$

where $\tau = Q^2/4M^2$. We determine the electric and magnetic form factors, G_E and G_M, at a given Q^2 by making several measurements at that fixed Q^2, varying the scattering angle θ (and the incident energy).

A second complication is the fact that this formula holds only in lowest order, while the measured cross section also involves mul-

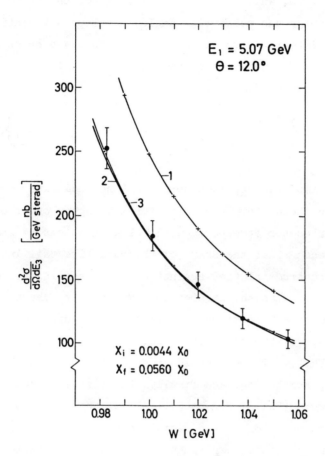

Fig. 1 Spectrum of scattered electrons from elastic e^-p scattering at fixed incident energy and scattering angle, plotted against the missing mass W in the final state (Ref. 3). This portion of the spectrum is between the elastic peak and the meson production threshold, and is accounted for by the radiative correction including multiple photon emission (curves 2 and 3).

tiple photon exchanges and bremsstrahlung. These radiative effects destroy the elastic kinematics; that is, the scattered electron energy spectrum, which would be a delta function in the case of fixed incident energy and scattering angle, develops a radiative tail. Even if we integrate over this tail, the observed cross section is modified by a factor which depends logarithmically on Q^2 and on the integration range ΔE:

$$\sigma_{obs} \approx \sigma_o \left\{ 1 - \frac{2\alpha}{\pi} \left[(\ln \frac{Q^2}{m_e^2} - 1)(\ln \frac{E}{\Delta E} - \frac{13}{12}) + \frac{17}{36} \right] \right\}, \qquad (5)$$

It is important to make sure that the radiative correction calculation has adequately accounted for any higher order effects. There are a number of experimental checks that can be made. Fig. 1 shows evidence[3] that the shape of the radiative tail in the scattered electron spectrum (plotted against missing mass instead of scattered energy) agrees with the prediction. We can also check (in Fig. 2) that the measured cross section, with the standard radiative correction, divided by the point cross section, is linear in $\tan^2 \frac{\theta}{2}$ at fixed Q^2 (Eq.4). The corrected cross sections should be the same for e^-p and e^+p scattering; and finally, the recoil proton polarization should be zero (Fig.3, for both). All of these checks have confirmed the expected behavior of the radiatively corrected cross section, although in each case, experimental difficulties oblige us to make the tests over a smaller Q^2 range than that covered by the simple cross section measurements. Anyway, let's assume that the nucleon form factors can be reliably derived from corrected measured cross sections using Eq.4.

The nucleon form factor data ranges in Q^2 up to about 25 GeV^2 and can be approximated within 10% (see Figs. 4,5) by the simple relations

$$G_E^p(Q^2) = G_M^p(Q^2)/\mu_p = G_M^n(Q^2)/\mu_n = 1/(1+Q^2/.71 \text{ GeV}^2)^2 \qquad (6)$$
$$G_E^n(Q^2) = 0 \;.$$

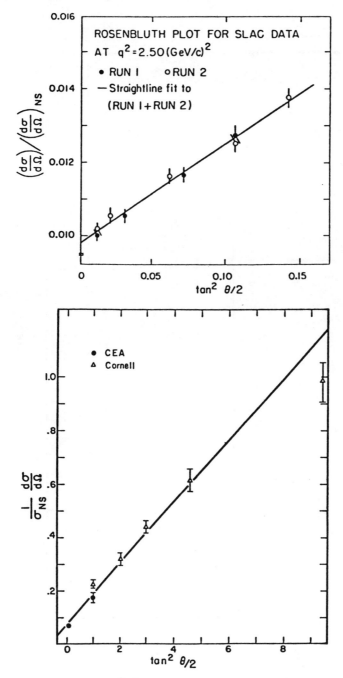

Fig. 2 Tests of the $\tan^2 \frac{\theta}{2}$ linearity of the elastic e^-p cross section, divided by the point cross section (Refs. 4,5,6).

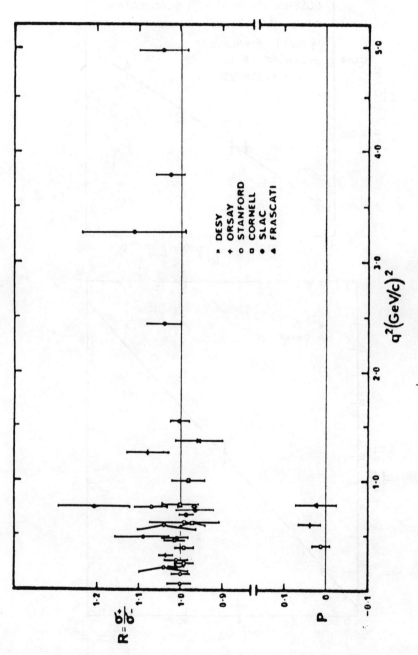

Fig. 3 The ratio R of e^+p and e^-p elastic cross sections, and the recoil proton polarization P in e^-p scattering (Ref. 7).

It is interesting that this simple "dipole" form is the Fourier transform of an exponential

$$\rho(r) = \rho_o \exp(-\sqrt{.71 \text{ GeV}^2}\ r), \qquad (7)$$

the proton charge distribution suggested by Hofstadter on the basis of the earliest low energy data.[1] The corresponding r.m.s. charge radius is $r_{rms} = \sqrt{12/.71 \text{ GeV}^2} = 0.8$ fm.

The immediate conclusions we can draw are (a) the proton and neutron are not elementary pointlike particles, and (b) the amplitude for the nucleon to remain a nucleon after the collision (Eq.2) drops rapidly with increasing momentum transferred. Just as in the case of electron-nucleus scattering, I take this as direct evidence that the proton and neutron are composite objects.

Before trying to get the data to tell us something about the constituents of the nucleon, I should mention an entirely different interpretation of the form factors. In the vector meson dominance model the virtual photon is assumed to couple to a neutral vector meson - ρ°, ω, ϕ, or whatever - which then couples to the nucleon. In the most naive form of the model each of the form factors is a sum over vector meson pole terms

$$F(Q^2) = \sum_V a_V \frac{m_V^2}{Q^2 + m_V^2}, \qquad (8)$$

with coefficients a_V depending on the V and VNN coupling constants. Such expressions can be made to agree with the experimental data tolerably well (Figs. 4,5), provided at least four vector mesons are included and the coefficients are adjusted to fit.

As Q^2 becomes larger and larger, more and more massive vector mesons have to be included in order to simulate the simple dipole Q^2 dependence with a sum of single-pole terms. Also the spacelike Q^2 in scattering gets farther and farther removed from the poles in the timelike region. So, although there is probably a lot of sense in

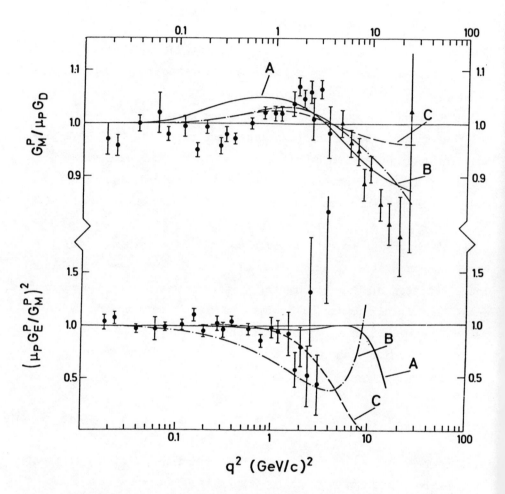

Fig. 4 Ratios of proton form factors and the dipole formula $G_D = 1/(1 + q^2/.71\ \mathrm{GeV}^2)^2$. The curves represent various vector dominance models; the data are from a compilation (Ref. 8).

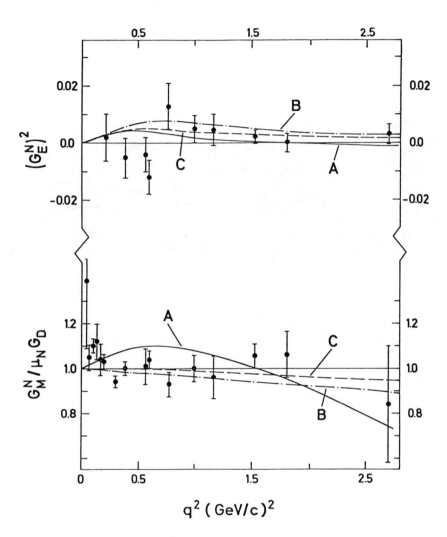

Fig. 5 As in Fig. 4, but with neutron form factors.

the vector meson dominance model applied to nucleon form factors, especially at low Q^2, it loses its predictive power at high Q^2. And it does not tell us why the vector meson masses and couplings should conspire to produce such a simple Q^2 dependence.

Suppose now that the nucleon is composed of several more-or-less pointlike constituents bound by some strong force. In order that the nucleon hold together in a high Q^2 elastic collision, some fraction of that Q^2 must be exchanged between the struck constituent and the other constituents. In the high Q^2 limit, each off-mass-shell constituent or exchanged field quantum will contribute a factor $1/Q^2$ to the amplitude and hence to the form factor. However, there will always be another Q^2 factor in the numerator for each exchange.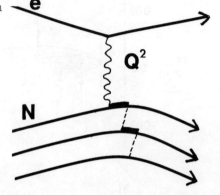
This follows[9,10] for any renormalizable field theory, such as spin-1/2 constituents bound by a scalar or vector field, or spin-0 constituents bound by a vector field. The net effect is that the form factor at high Q^2 will be proportional to $1/(Q^2)^{N-1}$, where N is the number of constituents. From the dipole behavior (Eq. 6 and Figs. 4,5) or from a direct plot of Q^4 times the measured form factor (Fig. 6), it is obvious that $2(N-1) = 4$ and $N = 3$. That is, we conclude from the high Q^2 behavior of the nucleon elastic form factors that the proton and neutron each consist of three particles.

DEEP INELASTIC SCATTERING

Figs. 7 and 8 show several examples of electron spectra in inelastic scattering - from atoms,[12] from a nucleus,[1] and from the proton.[13-15] In the atomic scattering the broad continuum below the elastic peak (Fig. 7, top) is understood as the quasielastic scat-

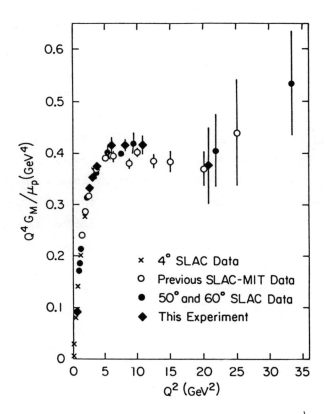

Fig. 6 Proton magnetic form factor, multiplied by Q^4, versus Q^2. Data compiled in Ref. 11.

tering from the individual electrons in the atom. In the nuclear case the broad maximum (Fig. 7, bottom) comes from scattering from the individual nucleons in the helium nucleus. This peak occurs near the energy for elastic scattering from free protons (the dashed peak), but is broadened and shifted by the effect of the Fermi momentum and binding energy of the nucleons. In some cases one also observes multiple peaks just below the elastic peak, corresponding to excitation of resonant final states of the atom or nucleus. The same features are obvious at higher energies in the spectrum from electron-nucleon scattering (Fig. 8), and it is natural to interpret the inelastic continuum as quasielastic scattering from the bound constituents of the nucleon.

In doing so, however, we have to avoid regions of the kinematic variables where the scattering is not dominated by constituent scattering. Near the elastic peak at low energy loss $\nu = E-E'$ and low Q^2 the scattering is dominated by resonance formation. Also at low Q^2, or more precisely low Q^2/ν, the virtual photon interacts peripherally with the whole nucleon through the vector meson dominance mechanism. To see why this is so, consider the distance Δs traveled by the virtual vector meson. It is related through the uncertainty principle to the energy mismatch between the vector meson and a mass Q^2 virtual photon having the same three-momentum:

$$\Delta s \approx 2\nu/(Q^2 + m_V^2). \tag{9}$$

If the intrinsic nucleon structure has a size scale Δr, then when $Q^2/2\nu < 1/\Delta r$ we will have $\Delta s > \Delta r$ and the vector meson will have a separate existence outside the nucleon; otherwise not. In fact, when $Q^2/2M\nu < 0.1$ the inelastic scattering cross section, divided by a virtual photon flux factor known from Q.E.D., is essentially the ordinary real photon cross section[16] modified by a vector meson propagator factor. So in the discussion of nucleon constituent effects I will avoid $Q^2/2M\nu < 0.1$.

EVIDENCE FOR CONSTITUENTS IN THE NUCLEON 43

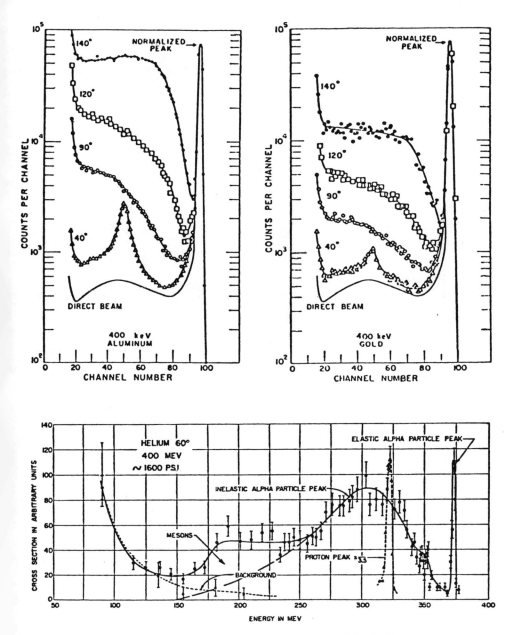

Fig. 7 Scattered electron energy spectra for 400 keV electrons scattered from aluminum and gold atoms (Ref. 12), and for 400 MeV electrons scattered from the helium nucleus (Ref. 1).

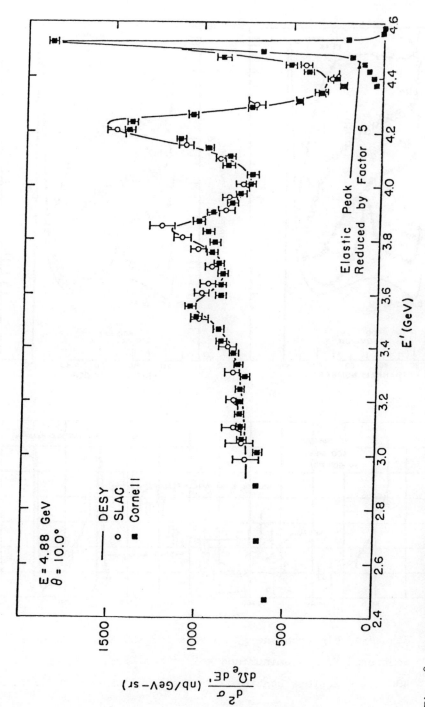

Fig. 8 Comparison of inelastic ep data from DESY (Ref. 13), SLAC (Ref. 14), and Cornell (Ref. 15).

Let me say a few words about the reliability of the data. First we note that different laboratories agree on the raw uncorrected measurements (Fig. 8). Fig. 9 shows a comparison of raw data, the corresponding radiatively corrected data, and the correction factor. Because the radiative tails for each contribution to the spectrum accumulate as you go to lower E' (higher W), they become appreciable, and the correction is often as much as a factor of two. Radiative corrections have been calculated independently by several groups, and there is general agreement among the experts, so there is a good chance that it is being done right. Our major conclusions will not be sensitive to errors of a factor of two, but it is important to remember that subtle effects in the data - small slopes, logarithmic dependences, and such - are quite vulnerable to errors in the radiative corrections.

The one-photon-exchange hypothesis for the corrected cross sections can be tested, just as in the case of elastic scattering. Again the cross section at fixed Q^2 and ν (or $Q^2/2M\nu$, or W, or some other measure of inelasticity) when divided by the point cross section should be linear in $\tan^2 \frac{\theta}{2}$. Actually, in the inelastic case, experimenters divide by a somewhat different factor and plot against the variable $\varepsilon = [1 + 2(1 + \nu^2/Q^2)\tan^2 \frac{\theta}{2}]^{-1}$ to make the linearity test (Fig. 10), but it's really the same test as in elastic scattering. The experiment is rather difficult, involving data runs at very different incident energies, scattering angles, and counting rates, and the accuracy is dominated by systematic errors in the relative normalization of different spectrometers, radiative corrections, and such. Within the errors the linearity is always confirmed. Fig. 11 shows the comparison of the cross sections for e^+p (or μ^+p) and e^-p (or μ^-p) scattering. Again, it is always easier experimentally just to measure cross sections than to carry out these checks, so we always have to use the one-photon-exchange hypothesis at higher Q^2 values than those where we have verified it.

The generalization of Eq. 4 to inelastic scattering involves two

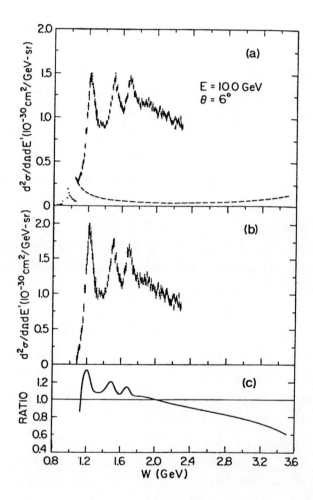

Fig. 9 Comparison of SLAC (a) raw data, (b) radiatively corrected data, and (c) ratio, in ep inelastic scattering (Ref. 17).

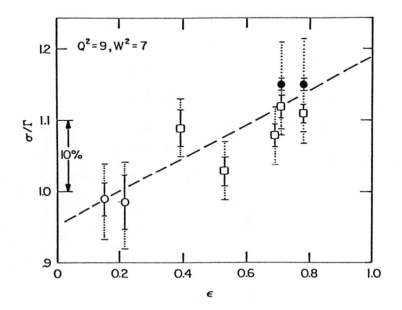

- ○ 50-60° DATA (ATWOOD)
- □ MIT−SLAC DATA
- ● NEW SLAC 20 GeV SPECTROMETER DATA

Fig. 10 Measured inelastic ep scattering cross section divided by the virtual photon flux factor Γ, at fixed $Q^2 = 9$ GeV^2 and $x = Q^2/2M\nu = 0.6$, plotted against the photon polarization parameter ε (see text). Data are from Refs. 18,19,11.

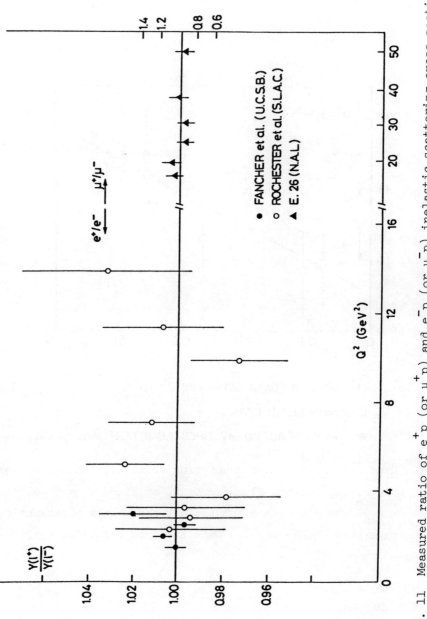

Fig. 11 Measured ratio of e^+p (or μ^+p) and e^-p (or μ^-p) inelastic scattering cross sections, versus Q^2. Data are from Refs. 20, 21, 22.

EVIDENCE FOR CONSTITUENTS IN THE NUCLEON

structure functions, depending on Q^2 and on one other variable which I will take to be Bjorken's $x = Q^2/2M\nu$. One of the many choices for structure functions leads to

$$\frac{d\sigma(Q^2,x,\theta)}{dQ^2 dx} = \left[\frac{4\pi\alpha^2}{Q^4} \frac{E'}{E} \cos^2\frac{\theta}{2}\right] \frac{\nu W_2(x,Q^2)}{x} \left\{1 + \frac{2(1+\tau/x^2)}{1+R(x,Q^2)} \tan^2\frac{\theta}{2}\right\}. \tag{10}$$

The expression in the square brackets is the cross section for elastic scattering from a point charge, as before. The second term in the curly brackets contributes only at large angles where the point cross section is very small, so it is usually negligible. The cross section is mainly determined by the structure function $\nu W_2(x,Q^2)$.

The most striking aspect of the deep inelastic data is its Q^2 dependence. In contrast to the behavior in elastic scattering, the ratio of the measured[23] inelastic cross section to the point cross section at fixed W (Fig. 12) decreases only slightly with increasing Q^2. This is just what you would expect to happen if the scattering were from pointlike constituents in the nucleon.

Another striking feature of the early data is the scaling phenomenon; that is $\nu W_2(x,Q^2)$ turned out to be mainly a function of x alone. When plotted against x (Fig. 13) the values for a wide range of Q^2 clustered around a single curve, or when plotted against Q^2 at a single x (Fig. 14) a flat dependence was observed. However, as more precise data over a wider kinematic range have become available, it is now clear that νW_2 does not really scale; it does depend separately on Q^2 as well as on x. At low Q^2, νW_2 has a trivial Q^2 dependence (Fig. 15); it must vanish at $Q^2 = 0$ in order that the virtual photon cross section connect to a finite real photoproduction cross section. But this scaling violation doesn't interest us, since it involves only low Q^2 where peripheral interactions dominate. At high Q^2, however, the Fermilab data[26-28] show systematic scaling violations. For $x < 0.2$, νW_2 increases with increasing Q^2, and when

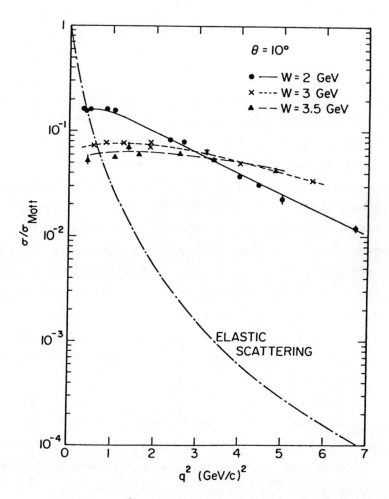

Fig. 12 Comparison of Q^2 dependence for inelastic and elastic ep cross sections (Ref. 23) divided by the point cross section.

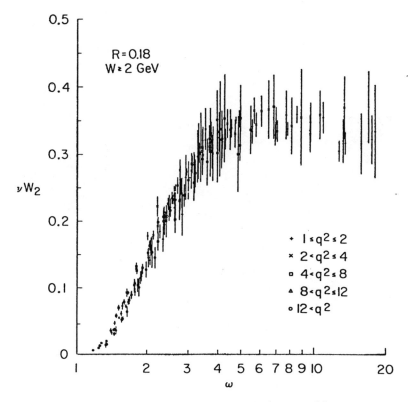

Fig. 13 The proton structure function νW_2 (Ref. 24) plotted against $\omega = 1/x$.

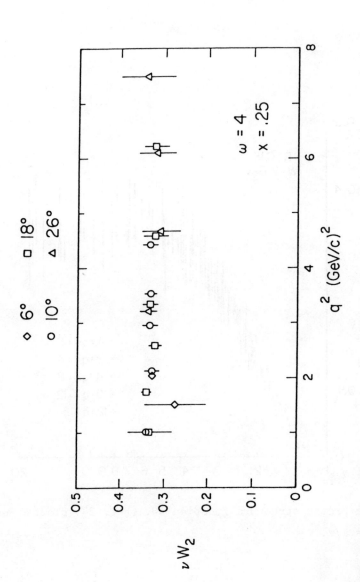

Fig. 14 The proton structure function νW_2 at fixed x = 0.25, plotted against Q^2. Data from Ref. 24.

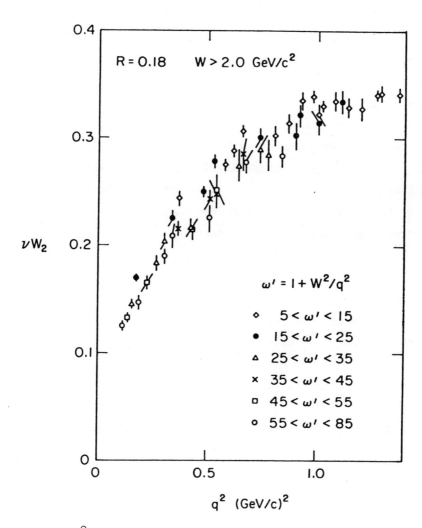

Fig. 15 Low Q^2 behavior of νW_2 for the proton. Data from Ref. 25.

x > 0.2 it decreases (Fig. 16). Thus as a function of x the structure function becomes more steeply peaked at small x, the higher the Q^2 value (Figs. 17,18). Note, however, that the Q^2 dependence is still only logarithmic, very much flatter than in the case of the elastic form factors.

Deep inelastic experiments with a deuterium target[29] tell us that the neutron scatters differently from the proton (Fig. 19), although the neutron νW_2 function also seems to scale, at least approximately.

How are we to interpret these data in terms of the constituents of the nucleon? We assume that the inelastic scattering cross section is just the incoherent sum of elastic scatters from the individual constituents, which for now we will take as free particles at rest.[30] Since we have elastic scattering from targets at rest, the Q^2 and the energy loss ν are not independent:

$$Q^2 = 2m_i \nu , \qquad (11)$$

where m_i is the mass of the i-th constituent. It therefore follows that the Bjorken x variable is constrained to be

$$x \equiv Q^2/2M\nu = m_i/M , \qquad (12)$$

the fractional mass of a constituent, which I will call x_i. The inelastic cross section is obtained by summing elastic cross sections for the constituents:

$$\frac{d\sigma_{inel}(Q^2,x,\theta)}{dQ^2} \sim \left[\frac{4\pi\alpha^2}{Q^4} \frac{E'}{E} \cos^2 \frac{\theta}{2}\right] \sum_{i \ni x_i = x} n_i q_i^2 F_i^2(Q^2), \qquad (13)$$

where n_i is the number of constituents of type i, q_i their charges, and $F_i(Q^2)$ their effective form factors.

Eq.(13) represents a discrete spectrum in E'; that is, the cros

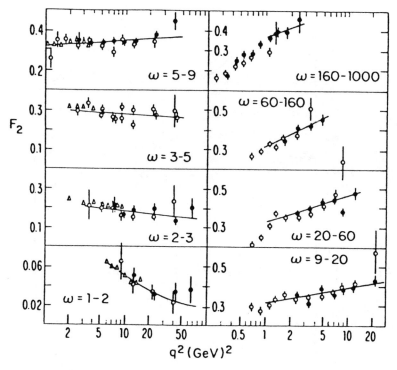

Fig. 16 Proton νW_2 as a function of Q^2 for various fixed $\omega = 1/x$. Data are from MIT-SLAC (triangles, Ref. 19) and Fermilab (Ref. 26).

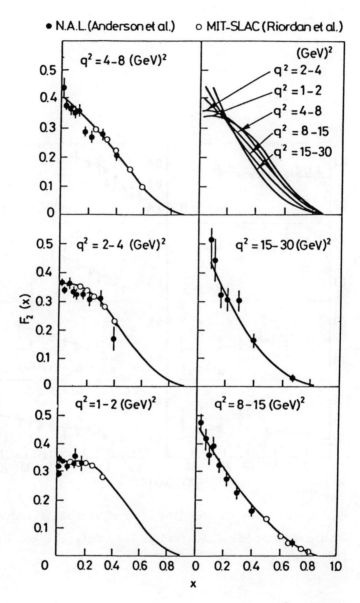

Fig. 17 Proton νW_2 versus x for various fixed Q^2. MIT-SLAC data are from Ref. 19; NAL data are from Ref. 27.

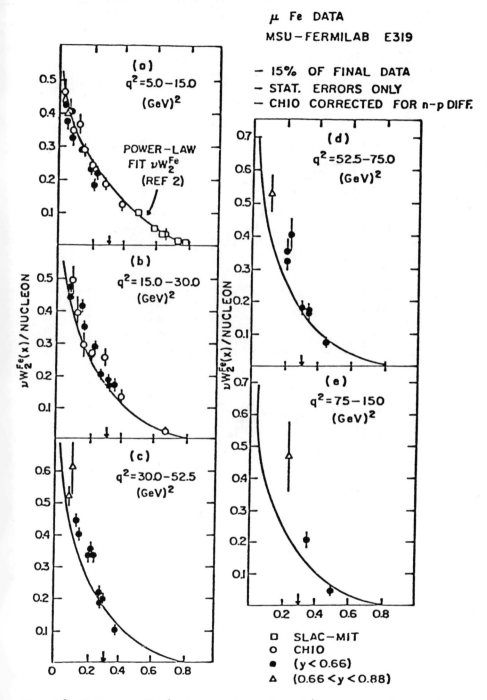

Fig. 18 Nucleon νW_2 (using an iron target) versus x for various fixed Q^2 (Ref. 28).

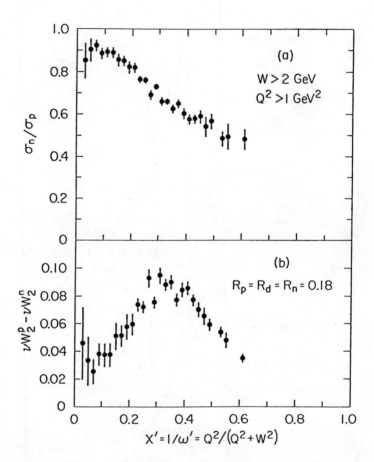

Fig. 19 Comparison of neutron and proton deep inelastic structure functions (Ref. 29).

section has contributions only where x equals one of the x_i. This, however, is an unrealistic consequence of our naive assumption that the constituents could be taken to be at rest. The effect of target motion is to smear the elastic kinematics; Q^2 is no longer equal to $2m_i\nu$, and the x spectrum of each constituent spreads out. Actually, if we scatter from a constituent with momentum k_{Li} in the direction of the virtual photon and transverse momentum k_{Ti} (in the rest frame of the initial nucleon), then x becomes

$$x = (\varepsilon_i - k_{Li})/M, \text{ where } \varepsilon_i^2 = m_i^2 + k_{Li}^2 + k_{Ti}^2 \quad (14)$$

(neglecting binding energy). Different x values in the scattered electron spectrum correspond to different longitudinal momenta of the struck constituents. We now rewrite Eq.(13), replacing the discrete n_i with a distribution $f_i(x)dx$:

$$\frac{d\sigma_{inel}(Q^2,x,\theta)}{dQ^2 dx} \sim \left[\frac{4\pi\alpha^2}{Q^4}\frac{E'}{E}\cos^2\frac{\theta}{2}\right]\sum_{\text{all } i} f_i(x)q_i^2 F_i^2(Q^2). \quad (15)$$

Comparing this with Eq.(10) and neglecting the $\tan^2\frac{\theta}{2}$ term, we get the relation

$$\nu W_2(x,Q^2) = x\sum_i f_i(x)q_i^2 F_i^2(Q^2). \quad (16)$$

The $f_i(x)$ are related to the momentum distributions of the constituents in the nucleon and hence to the unknown bound-state wave function.

Since the electron scattering from a constituent is proportional to its charge squared (Eq.(16)), we can hope to learn something about the charges of the constituents from the experimental data. In order to avoid the necessity of knowing the $f_i(x)$ distributions, we consider integrals over x:

1) the number integral, $\int \sum f_i(x) dx = N = 3$ (17)

   (the number of constituents, taken from elastic data);

2) the charge integral, $\int \sum f_i(x) q_i \, dx = \begin{array}{l} 1 \text{ for proton} \\ 0 \text{ for neutron;} \end{array}$ (18)

3) the energy integral, $\int x \sum f_i(x) dx = 1$ (19)

   (follows from $\sum \varepsilon_i = M$ and $\sum k_{Li} = 0$);

4) the data integral,

$$\int \nu W_2 \, dx = \int x \sum f_i(x) q_i^2 F_i^2(Q^2) dx = \begin{array}{l} .19 \text{ for proton}^{26} \\ .13 \text{ for neutron}^{29}. \end{array}$$ (20)

In order to eliminate the f_i I will make a few plausible assumptions. They are almost certainly not strictly true, but they should be close enough to the truth so as not to affect my main conclusions. First, I assume that the $f_i(x)$ distribution for each constituent is the same, i.e., independent of i. I fix Q^2 somewhere in the 2 to 8 GeV^2 range, and assume that the form factors $F_i(Q^2)$ are simply constants, the same for each constituent i. Finally, I take the neutron to be identical to the proton, except that one constituent, say i = 3, has a charge q_3-1 instead of q_3. With these assumptions the integrals (17)-(20) lead to three algebraic equation for the four unknown quantities q_1, q_2, q_3, and F^2:

$$q_1 + q_2 + q_3 = 1 \quad , \quad (21)$$

$$\frac{q_1^2 + q_2^2 + q_3^2}{3} F^2 = .19 \quad , \quad (22)$$

$$\frac{q_1^2 + q_2^2 + (q_3-1)^2}{3} F^2 = .13 \; . \quad (23)$$

Without more information we cannot find a unique solution, but we can plot the q_1, q_2, q_3 values against the value of F^2 assumed (see Fig. 20). The only solution with integral charges is at

EVIDENCE FOR CONSTITUENTS IN THE NUCLEON

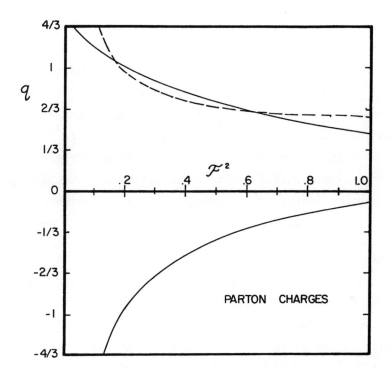

Fig. 20 The solutions of equations (21)-(23) for the proton constituent charges q_1, q_2, and q_3 (dashed) as a function of the assumed constituent form factor squared. For the neutron q_3 is replaced by $q_3 - 1$.

$F^2 = 0.2$, implying q_i = +1, -1, +1 for the proton, and +1, -1, 0 for the neutron. Another solution worth noting is at $F^2 = 0.6$: 2/3, -1/3, 2/3 for the proton and 2/3, -1/3, -1/3 for the neutron, the standard quark model. Of course, other nonintegral solutions are allowed.

The non-scaling behavior of νW_2 at high Q^2 comes from the constituent form factors $F(Q^2)$ in this picture. Looking at Fig. 16, the behavior of $\nu W_2(x,Q^2)$ at high Q^2 and $x > 0.2$ (avoiding the peripheral region) suggests a logarithmic decrease in $F(Q^2)$. This would be expected if the constituents did not act as bare point charges but were each surrounded by a cloud of virtual quanta of the binding field, which share the energy of the constituents and can be "radiated" in the collision. In fact, the behavior of $F(Q^2)$ implied by Fig. 16 is quite reminiscent of the effect of electromagnetic radiative corrections (Eq.(5)). One argument for the standard quark charge solution instead of the integral charge solution (Fig. 20) is that the former implied an F^2 closer to $F = 1$, as one would expect for an almost Q^2-independent (and therefore almost pointlike) form factor.[31] This is not proof, however.

The motivation for basing our analysis on the integral of νW_2 over x (Eq.(20)), was to minimize the effect of the low x region which, as we have seen, has more to do with the virtual vector meson components of the photon than the constituents of the nucleon.[32] Another way of avoiding this kind of background, without suppressing entirely the information contained in the x dependence of νW_2, is to look at the difference $\nu W_2^p - \nu W_2^n$ (Fig. 19). If the vector meson couplings are charge symmetric, their contributions should cancel, leaving an x spectrum more or less proportional to $xf(x)$. Since in the rest frame of the nucleon we can associate xM with $\varepsilon - k_L$ (Eq.(14)) the data then tell us the energy and momentum distribution of the constituents in the nucleon. The peak of the x distribution should occur at $<\varepsilon>/M$. The fact that it occurs at $x = 1/3$ (see Fig. 19) confirms our supposition that there are three constituents. The spread in x tells us the r.m.s. value of k_L and hence of the consti-

EVIDENCE FOR CONSTITUENTS IN THE NUCLEON

tuent momentum k. We get

$$k_{rms} \approx 280 \text{ MeV}. \qquad (24)$$

The other nucleon structure function also contains useful information. $R(x,Q^2)$ is obtained as the ratio of the slope to the intercept on the vertical axis in plots like that of Fig. 10, and is therefore fraught with all of the experimental difficulties which made the linearity test so difficult. The data for R (Fig. 21) have large errors and have changed in the course of time.

R is actually the ratio σ_L/σ_T of the absorption cross sections for longitudinally and transversely polarized virtual photons. If we for the moment assume that the nucleon constituents have no transverse momentum, we can use the conservation of the longitudinal component of angular momentum to conclude that R = 0 if the constituents have spin 1/2 and R = ∞ if they have spin 0. Transverse momentum introduces orbital angular momentum and therefore makes the R prediction less extreme. It turns out[33] that for spin 1/2, neglecting binding energy, one expects

$$R = 4 \frac{\langle k_T^2 \rangle + m^2}{Q^2}, \qquad (25)$$

which is about $0.6 \text{ GeV}^2/Q^2$ (using Eq.(24)). The average measured R, around 0.3, certainly favors spin 1/2 constituents, although there is not much more one can say from the data. The predicted $1/Q^2$ behavior is not confirmed and the magnitude prediction seems too small. Here I am afraid we have to hope for better data.

A more direct way to look for spin effects is to scatter longitudinally polarized electrons from longitudinally polarized protons. To see what we might expect, suppose we have a nucleon consisting of three spin 1/2 constituents. Let two of the constituents, called "a", have identical but unknown charges, and let the third, called "b", have a different unknown charge. A wave function

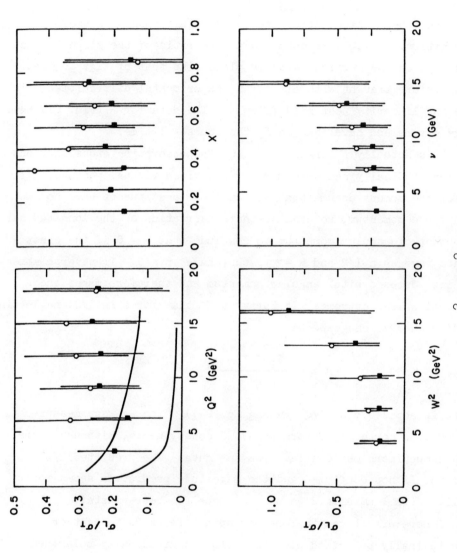

Fig. 21 Data for $R = \sigma_L/\sigma_T$ plotted against Q^2, x, W^2, and ν – compiled in Ref. 11. Open circles use data from Refs. 11 and 18; black squares include data from Ref. 19, as well.

EVIDENCE FOR CONSTITUENTS IN THE NUCLEON

for the nucleon in a pure spin state would be

$$|N\uparrow\rangle = (1/\sqrt{18})[2|a\uparrow b\downarrow a\uparrow\rangle + 2|a\uparrow a\uparrow b\downarrow\rangle + 2|b\downarrow a\uparrow a\uparrow\rangle$$
$$- |a\uparrow a\downarrow b\uparrow\rangle - |a\uparrow b\uparrow a\downarrow\rangle - |a\downarrow b\uparrow a\uparrow\rangle$$
$$- |b\uparrow a\downarrow a\uparrow\rangle - |b\uparrow a\uparrow a\downarrow\rangle - |a\downarrow a\uparrow b\uparrow\rangle]. \qquad (26)$$

High energy electrons do not change their helicity in scattering, so that in order that angular momentum be conserved, only electrons and constituents with opposite spins can scatter. From this we can derive the ep or en polarization asymmetry in terms of the constituent charges q_a and q_b:

$$A = \frac{\sigma_{\rightarrow\leftarrow} - \sigma_{\rightarrow\rightarrow}}{\sigma_{\rightarrow\leftarrow} + \sigma_{\rightarrow\rightarrow}} = \frac{4q_a^2 - q_b^2}{6q_a^2 + 3q_b^2} . \qquad (27)$$

We then have the following predictions:

For q_i = 2/3, 2/3, -1/3   A=5/9   (standard quark model proton)
         2/3, -1/3, -1/3   0   (standard quark model neutron)
         1, 1, -1   1/3   (integral charge solution for proton)
         1, 0, -1   ?   (integral charge solution for neutron)
         1, 0, 0   -1/3   (another model for the proton)

We now have experimental data[34] for the proton asymmetry A as a function of x (Fig. 22). If we again avoid low x, it is clear that the data favor A = 5/9 and thus the standard quark charge solution of Eqs.(21-23).

We can get even better confirmation of this from the nucleon anomalous magnetic moments, the Q^2 = 0 limits of the elastic magnetic form factors. Taking the same a, a, b model for the nucleon with the same spin wave function, and writing the nucleon magnetic moment as the sum of the Dirac moments of the constituents, we have

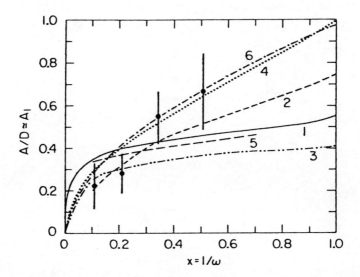

Fig. 22 The longitudinal polarization asymmetry in ep inelastic scattering, plotted against x (Ref. 34). The curves represent variations of the quark parton model.

$$\mu_N = \sum_i \frac{e\hbar}{2m_i c} q_i (2s_{zi}) = \frac{e\hbar}{2Mc} \frac{4q_a - q_b}{3} \frac{M}{m} . \qquad (28)$$

We therefore predict the following moments:

For q_i = 2/3, 2/3, -1/3 $\mu_p/(e\hbar/2Mc)$ = M/m ≈ 3

        2/3, -1/3, -1/3 μ_n "   = -2/3(M/m) ≈ -2

        1, 1, -1 μ_p "   = 5/3(M/m) ≈ 5

        1, 0, -1 μ_n "   = ?

        1, 0, 0 μ_p "   = -1/3(M/m) ≈ -1

The experimental values, $\mu_p/(e\hbar/2Mc)$ = 2.7928456 and $\mu_n/(e\hbar/2Mc)$ = -1.9130421 clearly establish the standard quark charge solution. In fact, the predicted ratio μ_n/μ_p = -2/3, independent of the effective quark mass m, is extremely close to the experimental ratio -0.685.

HADRON ELECTROPRODUCTION

As everyone knows, we don't see fractionally charged quarks knocked out of the nucleon. There must be some kind of final state interaction, involving the creation of new quark-antiquark pairs, which dresses the struck quark and the spectator diquark so that only the familiar hadrons appear. This second stage of the electroproduction process tends to mask what happens in the first state, and makes it difficult to infer from the final hadrons the identity and properties of the struck constituent of the target nucleon. It is not entirely hopeless, however. Martin and Osborne[35] have devised a clever scheme for extracting information from the four inclusive pion electroproduction processes,

$$\begin{aligned} ep &\to e\pi^+ \ldots \\ ep &\to e\pi^- \ldots \\ en &\to e\pi^+ \ldots \\ en &\to e\pi^- \ldots \end{aligned} \qquad (29)$$

Suppose again that the proton consists of two kinds of constituents, two of type "a" and one of type "b", and that the neutron has two "b" and one "a". Assume that the $f_a(x)$ and $f_b(x)$ for the proton are the same as the $f_b(x)$ and $f_a(x)$ for the neutron, respectively. The yield of π^+ produced from the proton into the forward hemisphere of the final-state hadron center-of-mass system is assumed to be given by an expression of the form

$$\pi_p^+ = f_a(x) q_a^2 D_a^+(z) + f_b(x) q_b^2 D_b^+(z), \qquad (30)$$

where $D_a^+(z)$, for instance, represents the probablity density for the struck "a" constituent to fragment into a π^+, as a function of whatever final state kinematic variables z you wish to use. Similar expressions describe the yields of the other three processes (29). The fragmentation functions are assumed to satisfy charge symmetry relations $D_a^\pm = D_b^\pm$.

The analysis[35] proceeds as follows. You form four sum and difference combinations of the four forward pion yields (as in Eq.(30)):

$$\begin{aligned}
A &= (\pi_p^+ + \pi_p^-) + (\pi_n^+ + \pi_n^-), \\
B &= (\pi_p^+ + \pi_p^-) - (\pi_n^+ + \pi_n^-), \\
C &= (\pi_p^+ - \pi_p^-) + (\pi_n^+ - \pi_n^-), \\
D &= (\pi_p^+ - \pi_p^-) - (\pi_n^+ - \pi_n^-).
\end{aligned} \qquad (31)$$

You then define three ratios of these quantities, which neatly separate the factors in Eq.(30):

$$r_1 = \frac{BC}{AD} = \frac{(q_a^2 - q_b^2)^2}{(q_a^2 + q_b^2)^2},$$

$$r_2 = \frac{BD}{AC} = \frac{(f_a(x) - f_b(x))^2}{(f_a(x) + f_b(x))^2} \qquad \text{(the f's apply to the proton)},$$

$$r_3 = \frac{CD}{AB} = \frac{D_a^+(z) + D_b^-(z)}{D_a^+(z) - D_b^-(z)} . \qquad (32)$$

From r_1 we can get the ratio $|q_a|/|q_b|$, called R_1 in Fig. 23. The data turn out to be independent of x (for x > 0.1) and consistent with the value 2 expected from the quark model. From r_2 we can solve for the ratio of majority quark to minority quark x distribution functions $f_a(x)/f_b(x)$, called u_v/d_v in Fig. 23. Above x = 0.1 the data are consistent (though barely) with the ratio 2 expected from the quark model. Finally, ratio r_3 tells us about the ratio of pion fragmentation functions averaged over the forward hemisphere, $\eta = <D_a^+>/<D_a^->$ = $<D_b^->/<D_b^+>$. This doesn't tell us anything about the constituents of nucleon, but the fact that η is experimentally independent of x confirms our two-step factorization of the electroproduction process. The conclusion of this analysis is that the inclusive forward pion electroproduction data demonstrate that the nucleon constituents indeed have charges 2/3, 2/3, -1/3 for the proton and 2/3, -1/3, -1/3 for the neutron.

Another way of relating the charges of the forward hadrons to the charge of the struck quark has been worked out by a DESY-Cornell collaboration.[36] This analysis has the advantage that one does not need to use data from neutron targets, but in order to apply it we have to make some assumption about what happens in the fragmentation process. The picture we use is as follows. The struck constituent moves away from the remaining two, dragging its binding field behind it. Eventually the potential energy in the field is high enough to allow the creation of a quark-antiquark pair. As the original struck quark and the created quarks keep moving, more quark pairs are created from the field until much of the original

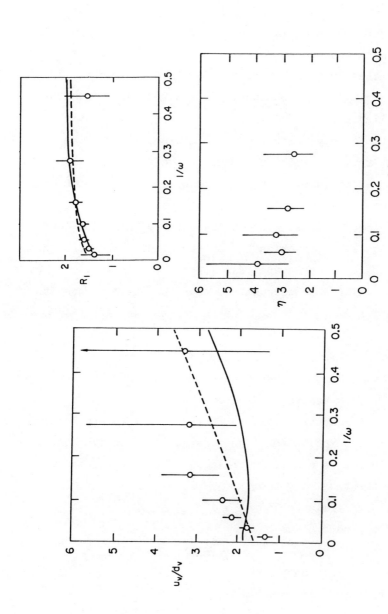

Fig. 23 The ratio $f_a(x)/f_b(x)$ of majority and minority quark distribution functions in the proton (called u_v/d_v here), the ratio of quark charges $R_1 = |q_a|/|q_b|$, and the ratio $\eta = \langle D_a^+ \rangle / \langle D_a^- \rangle$ of favored and unfavored fragmentation functions averaged over the forward hemisphere, all plotted against $x = 1/\omega$ (Ref. 35).

EVIDENCE FOR CONSTITUENTS IN THE NUCLEON

kinematic energy of the struck quark is converted to hadrons. If we now make some cut in longitudinal momentum and sum the charge of all hadrons forward of the cut, we will get a contribution q_i from the leading struck quark, with a probability weighted by q_i^2, another contribution from the trailing antiquark which we will assume has equal probability of being either kind of nucleon constituent (neglecting the creation of heavier quarks not present in the nucleon). All other quark charges between the leading and trailing ends cancel in pairs.

Again using q_a and q_b to denote the charges of the majority and minority constituents in the nucleon, we get the following expression for the average charge of the forward hadrons.

$$<c> = \frac{2q_a^3 + q_b^3}{2q_a^2 + q_b^2} - 1/2(q_a + q_b), \qquad (33)$$

and thus the following predictions:

For q_i = 2/3, 2/3, -1/3 $<c>$ = 7/18 = 0.39

   2/3, -1/3, -1/3   1/6

   1, 1, -1   1/3

   1, 0, -1   ?

   1, 0, 0   1/2

Fig. 24 shows measurements[36] of $<c>$ with a proton target. Above $x = 0.1$ the data are consistent with $<c> = 7/18$ as expected from the standard quark model.

CONCLUSIONS

Direct evidence from electron and muon scattering experiments alone have led us to the following conclusions about nucleon structure.
1) The nucleon is made up of three constituent particles.
2) These constituents are essentially pointlike, with slight departures from pointlike behavior rather similar to

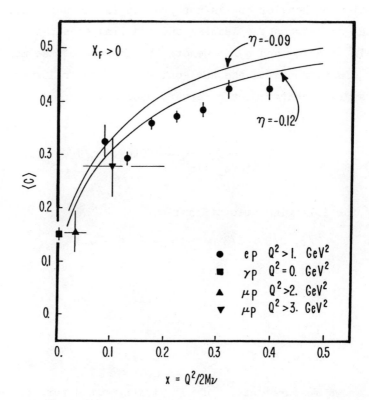

Fig. 24 Average charge per event of electroproduced hadrons, in the forward hemisphere of the hadron center-of-mass frame, plotted against x (see Ref. 36).

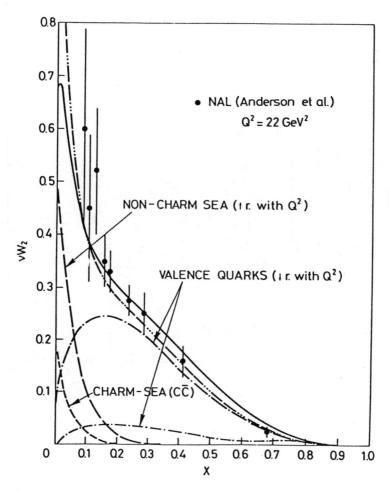

Fig. 25 Muon-proton scattering data (Ref. 27) compared with QCD fits (Ref. 38).

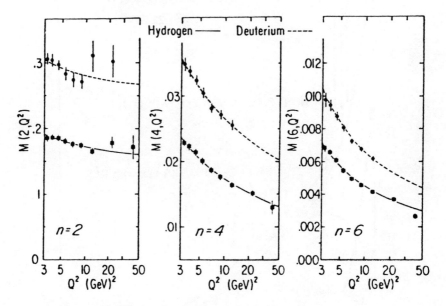

Fig. 26 Q^2 dependence of moments of $\nu W_2(x, Q^2)$ for the proton and for a deuterium target (Ref. 39). Curves are a best fit of the form predicted by QCD.

radiative corrections.

3) The charges of the constituents (in the usual units) are 2/3, 2/3, -1/3 in the proton and 2/3, -1/3, -1/3 in the neutron.[37]
4) The constituents have spin 1/2.
5) The root-mean-square momentum of the constituents in the nucleus is about 280 MeV.

These conclusions are in complete agreement with the quark model.

Over the years the once lowly quark model has developed into a sophisticated theory, quantum chromodynamics, involving the quarks and the colored gauge gluons which bind them. Fig. 25 is a sample comparison of the experimental νW_2 at a high Q^2 and a fit based on QCD. Fig. 26 shows a fit[39] to several moments of νW_2. From the fit one extracts the running coupling constant of QCD,

$$\alpha_s(Q^2) = \frac{12\pi}{33-2f} \frac{1}{\ln Q^2/\Lambda^2} \quad (f = \text{no. of quark flavors}), \tag{34}$$

with the result $\Lambda = 600 \pm 80$ MeV.

And of course the same ideas that have been so successful in interpreting the electron scattering data have also become the basis of our understanding of all high energy phenomena.

NOTES

1. R. Hofstadter, Rev. Mod. Phys. **28**, 214 (1956).
2. In the approximation where the nucleon is treated as a static potential; see J. Schwinger, Phys. Rev. **76**, 760 (1949). See also L.W. Mo and Y.S. Tsai, Rev. Mod. Phys. **41**, 205 (1969) for more accurate formulas.
3. W. Bartel et al., DESY preprint 71/47. See also W. Bartel et al., Nucl. Phys. **B58**, 429 (1973).

4. P.N. Kirk et al., Phys. Rev. D $\underline{8}$, 63 (1973). SLAC data.
5. K.W. Chen et al., Phys. Rev. Letters $\underline{11}$, 561 (1963). CEA data.
6. K. Berkelman et al., Phys. Rev. $\underline{130}$, 2061 (1963). Cornell data.
7. J.G. Rutherglen, Proceedings of the Fourth International Symposium on Electron and Photon Interactions at High Energies, Liverpool, 1969.
8. R. Felst, DESY preprint 73/56. See also W. Bartel et al., Nucl. Phys. $\underline{B58}$, 429 (1973).
9. S.J. Brodsky, G.R. Farrar, Phys. Rev. Letters $\underline{31}$, 1153 (1973) and Phys. Rev. D $\underline{11}$, 1309 (1975).
10. V. Matveev, R. Muradyan, A. Tavkhelidze, Lettere al Nuovo Cimento $\underline{7}$, 714 (1973).
11. M. Mestayer, SLAC report 214 (1978).
12. C.E. Dick and J.W. Motz, Phys. Rev. $\underline{171}$, 75 (1968).
13. W. Bartel et al., Phys. Letters $\underline{28B}$, 148 (1968).
14. E.D. Bloom et al., Proceedings of the Kiev Conference on High Energy Physics, 1970.
15. D.E. Andrews, Ph.D. thesis, Cornell University (1972).
16. Provided we ignore the effect of longitudinally polarized virtual photons.
17. E.D. Bloom et al., Phys. Rev. Letters $\underline{23}$, 930 (1969).
18. W.B. Atwood et al., Phys. Letters $\underline{64B}$, 479 (1976).
19. E.M. Riordan et al., Phys. Letters $\underline{52B}$, 249 (1974).
20. D.L. Fancher et al., Phys. Rev. Letters $\underline{37}$, 1323 (1976) and
21. L.S. Rochester et al., Phys. Rev. Letters $\underline{36}$, 1284 (1976).
22. K.W. Chen, Proceedings of the E.P.S. International Conference on High Energy Physics, Palermo, 1965.
23. M. Breidenbach et al., Phys. Rev. Letters $\underline{23}$, 935 (1969).
24. G. Miller et al., Phys. Rev. D $\underline{5}$, 528 (1972).
25. S. Stein et al., Phys. Rev. D $\underline{12}$, 1884 (1975).
26. B.A. Gordon et al., Phys. Rev. Letters $\underline{41}$, 615 (1978).
27. H.L. Anderson et al., Phys. Rev. Letters $\underline{38}$, 1450 (1977).
28. K.W. Chen et al., Michigan State preprint MSU-CSL-40 (1977).

29. J.S. Poucher et al., Phys. Rev. Letters $\underline{32}$, 118 (1974).
30. In the conventional parton model (R.P. Feynman, Photon-Hadron Interactions, Benjamin, p. 132) the nucleon is viewed in the infinite momentum frame. Although this leads to a more precise and unambiguous interpretation of the x variable (i.e., fractional parton momentum), I prefer for the present qualitative discussion to treat the nucleon in its rest frame.
31. A constituent form factor $F^2 = 0.6$ in my picture is equivalent to the statement in the conventional parton language that 60% of the nucleon momentum is carried by charged partons (quarks) and 40% by neutrals (gluons).
32. The wee partons or sea quarks of the conventional parton model are the peripheral vector meson effects in my picture. What I call constituents of the nucleon will turn out to be the valence quarks of the standard model.
33. R.P. Feynman, Photon-Hadron Interactions, Benjamin, p. 139.
34. M.J. Alguard et al., Phys. Rev. Letters $\underline{41}$, 70 (1978).
35. J.F. Martin and L.S. Osborne, Phys. Rev. Letters $\underline{38}$, 1193 (1977).
36. G. Drews et al., Phys. Rev. Letters $\underline{41}$, 1433 (1978) and R. Ericson (private communication).
37. The experimental data do not distinguish between fractionally charged quarks and integrally charged quarks which have time-averaged fractional charges.
38. A.J. Buras and K.J. Gaemers, Nucl. Phys. $\underline{B132}$, 249 (1978).
39. H.L. Anderson et al., Phys. Rev. Letters $\underline{40}$, 1061 (1978).

PHYSICS AFTER τ AND T

A. Pais

Rockefeller University

New York, New York 10021

The unification program for weak and electromagnetic interactions is in a relatively quiescent stage at this time. We have reached a curious state of affairs. The standard model, the SU(2) × U(1) gauge theory with four quarks, four leptons and one complex Higgs doublet is very successful insofar as the phenomenology of these quarks and leptons is concerned. At the same time it is obvious that the standard model is incomplete. We must incorporate at least one new charged lepton, τ, and its neutrino and at least one new quark. This is the bottom quark b with charge $Q = -1/3$. The existence of this fifth quark is strongly suggested[1] by the discovery of the T.

The fourth quark (c) had a reason to be there; it had a role to play in the suppression of strangeness-changing neutral transitions s ↔ d by means of the GIM mechanism.[2] The four-quark model has another welcome property: the suppression of c ↔ u transitions. It is much less clear what the existence of further quarks is telling us.

It is generally agreed that the next step in the developments must be such that the phenomenological implications of the standard model can be incorporated without substantial modifications. I shall refer to this requirement as "the standard constraint." It is not all that difficult to implement this constraint. A very simple way of doing this is to retain the $SU(2) \times U(1)$ gauge group and the single Higgs doublet but to introduce more than two left-handed quark doublets and, correspondingly, more right-handed quark singlets and to proceed likewise for leptons. As is well known, this can be done in such a way that the theory remains anomaly-free.

In such a theory the left-handed quark representations can be written as

$$\begin{pmatrix} U_i \\ D_i \end{pmatrix}_L, \quad D_i = \alpha_{ij} d_j, \quad \alpha^+ \alpha = 1 \tag{1}$$

Here $u_i = u, c, t, \ldots$ are the physical "up" states with $Q = 2/3$ and $d_i = d, s, b, \ldots$ are the physical "down" states with $Q = -1/3$. The structure of the unitary mixing matrix α is determined by the Yukawa couplings of the quarks and Higgs fields. One Higgs doublet suffices to unfold the quark spectrum. Regarding flavor changing neutral processes (FCNP) such a theory has the property that <u>all</u> FCNP are suppressed at essentially the level of the GIM suppressions for $s \leftrightarrow d$ and $c \leftrightarrow u$ in the standard model. This suppression holds not only in the gauge sector but also in the Higgs sector. Indeed, this theory fulfills trivially the criterion of Glashow and Weinberg:[3] if all fields with the same conserved quantum numbers and chirality occupy equivalent positions within equivalent representations of the gauge group then, in a natural way, FCNP of any kind are absent to $O(G)$ and suppressed in $O(G\alpha)$. Theories of this kind have two other important properties. First the presence of the b quark implies the existence of the top quark t. Secondly, for the case of more than two doublets, the matrix α is parametrized by angles <u>and</u> by non-trivial phases.[4] The latter allow for the possibility of natural CP violation in the gauge sector. These naturally occurring phases are not calculable. This CP option is interesting but not compelling.

Numerous extensions of the standard model to larger gauge groups have been studied in which again all FCNP are suppressed. (In attempts to calculate the Cabibbo angle in SU(2) × U(1) theories, or for other gauge groups, the Higgs sector is often enlarged. The impact of this enlargement on FCNP is not always considered.) Are there other ways to implement the standard constraint? Howard Georgi and I have found that the answer is yes.[5] We have discovered a new class of theories with signatures quite distinct from the generalized standard model described by Eq.(1). These differences show up already at the level of the particle content: these theories need not contain a t-quark. There are other qualitative differences, manifest to leading order in the Fermi constant G. Namely, these theories embody the suppression of s ↔ d and c ↔ u to the same level and with the same naturalness as in GIM. However, it is not true that all FCNP are equally supressed. In particular there are FCNP involving the b with amplitudes O(G). Before I indicate our line of reasoning, I wish to stress that we do not advocate these new models as being in any known way superior to models of the type Eq.(1). However, it does seem timely to point out the existence of hitherto unforeseen options for implementing the standard constraint.

The simplest example is the gauge group SU(3) × U(1) with charge operator $Q = T_3 + T_8/\sqrt{3} + S$ and with a representation content which exists exclusively of triplets and singlets. The theory contains two left triplets with S = 0:

$$\Psi_{iL} = \begin{pmatrix} U_i \\ D_i \\ B_i \end{pmatrix}_L , \quad i = 1,2 \qquad (2)$$

so that $Q = 2/3$ for U_i; $Q = -1/3$ for D_i, B_i. The U; D, B are linear combinations of mass eigenstates u,c; d,s,b,ℓ. ℓ is a new quark with $Q = -1/3$. The right-handed quark states are singlets with appropriate S-values. Since the D_i and B_i have the same Q, the Glashow-Weinberg criterion does not hold. The particle mixing has become two-dimensional. Not only is there "horizontal mixing" between D_1, D_2 and between B_1, B_2, much as in Eq.(1), but in general there is also vertical mixing between the D's and the B's:

$$\begin{pmatrix} D_1 \\ D_2 \\ B_1 \\ B_2 \end{pmatrix}_L = Y \begin{pmatrix} d \\ s \\ b \\ \ell \end{pmatrix}_L , \qquad (3)$$

where Y is in general an arbitrary 4×4 unitary matrix.

Imagine a superstrong first stage spontaneous symmetry breakdown from $SU(3) \times U(1)$ to the standard model $SU(2) \times U(1)$ gauge group such that the (U_i, D_i) are doublets and the B_i are singlets with respect to this $SU(2)$. Five vector bosons are massive at this stage. Their masses are supposed to be so large compared to the W- and Z-masses generated in the second stage breakdown $SU(2) \times U(1) \to U(1)$ that all but the effects of W, Z and photon exchange may be neglected. It has been shown in GP[5] that this can be achieved in such a way that the Z-boson is coupled to the current operator $T_3 - Q\sin^2\theta$, where θ is the usual Weinberg angle. The contributions of the ψ_{iL} to the W-current and to the Z-current are proportional to

$$\sum_i \overline{U}_i D_i \qquad (4)$$

and to

$$\sum_i \{(\overline{U}_i U_i - \overline{D}_i D_i) - \sin^2\theta \, (2\overline{U}_i U_i - \overline{D}_i D_i - \overline{B}_i B_i)/3\} \qquad (5)$$

respectively. (The shorthand $\overline{ab} \equiv \overline{a}\gamma_\mu(1 + \gamma_5)b/2$ has been used). In their generality, the expressions (4) and (5) spell disaster. The neutral current (5) will generate transitions $s \leftrightarrow d$ to $O(G)$ and that charged current (4) will generate $s \leftrightarrow d$ and $c \leftrightarrow u$ to $O(G)$ -unless precautions are taken which strongly restrict the form of Y in Eq.(3). These restrictions must be induced by the structure of the quark couplings to Higgs fields.

Even though we are dealing here with an effective $SU(2) \times U(1)$ theory, these Yukawa couplings must be $SU(3) \times U(1)$ invariant. Further constraints are necessary, however. We proceed as follows. The Higgs fields are taken to be three $S = 1/3$ triplets ϕ, ϕ_D, ϕ_B and one $S = -2/3$ triplet ϕ_U. The most general VEV's consistent with electromagnetic gauge invariance are

$$<\phi> = \begin{pmatrix} 0 \\ 0 \\ \lambda \end{pmatrix}, \quad <\phi_D> = \begin{pmatrix} 0 \\ \delta_1 \\ \delta_2 \end{pmatrix}, \quad <\phi_B> = \begin{pmatrix} 0 \\ \beta_1 \\ \beta_2 \end{pmatrix}, \quad <\phi_U> = \begin{pmatrix} \alpha \\ 0 \\ 0 \end{pmatrix}.$$

(6)

As is shown in GP,[5] the hierarchy $SU(3) \times U(1) \to SU(2) \times U(1) \to U(1)$ is established by taking $\lambda \gg$ the magnitudes of all other VEV's. The Yukawa couplings are taken to be

$$\sum_{i,j=1,2} \{A_{ij} \bar{\Psi}_{iL} \phi_U U_{jR} + C_{ij} (\bar{D}_{iR} \phi_D^\dagger + \bar{B}_{iR} \phi_B^\dagger) \Psi_{jL}\} + h.c. \qquad (7)$$

where $(D_1, D_2, B_1, B_2)_R \equiv (d, s, b, \ell)_R$. These are, of course, not the most general $SU(3) \times U(1)$ invariant couplings insofar as the $Q = -1/3$ particles are concerned. However, the expression (7) is strictly renormalizable all the same since it is the most general form compatible with the three discrete symmetries (which need not apply to the dimension two terms in the Higgs Lagrangian!)

$$\phi \to -\phi \; ; \quad \phi_D \to -\phi_D \, , \quad D_{iR} \to -D_{iR} \; ; \quad D_R \leftrightarrow B_R \, , \quad \phi_D \leftrightarrow \phi_B.$$

(8)

The couplings for $Q = 2/3$ states are as usual. Those for the $Q = -1/3$ states take the very special form given in Eq.(7), however, as a result of (8). It is this special form which prevents the potential disasters signaled earlier.

The $Q = -1/3$ mass terms are contained in the expression

$$(\bar{d}, \bar{s}, \bar{b}, \bar{\ell}) X \begin{pmatrix} D_1 \\ D_2 \\ B_1 \\ B_2 \end{pmatrix}_L + \text{h.c.} \qquad (9)$$

where the 4×4 matrix X is given by

$$X = \begin{vmatrix} \delta_1 C_{11}, & \delta_1 C_{12}, & \delta_2 C_{11}, & \delta_2 C_{12} \\ \delta_1 C_{21}, & \delta_1 C_{22}, & \delta_2 C_{21}, & \delta_2 C_{22} \\ \beta_1 C_{11}, & \beta_1 C_{12}, & \beta_2 C_{11}, & \beta_2 C_{22} \\ \beta_1 C_{21}, & \beta_2 C_{22}, & \beta_2 C_{21}, & \beta_2 C_{12} \end{vmatrix}$$

$$= \begin{vmatrix} C_{11} & C_{12} \\ C_{21} & C_{22} \end{vmatrix} \otimes \begin{vmatrix} \delta_1, & \delta_2 \\ \beta_1, & \beta_2 \end{vmatrix}. \qquad (10)$$

That is, X is a tensor product of a 2×2 matrix in the horizontal space (the first factor) and a 2×2 matrix in the vertical space (the second factor). The matrix Y in Eq.(3) must therefore factor correspondingly:

$$Y = \begin{vmatrix} c_1, & s_1 \\ -s_1, & c_1 \end{vmatrix} \otimes \begin{vmatrix} c_2, & s_2 \\ -s_2, & c_2 \end{vmatrix}, \qquad (11)$$

where $c_i = \cos \theta_i$, $s_i = \sin \theta_i$. Thus

$$\begin{vmatrix} C_{11} & C_{12} \\ C_{21} & C_{22} \end{vmatrix} = \begin{vmatrix} a & \cdot \\ \cdot & b \end{vmatrix} \cdot \begin{vmatrix} c_1, & -s_1 \\ s_1, & c_1 \end{vmatrix}, \qquad (12)$$

$$\begin{vmatrix} \delta_1 & \delta_2 \\ \beta_1 & \beta_2 \end{vmatrix} = \begin{vmatrix} c & \cdot \\ \cdot & d \end{vmatrix} \cdot \begin{vmatrix} c_2, & -s_2 \\ s_2, & c_2 \end{vmatrix}, \qquad (13)$$

and we have the mass relation

$$\frac{m_d}{m_s} = \frac{m_b}{m_\ell} \qquad (14)$$

which determines "which is which" regarding the b and ℓ states. ℓ is very heavy!

The special structure of the Yukawa couplings has reduced the possible number of angles on which Y depends from six to two. All phases in Y are removable, as in the standard model. Thus, CP violation is not generated in the gauge sector (but can be generated in the Higgs sector[6]).

It follows from Eqs.(2), (3) and (11) that the two triplets can be written as

$$\begin{bmatrix} u \\ c_1(c_2 d + s_2 b) + s_1(c_2 s + s_2 \ell) \\ c_1(c_2 b - s_2 d) + s_1(c_2 \ell - s_2 s) \end{bmatrix}_L, \begin{bmatrix} c \\ c_1(c_2 s + s_2 \ell) - s_1(c_2 d + s_2 b) \\ c_1(c_2 \ell - s_2 s) - s_1(c_2 b - s_2 d) \end{bmatrix}_L. \qquad (15)$$

Recall that the top two components are the doublets with respect to the effective $SU(2) \times U(1)$ group. We can now at once read off a number of implications of the model.

(I) There are no t-quarks.

(II) Flavor changes. Of the seven possible flavor changes, five are naturally suppressed in the gauge sector. These five include $s \leftrightarrow d$ and $c \leftrightarrow u$. We have therefore generalized the GIM mechanism. The two non-suppressed flavor changes are $b \leftrightarrow d$ and $s \leftrightarrow \ell$. The neutral current coupled to Z contains $\bar{b}d$, $\bar{\ell}s$ terms and their conjugates, see Eq.(5). Since the Higgs fields contain a number of real Higgs mesons, we must also ask what flavor changes are induced by these mesons. One finds that the situation is precisely the same as for the gauge bosons, all but $b \leftrightarrow d$ and $s \leftrightarrow \ell$ are naturally suppressed (see GP,[5] Eq.(2.21)). No constraints on

Higgs masses are involved. We have therefore generalized the Glashow Weinberg theorem.

(III) The amplitudes for $b \to u + W^-$, $b \to c + W^-$, $b \to d + Z$ are proportional to $c_1 s_2$, $s_1 s_2$, $c_2 s_2$ respectively. Thus $b \to u$ dominates over $b \to c$, independently of the magnitude of θ_2. $b \to u + W^-$ is of the same general order as $b \to d + Z$.

(IV) The chains $d \to b \to s$ and $d \to \ell \to s$ are forbidden. Therefore, the $\overline{K}K$ system remains unaffected by the presence of the large $d \leftrightarrow b$ and $s \leftrightarrow \ell$ mixings.

(V) There is strong mixing $\overline{b}d \leftrightarrow \overline{b}d$.

(VI) If the charged current contains the usual lepton terms, then it is of the form

$$\overline{u}c_2(c_1 d + s_1 s) + \overline{c}c_2(c_1 s - s_1 d) + \overline{\nu}_e e + \overline{\nu}_\mu \mu + \ldots \quad (16)$$

(I return below to the realization of this assumption). Then θ_1 is the Cabibbo angle, insofar as it is defined by the ratio of semileptonic d- and s-decays. Note also that the relative rates for non-leptonic charm decays are the same as in the standard model. Eq.(16) implies that θ_2 is constrained by the demand of Cabibbo universality Using the more generous errors allowed by the analyses of Shrock and Wang[7] one finds[5] that

$$s_2^2 \leq 0.014 \quad , \quad \theta_2 \leq 6.7° \quad (17)$$

With the help of Eq.(17) the lifetime of the b-quark is estimated[5] to be $\geq 10^{-13}$ sec.

(VII) Since $D° - \overline{D}°$ mixing is not observed, the $\Delta C = 2$ mass mixing must be smaller than the decay rate for these mesons. Since the chain $c \to \ell \to u$ is allowed and since ℓ is so massive, we must as if this mass mixing can be kept sufficiently small. With the help o Eqs.(14) and (17) it was found that this condition can be met.[5] Thu not only $\Delta S = 2$ but also $\Delta C = 2$ effects are naturally suppressed to requisite levels.

(VIII) Assuming that also the neutral current has the conventional lepton contribution one can estimate the branching ratios for various b-quark decay modes. The results are as follows.[5]

b-decay mode	Branching ratio (%)
$ud\bar{u}$	33
$u\nu e^-$, $u\nu\mu^-$	12 each
$us\bar{c}$	11
$dd\bar{d}$	10
$ds\bar{s}$	7
$d\nu\bar{\nu}$	7
$d\mu\bar{\mu}$, $de\bar{e}$	2 each
$us\bar{u}$	2

Let us summarize what has been done so far. First, the example of the six quark system in $SU(3) \times U(1)$ discussed above shows that there are new ways of satisfying the standard constraint in all its aspects. The methods described here can also be applied to larger quark contents of $SU(3) \times U(1)$ and to other gauge groups. Secondly, it has been shown that horizontal/vertical particle mixing leads to consequences which differ qualitatively from those obtained from the usual (purely horizontal) mixing insofar as quarks beyond u,d,s,c are concerned. One obvious problem remains to be considered. How can the six quark system treated above be made part of a full-fledged gauge theory? This question has been considered in some detail in GP. I shall conclude with a brief review of the main points, omitting all technical details.

The requirements to be met are: (1) Leptons are to be incorporated. (2) The introduction of any new fermions beyond the six quarks leads to new Yukawa couplings. These should be invariant under the three transformations (8), in order to maintain the quark mixing patterns established above. (3) The theory should be anomaly free. All these requirements can be met without introducing more

Higgs fields than the four triplets employed previously.

One naive way of doing this is to introduce three $S = 1/3$ triplets of right-handed leptons:

$$L_{aR} = \begin{pmatrix} \ell_a \\ \bar{\nu}_a \\ n_a \end{pmatrix}_R , \quad a = 1,2,3, \qquad (18)$$

where $\ell_1 = e^+$, $\ell_2 = \mu^+$, $\ell_3 = \tau^+$. The $\bar{\nu}_a$ are the corresponding antineutrinos, the n_a are massive neutral leptons. The left-handed leptons are SU(3) singlets with $S = Q$. The problem of giving appropriate masses to the leptons (and keep the neutrinos massless) via Higgs couplings which satisfy (8) can easily be solved. The lepton parts of the W- and Z- currents are normal in the sense defined earlier. However, the theory as it stands is not anomaly free. This can be remedied by introducing a third triplet, a $\bar{3}$, of quarks: $(D_3, U_3^1, U_3^2)_L$ with $S = 1/3$. (The corresponding R-states are singlets. In turn, these quarks can be given mass via Higgs couplings which again respect invariance under (8). These new quarks do not mix with the six introduced earlier, so the lightest of them is stable.

An alternative and more sophisticated method does not demand more than six quarks and also opens a way for incorporating the whole scheme in a super-unified gauge group, SU(6). This SU(6) contains $SU(3)_{flavor}$ in much the same way as the well known SU(5) theory[9] contains the $SU(2)_{flavor}$ of the standard model. The lepton content is as follows. Four 3's of right-handed fields:

$$L_{iR}^x = \begin{pmatrix} \ell_i^x \\ \nu_i^x \\ \nu_i^{-x} \end{pmatrix} , \quad x,i = 1,2; \qquad (19)$$

two $\bar{3}$'s of left-handed fields:

$$\Lambda_{iL} = (M_i^0, \ell_i^2, -\ell_i^1)_L , \qquad (20)$$

and two right-handed singlets M_{iR}^{o}. The four fields ℓ_{iR}^{x} correspond to e^+, μ^+, τ^+ and a new lepton T^+. The M_i^o are massive neutral leptons, the ν_i^x and $\nu_i^{\prime x}$ are neutrinos. All of the fermions, leptons and quarks, now carry a "horizontal index," $i = 1,2$. Thus, there are two "families" of fermions with the same gauge structure and it turns out that each family is anomaly free, (cf. GP[5] Section III). The lepton-Higgs couplings can again be made to obey (8). In addition we subject these couplings to a global symmetry which prevents the transitions μ or $T \leftrightarrow e$ or τ but not the transitions $\tau \leftrightarrow e$, $T \leftrightarrow \mu$.

Both lepton versions are in accord with the consequences (I) - (VIII). The second version[10] has two additional features. 1) There is a small branching ratio ($<6.10^{-4}$) for $\tau^- \to e^+e^-e^-$. 2) The relation (14) is extended, lepton and quark masses are related:

$$\frac{m_b}{m_d} = \frac{m_\ell}{m_s} = \frac{m_\tau}{m_e} = \frac{m_T}{m_\mu}. \tag{21}$$

The T is very heavy! It may be a flaw of this scheme that $m_\tau/m_e = m_b/m_d$ implies $m_d \simeq 1.3$ Mev. while current algebra suggests $m_d \simeq 7.5$ Mev.

The purpose of the work reported here was to show that the standard constraint does not necessarily imply the natural suppression of all flavor changing neutral processes.[11] It remains to be seen whether experiment will demand any such non-trivial generalizations of the GIM mechanism as were discussed here.

REFERENCES

1. S.W. Herb et al. Phys. Rev. Lett. <u>39</u>, 252 (1977); Ch. Berger et al. Phys. Lett. <u>76B</u>, 243 (1978); C. Darden et al. Phys. Lett. <u>76B</u>, 246 (1978).

2. S. Glashow, J. Iliopoulos and L. Maiani, Phys. Rev. $\underline{D2}$, 1285 (1970)
3. S.L. Glashow and S. Weinberg, Phys. Rev. $\underline{D15}$, 1958 (1977).
4. M. Kobayashi and T. Maskawa, Progr. Theor. Phys. $\underline{49}$, 652 (1973)
5. H. Georgi and A. Pais, "Generalization of GIM-horizontal and vertical flavor mixing," Harvard-Rockefeller Preprint No. HUTP-78/A059, COO-2232B-168, referred to in what follows as GP. The talk I gave in Coral Gables contained a number of technical details which I can omit from the present written version since I can refer the reader to the preprint just quoted. The title of the present paper has been borrowed from GP.
6. S. Weinberg, Phys. Rev. Lett. $\underline{37}$, 657 (1976); P. Sikivie, Phys Lett. $\underline{75B}$, 141 (1976).
7. R. Shrock and L.L. Wang, Phys. Rev. Lett. $\underline{41}$, 1692 (1978).
8. GP, Appendix B.
9. H. Georgi and S.L. Glashow, Phys. Rev. Lett. $\underline{32}$, 438 (1974).
10. For the embedding of this version in SU(6) see GP, Appendix A.
11. For other attempts in this direction see F. Gürsey, P. Ramond and P. Sikivie, Phys. Rev. $\underline{D12}$, 2166 (1975); Y. Achiman and B. Stech, Heidelberg preprint, HD-THEP-78-20 (1978).

PROTONS ARE NOT FOREVER

D. V. Nanopoulos*

Harvard University

Cambridge, Massachusetts 02138

PREFACE

The aim of this talk is to convince people that grand unified theories give a very nice and plausible explanation of a whole lot of different and at first sight unrelated phenomena, and they definitely have the merit and right to be taken seriously. A detailed analysis of many of the things that I mention here may be found in Ref. 1, which the interested reader may consult. I'll try to argue in simple words that a grand synthesis (unification) of strong-electromagnetic and weak interactions beyond aesthetic appeal and conceptual satisfaction may be a necessity if for example we want to understand charge quantization. Such a picture has made many successful predictions[1] such as $\sin^2\theta_W \simeq 0.20$, $m_s \simeq 0.5$ GeV and $m_b \simeq 5$ GeV and for the first time it is possible to explain the absence of appreciable accumulations of antimatter in the universe, and directly and successfully estimate the baryon to photon ratio in the universe in terms of observable parameters of elementary particle physics. Then a dramatic and unavoidable consequence is the instability of the proton,

---

*Research supported in part by the National Science Foundation under Grant Number PHY77-22864.

with a lifetime not far above present lower bounds. Since the calculations of the proton lifetime have the usual theoretical uncertainties, we would like to urge experimentalists to improve measurements on the proton lifetime, because we believe that the decay of the proton will be the most dramatic confirmation of a gran synthesis of all elementary particle forces.

I. INTRODUCTION

By now, it is a rather common belief that gauge theories of weak, electromagnetic and strong interactions are the right way to describe nature. Fortunately such a belief not only comes from the beauty and aesthetic appeal of gauge theories, but it also has quite a substantial experimental support.

Before getting to the main part of this talk which concerns grand unified theories, I would like to review the highlights of recent experimental data concerning the structure of weak, electromagnetic and strong interactions at present energies. As we will see, the simplest of all weak models, the standard SU(2) × U(1) Weinberg-Salam-GIM model[2] is in great shape, and we may even consider it as the theory of weak and electromagnetic interactions at least at present energies, as of course we may call quantum chromodynamics[3] the fundamental theory of strong interactions.

Then armed with this information, I will try to present an example of a grand synthesis of weak, electromagnetic and strong interactions, which beyond conceptual satisfaction predicts[1] the correct value of $\sin^2\theta_W \simeq 0.20$ as measured by recent experiments,[4] the correct masses for s and b quarks[6,7] $m_s \simeq 0.4-0.5$ GeV; $m_b \simeq$ 5 GeV, and of course explains the quantization of electric charge (the fact that the proton and the positron have the same charge).

II. WHY THE SU(2) × U(1) W-S-GIM MODEL AND QCD SEEM TO BE THE RIG CHOICES

    A. <u>SU(2) × U(1) - W-S-GIM</u>[2]

In this model all left-handed fermions come in SU(2) doublets and all right-handed *massive* fermions come in singlets. For simplicity let us define as a generation the following set:

$$\text{generation} \equiv \left\{ \begin{pmatrix} q_A \\ q_C \end{pmatrix}_L ; \quad q_{A_R} , \quad q_{C_R} \\ \begin{pmatrix} \nu_\ell \\ \ell^- \end{pmatrix}_L ; \quad \ell^-_R \right. \tag{1}$$

where $q_A = u,c,t,\ldots;$ $\quad q_C = d,s,b,\ldots$

$\nu_\ell = \nu_e, \nu_\mu, \nu_\tau, \ldots;$ $\quad \ell^- = e^-, \mu^-, \tau^-, \ldots$

Up to now we have "discovered" three generations. Now let us notice the following things:

1. Cancellation of Adler anomalies between quarks and leptons in each generation.
2. "Natural"[8] explanation of the absence of $\Delta F = 1$ ($F \equiv$ any flavor) neutral currents.
3. "Natural"[9] CP violation[10,11,12] consistent[12,13] with experiment. (Six quarks are the minimum number needed[10] for such an explanation[10,11,12] of CP violation.)

All these properties plus the beautiful agreement with experiment (see below) make this model almost unique in describing the weak and electromagnetic interactions at present energies (and maybe for our generation).

There are though some points that may be annoying and need more thought. Namely:

1. There seems to be someone missing from each generation, namely there is no ν_{ℓ_R}, which of course expresses the fact that the neutrino is massless.
2. There is no principle to tell us *a priori* how many generations exist.

3. There is no relation between quark and lepton masses as well as between quark and lepton charges.
4. The Weinberg angle (θ_W) is an arbitrary parameter that has to be taken from experiment.
5. Because of the existence of the U(1) factor, the electric charge is not quantized.

I have spelled out explicitly all these points in order to appreciate more the predictive fertility of a grand synthesis of all interactions.

Experimentally the "standard" model[2] is very healthy. All data in neutrino neutral current physics[14] are in agreement with the model: deep inelastic neutrino production, elastic $\nu(\bar{\nu})$-p scattering, exclusive and inclusive neutrino-pion production, purely leptonic $\nu(\bar{\nu})$-e scattering. The apparent disagreement between theory and experiment[15] on ν_μ-e scattering seems to be going away and new experimental data[5] support the "standard" model. In neutrino charge current physics the absence of any y-anomaly supports the left-handedness of the "standard" model while combined with Lederman's[6] $T(|Q| = 1/3)$[7] implies the existence of a third quark doublet, which is what is exactly needed for a natural explanation[10,11,12] of CP violation. At the same time the existence of a third quark doublet implies through the demand of cancellation of Adler anomalies (renormalizability!) the existence of a third lepton doublet. But it exists. It is Perl's[16] τ-lepton plus its accompanied neutrino (ν_τ). Lately, thanks to intense experimental effort, one of the big problems of tau-physics, namely the absence of $\tau \to \nu\pi$, has been resolved. It has been seen[17,18] and with the correct, predicted branching ratio (DELCO finds a B.R. = 0.080 ± 0.03 and PLUTO 0.087 ± 0.02; theory predicts 9%). In addition the Michel ρ-parameter has been measured ($\rho = 0.66 \pm 0.13$) and agrees with the V-A form of the τ-ν_τ coupling (the value of ρ for a V-A coupling after radiative corrections is 0.64). A new limit of $m_{\nu_\tau} < 0.25$ GeV has been found[18] and a more accurate determination of τ-mass has been achieved[18,19] (DELCO finds

$m_\tau = 1782^{+2}_{-7}$ MeV and DESY 1787^{+10}_{-18} MeV). Thus, it looks reasonable to assume that a third lepton doublet which is left-handed and with $m_{\nu_\tau} \simeq 0$ has been discovered. The most dramatic confirmation of the "standard" model came very recently from the observation[4] of parity violation in longitudinally polarized electron-deuteron (proton) scattering in a beautiful experiment[4] at SLAC. The predicted[20] asymmetry agrees both in sign and magnitude with experiment (for both targets deuteron and proton) for $\sin^2\theta_W \simeq 0.20$.

$$A_{deut} \equiv \frac{\sigma_R - \sigma_L}{\sigma_R + \sigma_L} = (-9.5 \pm 1.6)10^{-5} Q^2 (GeV^2) .$$

Notice that this asymmetry is very sensitive to $\sin^2\theta_W$, and maybe is the most accurate way to determine it. Also the aforementioned value of $\sin^2\theta_W$ is in the range determined[14] from neutrino physics. It is worth noticing that the most recent ν_μ-e scattering experiment[5] finds also $\sin^2\theta_W \simeq 0.20$, and that purely leptonic processes are the best laboratory to determine model parameters, since they are free from the usual hadronic-vertex assumptions, parton model, ... Parity violation in atomic physics seems to be experimentally contradictory. Different groups[21] have different results and the situation seems very cloudy. I hope that after the good news[4] from e-p(d) scattering the situation will be clearer.

I'll take as a working hypothesis that we have seen three fermion left-handed doublets (plus necessary singlets), which are described successfully by the "standard" model with $\sin^2\theta_W \simeq 0.20$. We notice in passing that a determination of the ratio

$$K \equiv \frac{M_W^2}{M_Z^2 \cos^2\theta_W}$$

is crucial,[22] since it depends on the Higgs content of the theory. Using recent experimental results one finds $K \simeq 0.98 \pm 0.05$, which is in beautiful agreement with the $I = \frac{1}{2}$ Higgs rule.[23]

B. <u>QCD</u>[3]

Here the situation is well known, and I will not repeat all the advantages of QCD. I would like to stress once more the remarkable agreement between theory[3] [SU(3) color as gauge group mediated by 8 *vector* gluons] and recent data on scaling violations in neutrino production from two groups[24] at CERN. The predicted[25] anomalous dimensions and the ones extracted from the data[24] literally coincide for the case of three colors and *vector* gluons. The arena of QCD applications[26] is big enough and the predictions are successful[25] when confronted with experiment,[24] that it does not take special courage, particularly these days, to say that QCD is the fundamental theory of strong interactions.

III. GRAND SYNTHESIS

After the successful attempt to unify weak and electromagnetic interactions via the gauge group SU(2) × U(1) and after the remarkable success of QCD as the theory of strong interactions, it is obvious and perhaps imperative to try to unify weak, electromagnetic and strong interactions via a gauge group. Perhaps this is a naive and simplistic attitude but let us see if it works.

Now let us see how we may choose the group that unifies all interactions (except gravity for now!).
(i) The gauge group has to be either simple or else admit a discret symmetry, so that it has a unique coupling constant which implies immediately *electric charge quantization* so that, if the scheme is successful, one of the big mysteries of almost a century of physics, i.e. why the proton and the electron have the same electric charge (in absolute value!) is resolved trivially. Also, having a unique coupling constant is what one may call *true unification*.
(ii) The gauge group has to contain $SU(3)_c$ × SU(2) × U(1), i.e., has to have at least rank 4. Let us concentrate for now on the groups of rank 4 that have just one coupling constant. There are 9

of them:

$[SU(2)]^4$, $[SO(5)]^2$, $[SU(3)]^2$, $[G_2]^2$, $SO(8)$, $SO(9)$, $S_p(8)$, F_4 and $SU(5)$.

The first two are out because they do not contain an SU(3) subgroup. The $SU(3)_c \times SU(2)$ representation content of a generation (1) is:

$$(3,2) + 2(\bar{3},1) + (1,2) + (1,1) \quad . \tag{2}$$

Now notice that the representation (2) is complex and only $[SU(3)]^2$ and $SU(5)$ among all the 9 groups admit complex representations. But the choice $[SU(3)]^2$ is impossible because leptons and quarks, having different color properties, must be in different representations, the electromagnetic charge is a generator of the weak SU(3) which does not allow fractional charges, and the sum of quark charge would have to be zero. Therefore the only possible rank 4 group is $SU(5)$.[27] (iii) Following the principle of economy and simplicity let us choose SU(5) as our gauge group that unifies everything, and anyway emerged *uniquely*[27] from all rank 4 groups. Let us see now, how a generation redistributes itself under SU(5). We observe that

$$5 = (3,1) + (1,2)$$
$$10 = (\bar{3},1) + (1,1) + (3,2) \tag{3}$$

where 5 is the fundamental representation of SU(5), 10 is the antisymmetric part of the product of two 5, $(5 \times 5 = 15 + 10)$, and we have exposed their $SU(3)_c \times SU(2)$ representation content. But a $\bar{5} + 10$ (3) is identical to (2), i.e., has exactly the $SU(3)_{col} \times SU(2)$ content that we are looking for! So we get[27] (R, Y, B are $SU(3)_c$ indices):

$$\bar{5}:\begin{pmatrix}\bar{d}_R\\\bar{d}_Y\\\bar{d}_B\\e^-\\\nu_e\end{pmatrix}_L \qquad 10:\frac{1}{\sqrt{2}}\begin{pmatrix}0 & \bar{u}_B & -\bar{u}_Y & -u_R & -d_R\\-\bar{u}_B & 0 & \bar{u}_R & -u_Y & -d_Y\\\bar{u}_Y & -\bar{u}_R & 0 & -u_B & -d_B\\u_R & u_Y & u_B & 0 & -e^+\\d_R & d_Y & d_B & e^+ & 0\end{pmatrix}_L \qquad (4)$$

Obviously for each generation we have a different $\bar{5} + 10$ representation. We make now the following comments:

(1) There are no Adler anomalies. The anomalies of the irreducible representations (4) ($\bar{5} + 10$) cancel. Actually, SU(5) is the only group of *any rank* with a ($\bar{5} + 10$) = 15 anomaly free representation with the correct $SU(3)_c \times SU(2) \times U(1)$ content! This is a very important statement, because it means that if nature chooses the generation (1) as the basic fundamental unit and all new quarks and new leptons follow the same pattern (new generations of the same origin) then SU(5) is *unique*.

(2) It is important that the neutrino is massless since otherwise we would have 16 fields and one could not put them on a $\bar{5} + 10$, which in turn means that we have either to go to higher representations which would imply a strange proliferation of particles, or to use peculiar SU(5) singlets. Both alternatives are very ugly and it seems satisfactory that Nature chooses the most economic way of the model, namely make the neutrino massless and use only $\bar{5} + 10$. In that way we get an algebraic insight into the first mystery of the $SU(2) \times U(1)$ model, i.e., why is someone missing from each generation!

(3) There is another gratifying observation: the electric charge operator Q has to be a generator of SU(5) [because of charge

quantization] and thus Q is traceless (Tr Q=0). This fact has the following consequences:

(i) Take the 5; then $(\text{Tr } Q)_5 = 0$ implies

$$-3Q_d + Q_{e^-} = 0 , \quad \text{i.e.} \quad \begin{aligned} Q_d &= \tfrac{1}{3} Q_{e^-} \\ Q_u &= \tfrac{2}{3} Q_{e^+} \end{aligned} \quad (5)$$

and in general if the color group was SU(n) then

$$Q_d = \frac{1}{n} Q_{e^-} ; \quad Q_u = \frac{n-1}{n} Q_{e^+} . \quad (6)$$

Thus there is a remarkable relation (I have not used at all the Gell-Mann-Nishijima formula) between the charge of quarks and the number of colors.

(ii) Since $(\text{Tr } Q)_{\text{quarks}} \neq 0$, and we need a traceless charge operator, it is a necessity to put quarks and leptons in the same irreducible representation. But then, we definitely create new kinds of exotic interactions (lepto-quark or di-quark currents), which in general violate (B) baryon and (L) lepton number conservation and *they make the proton unstable!*

(iii) The tracelessness of charge operator is essential in the cancellation of anomalies, i.e. keeping the theory renormalizable. So in a grand unified theory we find an interesting interrelation between charge quantization, having an anomaly-free theory (renormalizability), violation of baryon and lepton number (proton decay) and the fractional charge of the quarks. Next we come to the phenomenological consequences of the Georgi-Glashow SU(5) model.[27]

IV. PHENOMENOLOGY OF THE SU(5) MODEL

It is clear that at present energies the world does not seem to be described by the SU(5) gauge group. That means that SU(5) is badly broken at present and eventually will become an exact symmetry at very, very high energies. We may envision the breaking of SU(5) at two stages:

$$SU(5) \underset{24}{\rightarrow} SU(3)_c \times SU(2) \times U(1) \underset{5}{\rightarrow} SU(3)_c \times U(1) \tag{7}$$

where the first stage happens through a 24 of Higgs [transforming like the adjoint representation of SU(5)], and the second one through a 5 of Higgs [transforming like the fundamental representation of SU(5)]. At the first stage twelve gauge bosons get masses and twelve (8 gluons, 3 weak gauge bosons and the photon) remain massless. At the next stage the 3 weak gauge bosons get masses and so we are left with 8 massless gluons plus a massless photon, as we should. As we will see, since the "new" 12 gauge bosons mediate bizarre interactions that may lead to proton decay (disintegration of the world!), we better make them very heavy ($\sim 10^{16}$ GeV) which, in turn, means that the vacuum expectation value (v.e.v.) of the 24 has to be "astronomically" higher than the v.e.v. of the 5 which corresponds to usual value of ~ 250 GeV. Fortunately this turns out to be possible (more later). By looking at the representations (4) it is obvious that the only way that fermions can get masses is through the $\bar{5} \times 10$ and 10×10 coupling. Since $\bar{5} \times 10 = 45 + 5$ and $10 \times 10 = \overline{45} + \bar{5} + 50$, the Higgs multiplet(s) should belong to the I.R. {5,45} of SU(5). Now how do we choose which is (are) the relevant multiplet(s)? First, by invoking again the principle of economy and simplicity, we may choose the 5. Also notice that we need anyway a 5 of Higgs for the second stage of symmetry breaking and if we have it why do we not use it? [This corresponds to the successful use (see Section II) of a doublet of Higgs in the SU(2) × U(1) model to break the symmetry down to the U(1) and at the same time to give masses to the three weak gauge bosons and to fermions.] But there is yet another interesting reason. This involves the generalization[28] of natural suppression of flavor changing neutral currents to grand unified theories. In our case this principle tells us that all fermions get their masses from one Higgs multiplet. Then putting everything together, we have the obvious choice of a 5 of Higgs to give masses to everybody (except of course the superheavy gauge bosons, who get their masses through the 24 of Higgs).

In such a case it is easy to see the following things happening:

(1) From the $(\bar{5})_f \times (5)_{Higgs} \times (10)_f$ coupling only the q_c and ℓ^- get masses and actually they are equal, i.e.,

$$m_{q_c} = m_{\ell^-} \quad . \tag{8}$$

Had we used the 45 of Higgs then we would get[29]

$$m_{q_c} = 1/3\, m_{\ell^-} \quad . \tag{9}$$

(2) From the $(10)_f \times (\bar{5})_{Higgs} \times (10)_f$ coupling only the q_A get masses. This is rather unfortunate because we cannot get relations between q_A and q_c (or ℓ^-) masses. So Eq. (8) remains intact and the masses of $Q = 2/3$ quarks remain arbitrary parameters together with various Cabibbo angles.

It should be stressed that the above mass relations, as well as the equality of all coupling constants, are true at super-high energies, where SU(5) is supposed to be an exact symmetry. Then one may ask what is the connection between measurable (at present energies) quantities, like coupling constants and quark masses and the corresponding quantities at super-high energies. One even may worry about the whole approach since the strong (QCD) and weak/E.M. coupling constants are very different

$$\alpha_s(Q^2 = 10\text{ GeV}^2) \simeq 0.2\text{-}0.3 \quad , \tag{10}$$
$$\alpha = \frac{1}{137} \quad .$$

Also at present energies the $Q = -1/3$ quarks (q_c) are much heavier than charged leptons (ℓ^-). Fortunately there is a simple answer to all these problems. The renormalization group assures us that coupling constants[30] and quark masses[31] vary with the q^2 at which they are probed, and that asymptotic freedom suggests that the strong interaction coupling[30] decreases as q^2 increases; simultaneously a U(1) coupling increases slightly with q^2. The same is true for the masses.[31] So, maybe the dissimilar values (10) can be brought together at some sufficient high q^2 which would correspond to the M_X^2 of superheavy gauge bosons X which unify the strong, weak and electromagnetic interactions. To get from the grand unified mass (GUM) M_X to present energies μ, it is necessary to know how the SU(3), SU(2) and U(1) coupling constants (g_3, g_2, g_1) vary, assuming always $g_i \ll 1$, we get[30,31]

$$\frac{1}{4\pi \cdot \alpha_{GUM}} \quad \frac{1}{g^2_{GUM}} = \frac{1}{g_i^2(\mu)} + 2\beta_i \ \ell n \left(\frac{\mu}{M_X}\right) \qquad (11)$$

$$(i=1,2,3)$$

where the β_i are defined by

$$\mu \frac{\partial}{\partial \mu} g_i(\mu) \simeq \beta_i \ g_i^3(\mu) + O(g_i^5) \qquad (12)$$

and if we neglect Higgs contributions and finite-mass effects, then

$$\beta_1 = \frac{f}{24\pi^2}, \quad \beta_2 = \frac{-1}{16\pi^2}[\frac{22}{3} - \frac{2}{3} f], \quad \beta_3 = \frac{-1}{16\pi^2}[11 - \frac{2}{3} f] \qquad (13)$$

where f is the number of quark (lepton) flavors. (f=6 is favorable in this model, see below.) It is easy then to show that[30,31]

$$\ell n \left(\frac{M_X}{\mu}\right) = \frac{\pi}{11\alpha}[1 - \frac{\alpha}{\alpha_s(\mu)} \cdot \frac{8}{3}] \qquad (14)$$

and

$$\sin^2 \theta_W(\mu) = \frac{1}{6} + \frac{5}{9} \frac{\alpha}{\alpha_s(\mu)} \qquad (15)$$

(We use the notation $\alpha_i \equiv \frac{g_i^2}{4\pi}$, $i=1,2,3$ and $\alpha_3 = \alpha_s$.). At high energies, where SU(5) is exact, it is simple group theory that tells us[27]

$$\sin^2\theta_W = \frac{3}{8} \, . \tag{16}$$

Not only can we estimate coupling constant renormalization,[30] but also mass renormalization[1] using the anomalous dimension[31] of the fermion mass operator,

$$\gamma_m = \frac{-g_3^2}{2\pi^2} + \frac{3g_2^2}{64\pi^2} - \frac{3}{8\pi^2} g_1^2 \tag{17}$$

we find[1]

$$\ln\left[\frac{m_q(\mu)}{m_\ell(\mu)}\right] = \ln\left[\frac{m_q(M_X)}{m_\ell(M_X)}\right] + \frac{4}{11-\frac{2}{3}f} \ln\left[\frac{\alpha_s(\mu)}{\alpha_{GUM}}\right] + \frac{3}{2f} \ln\left[\frac{\alpha_1(\mu)}{\alpha_{GUM}}\right] \, . \tag{18}$$

So we have to determine M_X, $\sin^2\theta_W(\mu)$ and $m_q(\mu)$ by using (14), (15) and (18). As an input we need $\alpha_s(\mu)$ and α [as given from (10)], the muon (0.105 GeV) and τ-lepton (1.8 GeV) masses, and the value of $\frac{m_q(M_X)}{m_\ell(M_X)}$, which we take to be equal to one, as indicated by (8). So we find[1]

for
$$\left.\begin{array}{l}\alpha_s(\mu^2=10\text{ GeV}^2) = 0.19-0.32 \\ m_\mu = 0.105 \text{ GeV} \\ m_\tau = 1.8 \text{ GeV} \\ m_q(M_X) = m_\ell(M_X) \\ f = 6\end{array}\right\} \rightarrow \left\{\begin{array}{l}\alpha_{GUM} = 0.022 \\ M_X = (3-6)10^{14} \text{ GeV} \\ \sin^2\theta_W = 0.20-0.21 \\ m_s = (0.38-0.50) \text{ GeV} \\ m_b = (4.8-5.6) \text{ GeV}\end{array}\right. \tag{19}$$

I have over-simplified things and avoided detailed analysis of all phenomena that should be taken into account before arriving at

(19). The interested reader should consult Ref. 1 for a more careful exposition. The results of (19) are really impressive, they agree *completely* with experiment, both for $\sin^2\theta_W$,[4,5,14] and for quark masses[6,7] (see Section II).

A few things concerning the results (19) are worth mentioning:

(1) It is an obvious thing to say that if I change the input then necessarily the output will be changed. So it is remarkable that α_s ($\mu^2 = 10$ GeV^2), as determined by quarkonium physics or scaling violations in lepto-production, is in the right range as to give an output consistent with experiment. The same of course is true for the masses. The message is that m_s and m_b have to be 0.5 and 5 GeV because m_μ and m_τ happen to be 0.105 and 1.8 GeV respectively, or vice-versa!

(2) It is crucial that we chose[1] the 5 instead of the 45 of Higgs to give masses to fermions. If we had chosen (9) as a boundary condition instead of (8) then we could predict the mass of the ϕ at 350 MeV and the mass of the Y at 3.5 GeV, which is clearly nonsense. So Nature seems to prefer an economic, simplistic and "natural" way to express herself, i.e. the use of a 5 of Higgs to give masses to fermions and to usual gauge bosons.

(3) We see from (18) that the quark masses depend on the number of flavors, f. Up to now we have always used f=6, but we may change f and see how the masses change. We find[1] that m_b increases by 10% (30%) if f becomes 8 (10) and so if we believe this picture, more perhaps than we should, we may conclude[1] that *there are only 6 flavors*. Recent calculations[32] have shown that higher-order corrections (i.e. including two loops in β and γ-functions) to quark masses tend to increase them. Consequently, it strengthens the conclusion of Ref. 1, that six flavors is the maximal number of flavors allowed in the SU(5) model. For details see Ref. 32. So despite the fact that we do not have a fundamental reason of how many generation there are, consistency of theory with experiment demands *at mos*

6 flavors. Unfortunately these numbers seem to be rather insensitive to the mass of the t-quark mass, in the SU(5) model. We notice in passing two interesting facts:

a. "Natural" CP violation demands[10,11,12] at least 6 quarks.

b. In another standard model, the big-bang model of the universe which seems to be quite successful[33] (e.g. it predicted correctly the existence and characteristics of the microwave radiation background in the universe, found[34] in 1965, etc.), the amount of (primordial) He in the universe puts limits[35] on the number of massless (two component) stable neutrinos (like the ones used in our picture) that may exist. Interesting enough this number turns out[35] to be 3, which in our language means f=6, just as demanded by our model. We leave the reader to draw his own conclusions on this particular subject.

Finally, I'll discuss the most funny, amusing and perhaps the most fundamental consequences of grand unified theories. As we mentioned in section III, the existence of superheavy bosons (X) mediating bizarre interactions is bound to cause unconventional things happening, which are related with the beginning (absence of antimatter; number of baryons/number of photons) and with the end (proton decay) of the Cosmos.

A. <u>The Beginning ...</u>

A natural property of grand unified theories is that, in general, baryon and lepton numbers are not conserved, of course, in an experimentally acceptable way. One could argue that this is the least disturbing since no one, nowhere, never, understood why we had to have these global U(1) symmetries and also by the fact that the universe is maybe full of black holes, which have no "hair", i.e., they do not conserve baryon number (electric charge, mass, angular momentum and color are the only things that are recognized from these strange objects). On the contrary, having a model with baryon number violation and also "hard" CP violation,[10,11,12] [it comes from couplings in the Lagrangian of dimension 4 which do not disappear at very high energies (temperatures)] it seems conceivable[36,37] that we are able to explain why the universe is dominated by matter as opposed to antimatter and furthermore to evaluate that ratio of photons to baryons in the universe:

$$\frac{N_\gamma}{N_B} \simeq 10^8 - 10^{10} \qquad (20)$$

which is a quantitative estimate of the absence of antimatter from the universe. Recent calculations[37] show that even the quantitative verification of (20) is possible with the standard parameters of the model determined from elementary particle physics (see Eq. 19). If such an explanation turns out to be correct, then beyond its importance to cosmology, it would suggest that a grand synthesis is indeed necessary. Then one may see, in his (her) wild dreams the following picutre emerging. Since matter-antimatter annihilates, existential reasons demand the dominance of one of two kinds, which in turn, within our philosophy, will demand baryon-number violation and CP violation (which by CPT invariance means the necessary T violation to violate reciprocally). Then the most natural way to have baryon-number violation is to mix quarks with leptons and thus we come to grand unified theories and to "hard" CP violation which arises naturally[10,11,12] with 6 quarks. So the answer to the question *"Who needs the muon?"* is simple enough. The muon exists since the τ-lepton had to exist; *otherwise we would not exist*. So, all these wild, wild speculations indicate that cosmology and elementary particle physics have a lot to learn from each other.

B. <u>The End ...</u>

Since baryon and lepton numbers are violated, there is no more reason for the proton to remain stable. Actually, it is an unavoidable and necessary consequence of the whole world picture we envisage here, that the proton is not everlasting. A detailed calculation[1] of the proton lifetime, as mediated by super-heavy gauge bosons, yields

$$\tau_P \simeq \frac{1}{\alpha_{GUM}^2} \cdot \frac{M_X^4}{M_{Proton}^5} \simeq 10^3 \cdot \frac{M_X^4}{M_{Proton}^5} = 2 \cdot 10^{31} \cdot \left(\frac{M_X}{10^{15}}\right)^4 \text{ years} . \quad (21)$$

Using (19) we get[1,38,39,40] from (21)

$$\tau_{Proton} \simeq 10^{31\pm 2} \text{ years} . \quad (22)$$

[Very similar results are also expected for the decay of neutrons bound in atomic nuclei.]

There is a real possibility[38] that the proton lifetime may be even smaller than the one indicated in Eq. (22), if we properly include the contributions of the superheavy Higgs bosons that also mediate proton decay. In the SU(5) model the mass of the superheavy Higgs bosons tends[1] to be smaller than the mass of the superheavy gauge bosons (this property is relevant to the stability of the vacuum) and so despite the smallness of their couplings to fermions their contribution to proton decay may be important. Parenthetically we mention that "light" superheavy Higgs bosons play also an essential role in the explanation[37] of the baryon number in the universe.

A few words now about a very important issue, namely that of the branching ratios of the different decay modes. In the SU(5) model the *main decays are*:[38,40]

Protons

Decay Mode	X is mainly	B.R. (very rough estimate)
$p \to e^+ + X$	$\pi^0; \rho; \omega; \eta; \ldots$	$\sim 80\%$
$p \to \bar{\nu} + X$	$\pi^+; \rho^+; \ldots$	$\sim 15\%$
$p \to \mu^+ + X$	$K^0; \ldots$	$\sim 5\%$
$p \to e^-; \mu^- + X$	$\pi^+\pi^+; \ldots$	Forbidden

Neutrons

Decay Mode	X is mainly	B.R. (very rough estimate)
$n \to e^+ + X$	$\pi^-; \rho^-; \ldots$	$\sim 80\%$
$n \to \bar{\nu} + X$	$\pi^0; \rho^0; \eta; \omega; \ldots$	$\sim 15\%$
$n \to \mu^+ + X$	$\pi^-; \ldots$	$\sim 5\%$
$n \to e^-; \mu^- + X$	$\pi^+; \ldots$	Forbidden

It is important to note that the B.R. for modes involving muons are not very high and this fact should better be kept in mind, especially by those who are planning new detectors to search for proton decay. Please notice, *it is impossible*, despite common belief to the contrary, to put together in the first generation [see (1)]$\{u,d,\nu_\mu,\mu\}$, and so to interchange the roles of e and µ in the tables above, i.e. to create high B.R. to muons, because then using (8) we will get $m_d/m_s = m_\mu/m_e$, clearly nonsense! It is another nice property of the SU(5) model that "light" stuff likes to go with "light" stuff. The present experimental lower bound[41] on the proton lifetime ($\tau_p > 2 \cdot 10^{30}$ years),[41] involves muons and thus, after the discussion we had above it should be compared with ten times Eq. (22)

or
$$\tau^\mu_{Proton} \simeq 10^{32 \pm 2} \text{ years} \qquad (23)$$

It is interesting to notice that the present experimental lower bound on the proton lifetime immediately puts a lower limit on M_X, which belongs in the same range as determined by the demand of equal strong, weak and electromagnetic coupling constants [see Eq. (19)]. That was not really obliged to happen, and easily we could find ourselves in big trouble. Nevertheless, everything seems to work like a Swiss clock! Therefore, since this model seems to make sense by giving very reasonable results, we want to urge experimentalists to

give another try and provide more stringent bounds on the proton
lifetime. Maybe a big surprise is waiting ahead in the very near
future!

Before coming to the conclusions, a few words about the Higgs
mechanism in the SU(5) model. We are[1] the first to have shown that
it is possible to keep the hierarchy of the two symmetry-breaking
scales not only in the tree approximation but also when one-loop
corrections to the potential are taken into account and, at the same
time, that the $I = \frac{1}{2}$ Higgs rule[23] remains intact. It is rather ugly
but at least you can do it. So the alleged obstacle[42] for grand
unified theories concerning the hierarchy of symmetry breaking is
evaded.[1,43]

V. CONCLUSIONS

In this talk I have tried to show that a grand unification of
strong, electromagnetic and weak interactions, despite the fact that
it really happens at super-high energies, has rather interesting
implications at present energies, namely:

(1) The neutrino is massless (two component); since nature chooses
the $15 = 10 + \bar{5}$ reducible representation there is no room for
the other two components of the neutrino.

(2) Despite the fact that a fundamental principle is missing, for
the time being, to tell us how many generations exist, consistency of theory with experiment implies that there are 3
generations, which means 6 quarks or leptons, in accord also
with cosmological bounds[35] on the number of neutrinos.

(3) There is a relation[1] between $Q = -1/3$ quark masses and charged
lepton masses, coming from the fact that both quarks and leptons
belong to the same irreducible representations and actually it
seems to be the correct one ($m_s \simeq 0.5$ GeV; $m_b = 5$ GeV) if we
want to make SU(5) natural,[28] i.e. choose only the 5 of Higgs
to give mass to ordinary particles (not the super-heavies).
There is also a nice relation between the number of colors and

the charge of the quarks [see (5), (6)]. So we understand why the quarks have fractional charges with respect to the electron charge.

(4) The Weinberg angle (θ_W) is successfully predicted[1] to be $\sin^2\theta_W \simeq 0.20$ in accord with recent experimental results.[4,5]

(5) The electric charge is quantized (we understand why $Q_p = Q_{e^+}$).

Notice that these implications of SU(5) model, are exactly what in section II were mentioned as desirable but impossible to find if we stuck inside the framework of the "standard model" [SU(3)$_c$ × SU(2) × U(1)]. There are some weak points that should be mentioned:

a. There is no relation between the Q = 2/3 quark masses and the Q = - 1/3 quark (or lepton) masses; thus the Q = 2/3 quark masses and the different Cabibbo angles are arbitrary. So despite the efforts during this year at Harvard and CERN from groups with common interests and ambitions, it has been impossible so far to pin down the mass of the t-quark.[47]

b. There is perhaps a more serious problem. Equation (8) predicts $m_d/m_s = m_e/m_\mu \simeq 1/200$ instead of $1/20$ (m_d, m_s here refer to "current algebra" masses) that everybody wants.[45] There seem to exist some complications with the very-light quark masses and definitely more thought and work is needed. Still there is hope in the O(10) model[44].

c. The hierarchy of symmetry breaking is possible but ugly. It is interesting though to notice that the super-breaking is of the order of the Planck mass ($M_{P\ell} \simeq 1.2 \times 10^{19}$ GeV $\simeq 2 \times 10^{-5}$ gr) at which, as is well known, quantum effects become important in gravity $\Delta g_{\mu\nu}/g_{\mu\nu} \sim 1$) and we can no longer use a flat Lorentzian spacetime manifold to describe physics. Maybe it is a coincidence or else there is a relation between quantum gravity and the super-breaking which we miss, at present, and thus the hierarchy of symmetry breaking looks so ugly. Actually the relation $M_X \sim \alpha_{GUM}^2 M_{Planck}$ may play a

significant role[37] at the early stages of the universe if
the explanation that was mentioned before about the ratio
of the number of photons to the number of baryons turns
out to make sense.

Recently it has been observed[46] that grand unified theories
improve the motivation for weak symmetry breaking by radiative
corrections, which would imply that[46]

$$m_{Higgs} \sim 10 \text{ GeV} \qquad (24)$$

for $\sin^2\theta_W \sim 0.20$.

A systematic study[46] of the phenomenology of a Higgs boson in
this mass range shows that it may be observable in T decays and pp
collisions, as well as in toponium decays. The significance of
finding a Higgs boson in the 10 GeV mass region can hardly be over-
estimated. Another challenging hunt for experimenters!

Now that it is pretty clear that unified gauge theories of weak-
electromagnetic interactions are the correct approach and QCD is very
healthy, grand unified theories seem to be our next and perhaps
immediate field of interest.

So experimentalists are urged to think again and improve, if
possible, the present limits on the proton lifetime, which is going
to be a crucial test of the whole, beautiful but far from established
idea, that strong, electromagnetic and weak forces arise from a
single fundamental interaction based on a simple gauge group like
SU(5), O(10), etc.

NOTES AND REFERENCES

1. A. Buras, J. Ellis, M. K. Gaillard and D. V. Nanopoulos, Nucl. Phys. **B135**, 66 (1978).
2. S. Weinberg, Phys. Rev. Lett. **19**, 1264 (1967); A. Salam, in Proc. 8th Nobel Symposium, Stockholm, ed. N. Svartholm (Almquist and

Wiksells, Stockholm, 1968) p. 367; S. L. Glashow, J. Iliopoulos and L. Maiani, Phys. Rev. D2, 1285 (1970); S. L. Glashow, Nucl. Phys. 22, 579 (1961).

3. H. Fritzsch, M. Gell-Mann and H. Leutwyler, Phys. Lett. 47B, 365 (1973); S. Weinberg, Phys. Rev. Lett. 31, 494 (1973); D. V. Nanopoulos, Nuovo Cimento Letters 8, 873 (1973).
4. C. Prescott et al., Phys. Lett. 77B, 347 (1978).
5. A. M. Cnops et al., Phys. Rev. Lett. 41, 357 (1978).
6. S. Herb et al., Phys. Rev. Lett. 39, 252 (1977); W. R. Innes et al., Phys. Rev. Lett. 39 1240 (1977); see also J. H. Cobb et al., Phys. Lett. 72B, 273 (1977).
7. Ch. Berger et al., Phys. Lett. 76B, 243 (1978); C. W. Darden et al., Phys. Lett. 76B, 246 (1978).
8. S. L. Glashow and and S. Weinberg, Phys. Rev. D15, 1958 (1977); E. A. Paschos, Phys. Rev. D15, 1966 (1977).
9. B. W. Lee, Phys. Rev. D15, 3394 (1977).
10. M. Kobayashi and K. Maskawa, Prog. Theor. Phys. 49, 652 (1973).
11. L. Mainani, Phys. Lett. 62B, 183 (1976); S. Pakvasa and H. Sugawara, Phys. Rev. D14, 305 (1976).
12. J. Ellis, M. K. Gaillard and D. V. Nanopoulos, Nucl. Phys. B109 213 (1976).
13. J. Ellis, M. K. Gaillard, D. V. Nanopoulos and S. Rudaz, Nucl. Phys. B131, 285 (1977).
14. See for example, L. F. Abbott and R. M. Barnett, Phys. Rev. D18 3214 (1978); J. J. Sakurai, UCLA 78-TEP-9 preprint and references therein; L. M. Sehgal, TH-Aachen preprint (PITHA-78-102) and references therein.
15. P. Alibran et al., Phys. Lett. 74B, 422 (1978).
16. M. L. Perl et al., Phys. Rev. Lett. 35, 1489 (1975).
17. G. Knies, talk given at the Trieste Conference, June 1978.
18. J. Kirkby, SLAC-PUB-2127 (June 1978).
19. W. Bartel et al., DESY 78/24 preprint (May 1978).
20. A. Love, D. V. Nanopoulos and G. G. Ross, Nucl. Phys. B49, 513 (1972); D. V. Nanopoulos, "Tests of Unified Weak and Electro-

magnetic Gauge Theories", D. Phil. thesis, University of Sussex (1973). For a modern, handy version see R. N. Cahn and F. J. Gilman, Phys. Rev. $\underline{D17}$, 1313 (1978) and references therein.

21. L. M. Barkov and M. S. Zolotarev, JETP Lett. $\underline{26}$, 379 (1978). This paper reports parity violation in atomic physics in accord with the W-S model. P. Baird et al., Nature $\underline{264}$, 528 (1976). This paper reports parity "conservation" in atomic physics.

22. V. Barger and D. V. Nanopoulos, Nucl. Phys. $\underline{B124}$, 426 (1977).

23. D. A. Ross and M. Veltman, Nucl. Phys. $\underline{B95}$, 135 (1975).

24. P. C. Bosetti et al., Nucl. Phys. $\underline{B149}$, 1 (1978) (BEBC group) J. G. H. de Groot et al., CERN preprint (1978)(CDHS group).

25. D. Bailin, A. Love and D. V. Nanopoulos, Nuovo Cimento Lett. $\underline{9}$, 501 (1974); H. D. Politzer, Phys. Rpt. $\underline{14C}$, 129 (1974); D. G. Gross and F. Wilczek, Phys. Rev. $\underline{D8}$, 3633 (1973), $\underline{D9}$, 980 (1974); H. Georgi and H. D. Politzer, Phys. Rev. $\underline{D9}$, 416 (1974).

26. J. Ellis, SLAC-PUB-2121 (1978).

27. H. Georgi and S. L. Glashow, Phys. Rev. Lett. $\underline{32}$, 438 (1974).

28. M. S. Chanowitz, J. Ellis and M. K. Gaillard, Nucl. Phys. $\underline{B128}$, 506 (1977).

29. This was also known to J. Ellis and P. Ramond.

30. H. Georgi, H. R. Quinn and S. Weinberg, Phys. Rev. Lett. $\underline{33}$, 451 (1974).

31. D. V. Nanopoulos and G. G. Ross, Phys. Lett. $\underline{B56}$, 279 (1975); H. Georgi and H. D. Politzer, Phys. Rev. $\underline{D14}$, 1829 (1976); H. D. Politzer, Nucl. Phys. $\underline{B117}$, 397 (1976).

32. D. V. Nanopoulos and D. A. Ross, CERN preprint TH-2536 (1978).

33. For a nice exposé, see S. Weinberg in Gravitation and Cosmology (Wiley, New York, 1972) chapter 15.

34. A. A. Penzias and R. W. Wilson, Astrophys. J. $\underline{142}$, 419 (1965).

35. V. F. Shvartsman, JETP Lett. $\underline{9}$, 184 (1969); G. Steigman, D. N. Schramm and J. E. Gunn, Phys. Lett. $\underline{B66}$, 202 (1977); J. Yang, D. N. Schramm, G. Steigman and R. T. Rood, Enrico Fermi Institute preprint 78-26.

36. M. Yoshimura, Phys. Rev. Lett. $\underline{41}$, 381 (1978); S. Dimopoulos and L. Susskind, SLAC-PUB-2126 (1978); D. Toussaint et al., Princeton preprint (1978); D. Toussaint and F. Wilczek, Princeton preprint (1978).

37. J. Ellis, M. K. Gaillard and D. V. Nanopoulos, Phys. Lett. $\underline{80B}$, 360 (1979), and CERN preprint in preparation; S. Weinberg, Harvard preprint HUTP-78/A040 (1978).

38. J. Ellis, M. K. Gaillard, S. L. Glashow and D. V. Nanopoulos, paper in preparation. There a detailed analysis is given of the proton lifetime involving all possible contributions and the different branching ratios of all the principal decay modes are calculated in grand unified models.

39. D. A. Ross, Nucl. Phys. $\underline{B140}$, 1 (1978).

40. C. Jarlskog and F. J. Yudurain, CERN preprint TH-2556 (1978).

41. F. Reines and M. F. Crouch, Phys. Rev. Lett. $\underline{32}$, 493 (1974).

42. E. Gildener, Phys. Rev. $\underline{D14}$, 1667 (1976); K. T. Mahanthappa and D. G. Unger, Phys. Lett. $\underline{78B}$, 604 (1978); K. T. Mahanthappa, M. A. Sher and D. G. Unger, preprint COLO-HEP-8 (1978).

43. I. Bars and M. Serdazoglou, Yale report COO-3075-188 (1978); R. N. Mohaptra and G. Senjanovic, CCNY-HEP-78/6 (1978); S. Weinberg, Harvard preprint HUTP-78/A060 (1978).

44. H. Georgi and D. V. Nanopoulos, Harvard preprints HUTP-78/A039 (1978) and HUTP-79/A001 (1979).

45. See for example, S. Weinberg in <u>Transactions of the New York Academy of Sciences</u>, Series II, Vol. 38, p. 185 (1977) and references therein.

46. J. Ellis, M. K. Gaillard, D. V. Nanopoulos and C. T. Sachrajda, CERN-preprint TH-2634 (1979).

47. We are proud to announce that very recently a simple extension of $SU(5) \to O(10)$, while keeping intact almost everything I have said here, does the trick to predict[44] the t-quark mass ~ 14 GeV and all the three Cabibbo K-M[10] angles, in agreement with experiment.

GAUGE HIERARCHIES IN UNIFIED THEORIES[†]

Itzhak Bars[*][‡]

Institute for Advanced Study

Princeton, New Jersey 08540

The problem of gauge hierarchies is reviewed. A distinction is made between gauge hierarchy in each order of perturbation theory and gauge hierarchy in the full theory. It is shown that, in the full theory, in all cases, it is possible to arrange for a phenomenologically adequate hierarchy provided certain effective mass parameters are restricted to be small or zero. The problem of hierarchy in each order of perturbation theory is clarified and the necessary conditions for the existence of such models are discussed.

I. INTRODUCTION

Unified gauge theories which attempt to describe the strong, weak, and electromagnetic interactions classify all elementary fields with a unified simple gauge group G. It is supposed that at some very large mass scale $\lambda \sim 10^{16}$ GeV the symmetry is broken spontaneously down to $SU(3) \times SU(2) \times U(1)$, and that at a much lower mass scale $\sim 10^2 - 10^3$ GeV a further spontaneous breaking occurs, giving

---

[†]Research sponsored by the Department of Energy under Grant #EY-76-S-02-2220.
[*]Alfred P. Sloan Foundation Fellow.
[‡]On leave of absence from Yale University.

the appropriate masses to W± and Z, and leaving unbroken the exact gauge symmetry SU(3) x U(1) of color and electromagnetism.

$$G \xrightarrow{\lambda} SU(3) \times SU(2) \times U(1) \xrightarrow{v} SU(3) \times U(1)$$

$$|\lambda| \sim 10^{16} \text{ GeV}, \quad |v| \sim 10^2 - 10^3 \text{ GeV}.$$

The mass scale $\sim 10^{16}$ GeV is arrived at by studying the effective strengths of the coupling constants for the factor groups in SU(3) x SU(2) x U(1) as a function of mass scales.[1] It is found that while at laboratory energies the strong, weak, and electromagnetic coupling constants differ from each other widely, at 10^{16} GeV they unify into the same dimensionless coupling strength of the overall group G. To this, one can add arguments such as proton stability, the explanation of net baryon number in the universe[2] etc. for requiring a very large mass scale, such as $\gtrsim 10^{16}$ GeV, within certain unified gauge theories. While the successes of the SU(2) x U(1) Weinberg-Salam model[2] lend support to the gauge theory point of view, it is worth mentioning that a unified theory of the type described above requires a remarkably bold extrapolation of our successful ideas from present laboratory energies by many orders of magnitude up to the Planck mass.

Assuming that this picture holds, theorists have investigated models that produce widely different mass hierarchies through spontaneous symmetry breaking in theories involving explicit Higgs scalars. It is widely known that in the tree approximation it is trivial to arrange for the desired hierarchy by approximation choosing the values of the parameters in the potential $V_o(\Phi)$, where

$$L = \text{K.E.} + V_o(\Phi) + \text{counterterms.} \tag{1.1}$$

However, on the technical side, it has been suggested[3] that radiative corrections may destroy the gauge hierarchy established in lowest order and restrict the ratio v/λ to

$$|v/\lambda| \sim \sqrt{\alpha}$$

rather than the required tiny value of $\sim 10^{-1}$. Since this observation several papers have been written,[4-9] some disagreeing[4,5,6,8,9] and some agreeing[3,7] with the view that gauge hierarchies are limited in renormalizable models with explicit Higgs mesons. This literature is somewhat confusing. Various authors mean different things even when they seem to agree on the view that gauge hierarchies are not limited as originally suggested.

In this talk I will try to clarify the problem, sharpen the question and give an answer. I will distinguish between two types of hierarchies:

(1) Hierarchy in each order of perturbation theory and

(2) Hierarchy in the full theory.

If in a given theory (1) fails but (2) holds, this only means that perturbation theory breaks down and definitely not that there is a limit on the gauge hierarchy. Unfortunately it is the misinterpretation of this fact and the lack of attention to the full theory (except in refs. 5,9) that accounts for most of the confusion in the literature. Here I will mainly discuss the related approaches of myself and M. Serdaroglu[5] and of S. Weinberg[9], and in the course of the discussion I will emphasize the various contributions of other authors. The most recent result arrived at by Weinberg is that, with appropriate restrictions on dressed parameters defined through the full theory[5], it is possible in all cases to arrange for the desired hierarchy. In some cases the hierarchy is unlimited while in other (perhaps more attractive) cases there is a limit which, however, is more than adequate for phenomenological purposes. While this settles the question of hierarchy in each order of perturbation theory, no examples of theories which possess hierarchy _naturally_ in each order are known. This type of theory, if it exists, is likely to have additional symmetries and bound to be very interesting.

II. AN EXAMPLE

To illustrate the problem and discuss the nature of the solution it is useful to investigate the simple case of an $O(n)$ gauge model with two Higgs multiplets $\vec{\chi}$ and $\vec{\eta}$ in the n-dimensional vector representation.[3] With two reflection symmetries $\vec{\chi} \to -\vec{\chi}, \vec{\eta} \to -\vec{\eta}$ the potential takes the form

$$V_o = -\frac{1}{2} m_1^2 \vec{\chi}^2 - \frac{1}{2} m_1^2 \vec{\eta}^2 + \frac{1}{4} f_1 (\vec{\chi}^2)^2 + \frac{1}{4} f_2 (\vec{\eta}^2)^2$$

$$+ \frac{1}{2} f_3 \vec{\chi}^2 \vec{\eta}^2 + \frac{1}{2} f_4 (\vec{\chi} \cdot \vec{\eta})^2 \quad . \qquad (2.1)$$

The parameters can always be arranged so that in the tree approximation the absolute minimum occurs when $<\vec{\chi}> \cdot <\vec{\eta}> = 0$ and a gauge hierarchy can result. One can always choose

$$<\vec{\chi}> = \begin{pmatrix} \lambda \\ 0 \\ 0 \\ 0 \\ \cdot \\ \cdot \\ \cdot \end{pmatrix} \qquad <\vec{\eta}> = \begin{pmatrix} 0 \\ v \\ 0 \\ 0 \\ \cdot \\ \cdot \\ \cdot \end{pmatrix}$$

with $|\lambda| \gg |v|$. Then one finds $(n-2)$ heavy gauge bosons with masses $M^2 = g^2 \lambda^2$, another $(n-2)$ light gauge bosons with masses $\mu^2 = g^2 v^2$, and 1 heavy gauge boson with mass $g^2(\lambda^2 + v^2)$. Of the 2n original scalars, 2n-3 get absorbed by the gauge bosons, leaving 3 physical scalars which can be identified by χ, η, and σ, where

$$\chi = \chi_1 - \lambda$$

$$\eta = \eta_2 - v$$

$$\sigma = (\lambda \eta_1 + v \chi_2) / \sqrt{\lambda^2 + v^2} \quad . \qquad (2.2)$$

GAUGE HIERARCHIES IN UNIFIED THEORIES

To sit at a minimum these scalars must have a positive mass matrix (positive eigenvalues)

$$\begin{pmatrix} m_\chi^2 & m_{\chi\eta}^2 \\ m_{\chi\eta}^2 & m_\eta^2 \end{pmatrix} > 0 \quad \text{and} \quad m_\sigma^2 > 0 \qquad (2.3)$$

which is obtained from $\partial^2 V_o/\partial \chi_i \partial \chi_j \big|_{\lambda,v}$, $\partial^2 V_o/\partial \chi_i \partial \eta_j \big|_{\lambda,v}$, $\partial^2 V_o/\partial \eta_i \partial \eta_j \big|_{\lambda,v}$. Note that $m_{\chi\eta}^2$ could be positive or negative but should satisfy $m_{\chi\eta}^4 < m_\chi^2 m_\eta^2$ with $m_\chi^2 > 0$, $m_\eta^2 > 0$. We must also satisfy the usual conditions $\partial V_o/\partial \chi_i \big|_{\lambda,v} = 0 = \partial V_o/\partial \eta_i \big|_{\lambda,v}$.

Thus, in lowest order one arrives at the following relations between the parameters in V_o and the masses of the Higgs bosons and gauge bosons after spontaneous breaking:

$$f_1/g^2 = (m_\chi^2/2M^2)_o$$

$$f_2/g^2 = (m_\eta^2/2\mu^2)_o$$

$$f_3/g^2 = (m_{\chi\eta}^2/2M\mu)_o \qquad (2.4)$$

$$f_4/g^2 = (m_\sigma^2/M^2+\mu^2)_o$$

$$m_1^2 = \tfrac{1}{2}(m_\chi^2 + \tfrac{\mu}{M} m_{\chi\eta}^2)_o = (\tfrac{f_1}{g^2} M^2 + \tfrac{f_2}{g^2} \mu^2)_o$$

$$m_2^2 = \tfrac{1}{2}(m_\eta^2 + \tfrac{M}{\mu} m_{\chi\eta}^2)_o = (\tfrac{f_2}{g^2} \mu^2 + \tfrac{f_3}{g^2} M^2)_o$$

where the index o under the parantheses $(\ldots)_o$ is included to emphasize that these are the zeroth order masses calculated from V_o.

There are 6 parameters in the potential $(f_1, f_2, f_3, f_4, m_1^2, m_2^2)$ and 6 elements in the resulting mass matrix $(M^2, \mu^2, m_\chi^2, m_\eta^2, m_{\chi\eta}^2, m_\sigma^2)_o$.

Both of these sets are physically measurable and one can choose one set over the other as an independent set in order to parametrize the theory. I emphasize that the second set is superior to discuss the problem of hierarchies. Indeed, except for the positivity of the mass matrix there are no conditions on the masses and any hierarchy such as

$$|\mu/M| \sim 10^{-14}$$

can simply be chosen. Since one has in mind a perturbation scheme the quartic coupling constants should satisfy $f_i(\kappa) \ll 1$ at an appropriate renormalization point κ. This implies (from Eq. 2.4) that m_χ^2/M^2, m_η^2/μ^2, $m_{\chi\eta}^2/M\mu$ and $m_\sigma^2/M^2+\mu^2$ should remain smaller than ~ 1. This allows a very large range for the mass parameters which is consistent with a hierarchy. To my mind, it is also attractive to take all quartic coupling constants of the same order of magnitude at the unification scale $|\lambda| \sim 10^{16}$ GeV. Suppose one takes $f_i(\lambda) \sim g^2(\lambda)$ (within factors of order 1 for different i = 1,2,3,4). Then from eq. (2.4) it is seen that the Higgs and gauge bosons fall naturally into light and heavy sectors, with χ and σ very heavy and η light (except for small mixing). Note that this happens without any of the coupling constants $f_i(\lambda)$ becoming large or vanishingly small. Note further that m_1^2 and m_2^2 are both large and of similar order of magnitude.

Now consider the one loop effective potential

$$V(\vec{\chi},\vec{n}) = V_0(\vec{\chi},\vec{n}) + V_1(\vec{\chi},\vec{n}) \quad . \tag{2.5}$$

where, after all renormalizations have been done, the parameters $f_i(\kappa)$ are defined through fourth derivatives of $V(\vec{\chi},\vec{n})$ evaluated at an appropriate subtraction point κ. Their numerical value is identified with their "physical value" that occurs in V_0 provided they are both taken at the same scale κ.

The correction $V_1(\vec{\chi},\vec{n})$ is of order $g^4(\kappa)$ and $f_i^2(\kappa)$ and shifts the minimum of the potential. Thus, the relation between the two sets of physical observables $(f_1, f_2, f_3, f_4, m_1^2, m_2^2)$ and $(\mu^2, M^2, m_\chi^2, m_\eta^2, m_{\chi\eta}^2, m_\sigma^2)$ will take a new form. In order to properly control large logarithms which could invalidate perturbation theory[10] one should solve the renormalization group equations for the parameters in V_0 and determine an "improved" $V_1(\vec{\chi}, \vec{\eta})$. This has not been tackled in the literature as the relevant equations are quite complicated even in the toy model under consideration. Accordingly, I will exhibit the relation between the two sets of physical parameters in a symbolic form, which is sufficient for the present discussion. Defining the mass parameters through second derivatives of the potential evaluated at the new minimum (λ', v') one finds the relations (with $(M^2)_1 = g^2(\lambda')^2$, $(\mu^2)_1 = g^2(v')^2$)

$$f_1/g^2 = (m_\chi^2/2M^2)_1 + a_1 \alpha$$

$$f_2/g^2 = (m_\eta^2/2\mu^2)_1 + a_2 \alpha$$

$$f_3/g^2 = (m_{\chi\eta}^2/2M\mu)_1 + a_3 \alpha$$

$$f_4/g^2 = (m_\sigma^2/M^2 + \mu^2)_1 + a_4 \alpha \qquad (2.6)$$

$$m_1^2 = \tfrac{1}{2}(m_\chi^2 + \tfrac{\mu}{M} m_{\chi\eta}^2)_1 + (b_1 M^2 + c_1 \mu^2)_1 \alpha$$

$$= [\tfrac{f_1}{g^2} + (b_1 - a_1)\alpha](M^2)_1 + [\tfrac{f_3}{g^2} + (c_1 - a_1)\alpha](\mu^2)_1$$

$$m_2^2 = \tfrac{1}{2}(m_\eta^2 + \tfrac{M}{\mu} m_{\chi\eta}^2)_1 + (b_2 \mu^2 + c_2 M^2)_1 \alpha$$

$$= [\tfrac{f_2}{g^2} + (b_2 - a_2)\alpha](\mu^2)_1 + [\tfrac{f_3}{g^2} + (c_2 - a_3)\alpha](M^2)_1$$

where the index 1 under the parantheses $(\ldots)_1$ is used to emphasize that the masses are calculated at the minimum of $V = V_0 + V_1$. $= g^2/4\pi$ is taken of the order of the fine structure constant $\sim 10^{-2}$

The parameters $(a_1,a_2,a_3,a_4,b_1,b_2,c_1,c_2)$ are functions of logs of ratios of masses and the renormalization point κ, as well as pure numbers. These parameters are typically of order 1 for appropriate values of κ, as can be inferred from detailed calculations.[11] Note the factors of the large mass M^2 in the expression for m_1^2 and m_2^2 that were pulled out in defining the parameters b_1,b_2,c_1,c_2. Such large mass shifts are typical of scalar theories and reflect the quadratic divergences that occur in quadratic terms. It is this large mass shift which is at the root of the problems with hierarchies in each order of perturbation theory as will be illustrated now.

Suppose we take $\kappa \simeq \lambda$ and also assume $f_i(\lambda)/g^2(\lambda) \sim O(1)$. Then comparing eqs. (2.4, 2.6) it appears at first sight that the terms of order α are a small correction, so that the masses $(\mu^2)_1$, $(M^2)_1$, etc. should be of the same order of magnitude as $(\mu^2)_o$, $(M^2)_o$, etc. If this were true then we would say that the tree approximation hierarchy is consistent with the 1-loop hierarchy. In that case the numerical values of the parameters could be chosen so as to establish a hierarchy at the tree diagram level, since the hierarchy would be maintained except for small corrections. Unfortunately, this conclusion is too hasty, as the analysis below will show.

Following Gildener[3] we consider the combination of parameters

$$\varepsilon = \frac{f_1 m_2^2 - f_3 m_1^2}{f_2 m_1^2 - f_3 m_2^2} \quad . \tag{2.7}$$

Let us evaluate this parameter in terms of the calculated Higgs and gauge boson masses by substituting equations (2.4, 2.6) in eq. (2.7) One finds

$$\left(\frac{\mu^2}{M^2}\right)_o = \varepsilon = \left(\frac{\mu^2}{M^2}\right)_1 \frac{1 + \alpha A + \alpha B (M^2/\mu^2)_1}{1 + \alpha C + \alpha D (\mu^2/M^2)_1} \tag{2.8}$$

where A,B,C,D are functions of f_i/g^2 and a_i,b_i,c_i, and with our

GAUGE HIERARCHIES IN UNIFIED THEORIES

assumptions they are typically of order 1. Now we see that there could be a conflict between hierarchies in the tree approximation and 1-loop. Thus, if one demands a large gauge hierarchy in lowest order then the left-hand side of this equation fixes

$$\varepsilon = (\mu^2/M^2)_o \sim 10^{-28} \quad .$$

In order to agree with this constraint on ε it is seen that the right-hand side must yield

$$(\mu^2/M^2)_1 \sim O(\alpha)$$

so that the numerator of the expression above can be made vanishingly small. On the basis of a similar (but much less detailed) argument Gildener concluded that there must be a limit of order $\sqrt{\alpha}$ on gauge hierarchies.

However, it was later pointed out[4,5,6] that if $B \sim 0$, which could be arranged by an unnatural condition on the parameters of V_o, then the hierarchy could be maintained at least through 1-loop. Note that the condition $B \sim 0$ corresponds in our notation to demanding a relation between the large mass shifts in the quadratic terms in eq. (2.6).

$$B \sim 0 \not\gtrless \frac{1}{g^2}[f_1(c_2 - a_3) - f_3(b_1 - a_1)] \sim O(\mu^2/M^2)_1 \quad . \qquad (2.9)$$

This is easy to satisfy, but now one must consider 2-loop, 3-loop corrections etc. and determine whether similar conditions are required to maintain the hierarchy. It is not clear that this unnatural process could be continued without generating inconsistent conditions on the parameters, but it remains as a possibility.

On the other hand, it was not necessary to demand that $\mu^2/M^2)_o \sim 10^{-28}$. What should be true is that (μ^2/M^2) calculated for the full theory must be vanishingly small. To illustrate this point of view which was emphasized by Bars and Serdaroglu[5] let us pretend that $V_0 + V_1$ is the full theory. This theory can be parametrized by the effective dressed parameters $(\mu^2, M^2, m_\chi^2, m_\eta^2, m_{\chi\eta}^2, m_\sigma^2)_1$ (with subscript 1). These can be taken as the free parameters of the theory and they can be allowed to take arbitrary values while the parameters $(f_1, f_2, f_3, f_4, m_1^2, m_2^2)$ of V_0 are regarded as dependent on these mass parameters. We can thus certainly have an arbitrary hierarchy for $(\mu^2/M^2)_1$ and the numerical values of f_i and m_i^2 can be chosen in accordance with this hierarchy rather than the lowest order hierarchy. There will again be heavy and light sectors if f_i are of similar orders of magnitude as before. Returning to eq. (2.8) we see now that the parameters of V_0 should be such that

$$(\mu^2/M^2)_0 = \varepsilon \sim O(\alpha)$$

rather than 10^{-28} as it would have been if one demanded a hierarchy in lowest order. This represents a breakdown of perturbation theory since large readjustments of ε are required when the 1-loop correction is included. Of course, $V_0 + V_1$ is not the full theory, and the numerical values of the parameters should not be chosen at the 1-loop level, but rather should be chosen only after the full effective potential is known. It is, of course, practically impossible to compute the full potential. Nevertheless, this argument shows that it is quite possible that the full theory will have a hierarchy while various orders of perturbation theory do not have it.[5] If this is the situation for a given theory, the interpretation should be that naive perturbation theory breaks down and certainly not that the gauge hierarchy is limited as originally suggested by Gildener. One should find then a better approximation method to estimate the functional relationship between the parameters of V_0 and the resulting masses after spontaneous breakdown in the full theory.

III. AN ANALYSIS OF THE FULL THEORY

Weinberg[9] has recently given an anlysis to determine the nature of the hierarchies in the full theory. He uses an approximation

GAUGE HIERARCHIES IN UNIFIED THEORIES

scheme which is based on

(i) Constraints on the mass matrix of the full theory, (analogous to the previous section).

(ii) The Appelquist-Carazzone theorem.[12]

(iii) The renormalization group of Gell-Mann and Low.

Consider the full potential in any theory $V(\Phi)$, where Φ is, in general, a reducible representation of the full group G (e.g., in the O(n) theory $\Phi = (\vec{\chi},\vec{\eta})$). Let $<\Phi_i> = \lambda_i$ be a stationary point at which the symmetry is broken down to a subgroup $G_1 \subset G$. We are searching for a nearby absolute minimum of the potential at $<\Phi_i> = \lambda_i + v_i$ with $|v| \ll |\lambda|$ which will give rise to a hierarchy. To do this we may expand the full potential around the stationary point $<\Phi_i> = \lambda_i$ by substituting $\Phi_i \to \lambda_i + \Phi_i$. Since we are interested in exploring the region close to the stationary point λ, Weinberg proposes to use an effective renormalizable field theory for this purpose. Basing his arguments on the Appelquist-Carazzone theorem, the only fields that are included in the effective renormalizable theory are the massless gauge bosons associated with G_1, any non-Goldstone Higgs bosons that are left massless or with small mass at $\Phi = \lambda$ plus any fermions with small or zero mass at $\Phi = \lambda$. We recall that the essence of the Appelquist-Carazzone theorem is that, in theories with heavy and light sectors, at mass scales much smaller than the heavy sector, the light sector decouples from the heavy sector provided the fields of the light sector correspond to a renormalizable field theory. The effect of the heavy sector is simply a renormalization of the parameters of the light sector. The effective theory may be used to estimate the terms in the expansion of the full potential around the point λ. For example, in the O(n) case, at $<\chi_1> = \lambda$ and $<\eta_i> = 0$, χ_1 and η_1 are superheavy and χ_i i = 2,...n are Goldstone bosons absorbed by the n-2 heavy gauge bosons. The fields that are included in the effective theory are the massless gauge bosons of the unbroken O(n-1) subgroup and η_i = 2,...,n with a common mass m^2 which is taken 0 or small.

More generally the light scalars are identified by examining the mass matrix at $\Phi = \lambda$

$$m_{ij}^2(\lambda) = (\partial^2 V(\Phi)/\partial\Phi_i \partial\Phi_j)_{\Phi=\lambda} \quad . \tag{3.1}$$

They may be chosen as

$$\phi_a = u_a^i(\Phi_i - \lambda_i) \quad , \tag{3.2}$$

where u_a^i are orthonormalized eigenvectors of μ^2 with small eigenvalues

$$m_{ij}^2 u_a^j = m_a^2 u_a^i \quad ,$$

$$u_a^i u_b^i = \delta_{ab} \quad , \tag{3.3}$$

$$|m_a| \ll |\lambda| \quad .$$

The smallness of some of the eigenvalues m_a^2 is achieved by appropriately choosing the parameters of V_o. It is perhaps helpful to try to understand this statement from the point of view of perturbation theory. Usually, in perturbation theory one takes the point of view that a given set of parameters such as f_i, etc. in V_o are known and any other physical quantity such as masses M^2, μ^2, m^2, etc. are calculated order by order in terms of the original parameters. Thus, if the original parameters are assigned a fixed numerical value, the the numerical value of the masses change in each order as new terms of the perturbation expansion are added in. This procedure makes sense when the shifts in physical quantities are small in each order. However, as we have seen in the previous section the mass terms can acquire large mass shifts. For example, even though $(m_\eta^2)_o$ was arranged to be small in lowest order $(m_\eta^2)_1$ (eq. 2.6) could be large if the original parameters were held at a fixed value. For this reason one may take a different point of view. After all, the mass parameters themselves are physical quantities and the full theory specifies a definite functional relationship between the masses and the parameters of V_0. This functional relationship is only partially known if one stops at a given order of perturbation theory. In the full theory we expect to have the freedom to pick some desired value at least for some of the masses and adjust the parameters of V_0 so as to produce the desired numerical value for the given mass. Thus one could demand that a given mass such as m_η^2 must have a fixed numerical value as the outcome of the theory. In order to satisfy such a physical requirement in perturbation theory (i.e., a fixed value $m_\eta^2 = (m_\eta^2)_0 = (m_\eta^2)_1 =$ etc.) one would have to readjust in each order the numerical values of the parameters of V_0, as the functional relationship between m_η^2 and V_0 gets discovered order by order. This procedure is as physically sensible as the ordinary one. If the readjustments in the parameters of V_0 were small in each order then perturbation theory would work. However, just the opposite was discovered in the last section, i.e., the parameter ε had to be readjusted by many orders of magnitude. So, we continue to hold on to the second point of view expressed here for the full theory, and emphasize that the perturbative expansion breaks down. We see, therefore, that Weinberg's proposal is in the same spirit as ref. 5 as explained in the previous section. Thus, for convenience we will take the mass parameters m_a^2 as an effective parametrization of the theory, replacing some of the original parameters.

The effective theory may be written by omitting the heavy fields as

$$V(\lambda + \phi_a u_a) = V(\lambda) + \frac{1}{2} m_a^2(\lambda) \phi_a^2 + \frac{1}{4!} f_{abcd}(\lambda,\kappa) \phi_a \phi_b \phi_c \phi_d + \cdots$$

$$+ \cdots \cdots \quad (3.4)$$

where the effective coupling constant is evaluated at a renormalization point κ. There are no linear terms because λ is a stationary point and it is assumed that no trilinear terms exist due to some reflection (or other) symmetry (as in the O(n) model). The remaining terms of the full potential (...) are to be estimated by calculating radiative corrections with the above effective theory of the light fields.

First examine the case of $m_a^2 = 0$, later $m_a^2 \neq 0$ will be taken into account. The 1-loop corrections must be included to find a nontrivial minimum. For this perturbative scheme to make sense it is necessary to choose κ at a value κ_o such that[14]

$$\min_{n^2=1} \{f_{abcd}(\kappa_o) n_a n_b n_c n_d\} = 0 \qquad (3.5)$$

For the O(n) example there is a single quartic coupling constant in the effective theory which now is chosen so that $f(\kappa_o) = 0$. This condition can be viewed as an exchange of one coupling constant for the parameter introduced at the renormalization point κ_o, thus keeping the number of parameters the same.[10] The one loop correction which now dominates the expansion can be calculated in order to determine the value of $<\phi>$ at the absolute minimum. Assuming that the minimum is attained along a direction n_a^o in eq. (3.5) we can write $\phi_a = n_a^o \phi$, in terms of which the 1-loop potential takes the form

$$V(\lambda + u^a n_a^o \phi) = V(\lambda) + A \phi^4 [\ln \frac{\phi^2}{\kappa_o^2} - \frac{25}{6}] \qquad (3.6)$$

where A is of order $g^4(\kappa_o)$, $f^2(\kappa_o)(=0)$ and $h^4(\kappa_o)$ (with h = Yukawa coupling). The minimum occurs at

$$<\phi> = v = \kappa_o e^{11/6} \qquad (3.7)$$

Now one must determine κ_o from eq. (3.5), since it sets the scale of the hierarchy v/λ. For this purpose one must solve the renormalization group equations for $f_{abcd}(\kappa)$. To get a feeling, conside simple cases (like the O(n) example) with a single quartic coupling constant $f(\kappa)$ and a gauge coupling constant $g(\kappa)$. Then, the equations can be put in the form[13]

$$16 \pi^2 \kappa \frac{d}{d\kappa} g(\kappa) = b\, g^3(\kappa)$$

$$\frac{16 \pi^2}{g^2(\kappa)} \kappa \frac{d}{d\kappa}\left(\frac{f(\kappa)}{g^2(\kappa)}\right) = \beta\left(\frac{f(\kappa)}{g^2(\kappa)}\right) \qquad (3.8)$$

$$\beta\left(\frac{f(\kappa)}{g^2(\kappa)}\right) = a\left(\frac{f(\kappa)}{g^2(\kappa)}\right)^2 + e\left(\frac{f(\kappa)}{g^2(\kappa)}\right) + d$$

where a, b, e, d are numerical constants which satisfy $a > 0$, $b < 0$, $d > 0$ and the sign of e depends on the details of the theory In figure 1 the function β is plotted against f/g^2 for various signs and magnitudes of e. The arrows show the shift in f/g^2 as κ decreases. The direction of the arrow is determined by the sign of $\beta(f/g^2)$ at the appropriate value of f/g^2, as given by the renormalization group equations.

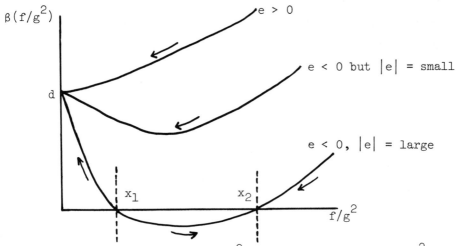

Fig. 1. Plot of β against f/g^2 showing the flow of f/g^2 as κ decreases.

We assume that $f(\lambda)$ and $g(\lambda)$ are given at the unification scale $\lambda \sim 10^{16}$ GeV, so that these provide boundary conditions for the differential equations. Thus, if $f(\lambda)/g^2(\lambda)$ is smaller than the positive root x_1 (if any) at which $\beta(x_1) = 0$, then $f(\kappa)$ will necessarily vanish at some smaller scale $\kappa = \kappa_o$, $f(\kappa_o) = 0$. This point is determined by the solution of eqs. (3.8) which give

$$\kappa_o = \lambda \exp\left[-\frac{16\pi^2}{g^2} F\left(\frac{f(\lambda)}{g^2(\lambda)}\right)\right] ,$$

$$F(x) = \frac{1}{2b}\{-1 + \exp[2b \int_0^x \frac{dt}{\beta(t)}]\} .$$

(3.9)

With $f(\lambda)/g^2(\lambda) \sim O(1)$ ($<x_1$), one gets $F(x) \sim 1$ and κ_o/λ or v/λ is tiny, allowing a limited but adequate gauge hierarchy. Weinberg estimates for the SU(2) x U(1) model (embedded in a larger model) that it is quite reasonable to have a guage hierarchy if $f(\lambda)/e^2(\lambda) \sim O(2-3)$, where e is the electromagnetic coupling constant and f is the quartic coupling constant of the Higgs scalar.

Now return to $m_a^2 \neq 0$, and for simplicity assume a common mass $m_a^2 = m^2$ (as in the O(n) model). If $m^2 < 0$ and if $f(\kappa)/g^2(\kappa)$ remains large as κ decreases from $\kappa = \lambda$, then the corrections in eq. (3.4), which are of order $g^4(\kappa)$ and $f^2(\kappa)$, will be unimportant in determining the minimum (provided, of course, $f(\kappa)$ is still small enough for perturbative expansions). This will be the case if $f(\lambda)/g^2(\lambda) > x_1$ in fig. 1 and x_2 is large enough. Now minimizing V one obtain $\langle\phi\rangle = v$

$$v^2 = 6(-m^2)/f(v) , \qquad (3.10)$$

where $\kappa = v$ has been chosen. In this case the hierarchy v/λ appear to be unlimited since m^2 could be chosen arbitrarily.

If in fig. 1 the value of e is so that $\beta(f/g^2)$ is always positive (i.e., e > 0, or e < 0 but $|e|$ = small) or if $f(\lambda)/g^2(\lambda)$ is less than x_1, then at a small enough κ we will have $f(\kappa) \ll g^4(\kappa)$.

GAUGE HIERARCHIES IN UNIFIED THEORIES

In this case the 1-loop correction becomes important again. Thus, choosing again $\kappa = \kappa_o$ so that $f(\kappa_o) = 0$, the effective potential including the 1-loop correction takes the form

$$V \simeq V(\lambda) + \frac{1}{2} m^2 \phi^2 + A\phi^4 [\ln(\phi^2/\kappa_o^2) - 25/6] \qquad (3.11)$$

The value of $<\phi> = v$ at the minimum can be seen to be a decreasing smooth function of $(-m^2)$ as m^2 moves from negative values through 0 to positive values. The case of $m^2 = 0$ has already been analyzed. As $m^2 > 0$ gets larger, at some point $V(\phi)|_{\phi=v}$ will exceed $V(\lambda)$ so that at that point the minimum will jump to $v = 0$ restoring the G_1 symmetry. Just before this happens v attains its minimum non-zero value which corresponds to the maximum hierarchy v/λ (except for the previous possibility of unlimited hierarchy). This point, however, could be unstable under a rise in the temperature.

IV. DISCUSSION

In this talk I argued that it is possible to arrange the parameters of V_o so that the full theory displays a hierarchy[5,9] $v/\lambda \sim 10^{-14}$. As Weinberg has shown the hierarchy may be limited in some cases and unlimited in others. But in all cases it is adequate as desired by phenomenology in unified models.

On the technical side, it is now clear that hierarchies depend on the smallness of the effective mass parameters of some scalars, defined at $<\Phi> = \lambda$. As argued here it is possible to fulfill this requirement in the full theory. The question before us now is whether there are any models that satisfy this condition in <u>every order</u> of perturbation theory in a <u>natural</u> way. Such a model, if it exists, is likely to have additional symmetries (supersymmetry?) which will make it quite interesting. A symmetry can be responsible for making $m_a^2 = 0$ for some scalars at $<\Phi> = \lambda$ in <u>each order</u> of perturbation theory. It is worth noting that in such a model the masses of the $W^\pm$ and Z bosons could be calculable in terms of the unification mass $\lambda \sim 10^{16}$ via formulas such as eq. (3.9).

NOTES AND REFERENCES

1. H. Georgi, H. Quinn, and S. Weinberg, Phys. Rev. Lett. <u>33</u>, 451 (1974).
2. For a review see D. V. Nanopoulos, previous talk in this conference.
3. E. Gildener, Phys. Rev. <u>D14</u>, 1667 (1976).
4. A. J. Buras, J. Ellis, M. K. Gaillard, and D. V. Nanopoulos, Nucl. Phys. <u>B135</u>, 66 (1978).
5. I. Bars and M. Serdaroglu, Yale preprint COO-3075-188 (revised).
6. R. N. Mohapatra and G. Senjanovic, CCNY preprint CCNY-HEP-78/6 (revised).
7. K. T. Mahanthappa and D. G. Unger, Phys. Lett. <u>78B</u>, 604 (1970); K. T. Mahanthappa, M. A. Sher, and D. G. Unger, preprint COLO-HEP-8.
8. O. K. Kalashnikov and V. V. Klimov, Lebedev preprint #110.
9. S. Weinberg, Harvard preprint HUTP-78/A060.
10. S. Coleman and E. Weinberg, Phys. Rev. <u>D7</u>, 1888 (1973), see also S. Weinberg, Phys. Rev. <u>D7</u>, 2887 (1973).
11. The large logarithms that occur in such calculations (e.g., refs 6,7) must be controlled by using the renormalization group as in ref. 10.
12. T. Appelquist and J. Carazzone, Phys. Rev. <u>D11</u>, 2856 (1975).
13. For detailed formulas see e.g., T. P. Cheng, E. Eichten, and L. F. Li, Phys. Rev. <u>D9</u>, 2259 (1974).
14. E. Gildener and S. Weinberg, Phys. Rev. <u>D13</u>, 3333 (1976).

ANOMALIES, UNITARITY AND RENORMALIZATION[†]

Paul H. Frampton

Department of Physics

The Ohio State University

Columbus, Ohio 43210

In these remarks, I wish to indicate how the renormalizability of a gauge theory with γ_5 couplings to fermions can impose additional dynamical constraints to those already well known in connection with the triangle anomaly. Here, I can only outline how I differ from previous workers; for more details, see Refs. 1,2.

Without γ_5 couplings, we know a spontaneously broken gauge theory is renormalizable by using 't Hooft's gauge.[3,4]

With γ_5 couplings, there is the additional complication of the anomaly problem. Here, I have a strong conjecture that I hope can be elevated to the status of a theorem.

Recall that for the lowest-order triangular diagram with two external vectors (couplings γ_μ, γ_ν; outgoing momenta p_1, p_2) and one external axial-vector (coupling $\gamma_5 \gamma_\lambda$; incoming momentum $q = p_1 + p_2$) the vertex satisfying

$$p_{1\mu} \Gamma_{\mu\nu\lambda} = p_{2\nu} \Gamma_{\mu\nu\lambda} = 0 \tag{1}$$

[†]This work is supported in part by the U.S. Department of Energy.

has an anomalous axial Ward identity[5]

$$q_\lambda \Gamma_{\mu\nu\lambda} = \frac{1}{4\pi^2} \epsilon_{\mu\nu}(p_1 p_2) \qquad (2)$$

which has been successfully related to the $\pi^0 \to 2\gamma$ decay rate.

Soon after 't Hooft's work, it was pointed out by several people[6] that this anomaly must be canceled in flavor dynamics to preserve renormalizability. In the general non-abelian case, let the flavor matrices T^a, T^b, T^c be associated with the vertices γ_μ, γ_ν, $\gamma_5\gamma_\lambda$ respectively and let the T^a contain the sign (± 1) corresponding to the fermion helicity. Then including the crossed diagrams and summing over flavors gives, to lowest order,

$$q_\lambda \Gamma_{\mu\nu\lambda}^{abc} \sim \text{Tr}(\{T^a, T^b\} + T^c) \frac{1}{4\pi^2} \epsilon_{\mu\nu}(p_1 p_2) \qquad (3)$$

and hence we require that

$$\text{Tr}(\{T^a, T^b\} + T^c) = 0 \quad . \qquad (4)$$

In an $SU(2) \otimes U(1)$ theory with only left-handed doublets and right-handed singlets, Eq.(4) reduces to $\sum Q_L = 0$, which is satisfied separately for each quark-lepton generation: $g_1 = (u,d,e)$, $g_2 = (c,s,\mu)$, $g_3 = (t,b,\tau)$ since $3(\frac{2}{3} - \frac{1}{3}) - 1 = 0$.

What was shown in Ref. 6 is that if Eq.(4) is <u>not</u> satisfied then there is a disaster: there remains the question of the proof of renormalizability when Eq.(4) <u>is</u> satisfied.

Concerning such a proof, with the usual anomaly cancellation, there is the two-step prescription:[7]

1. Dimensionally regularize meson loops
2. Handle the closed fermion loops one by one, performing the Dirac trace in four dimensions and then regularizing.

My point is that during step (1), unless the triangular vertex has been defined for generic dimension, n, there is a danger of violating unitarity. This is especially so when the triangle is involved in an overlapping divergence, as in Fig. 1. In such a case, the $(4-n)^{-1}$ in the overlapping divergence may give a finite contribution with the $(4-n)$ term from the triangle. To throw away such a $(4-n)/(4-n)$ piece of the Feynman amplitude can lead to violation of unitarity, in particular of the cutting rules.[8]

With this motivation, we examine the expansion of the triangle in $(4-n)$. In general, the fermion masses m_α, m_β, m_γ (corresponding to the flavor labelling $T^a_{\alpha\beta}$, $T^b_{\beta\gamma}$, $T^c_{\gamma\alpha}$) may be different for the three sides of the triangle. After straightforward algebra, one finds

$$q_\lambda \Gamma^{abc}_{\mu\nu\lambda} = \frac{1}{4\pi^2} \varepsilon_{\mu\nu}(p_1 p_2) \times$$

$$\sum_{\text{flavors}} T^a_{\alpha\beta} T^b_{\beta\gamma} T^c_{\gamma\alpha} \left[1 + (4-n) f\left(m_\alpha, m_\beta, m_\gamma, p_1^2, p_2^2, q^2\right) + \cdots \right] \quad (5)$$

after ensuring that the vector identities are satisfied. The right-hand side of Eq.(5) should be zero for all energies p_1^2, p_2^2, q^2 and the necessary and sufficient condition for this is mass degeneracy within each quark-lepton generation. This is sufficient to kill all terms including higher powers $(4-n)^p$; my conjecture is that it is actually necessary for renormalizability.

Thus, with such mass degeneracy the proof of renormalizability seems complete but without this constraint the proof is at best incomplete and may not even hold.

The conventional wisdom appears to be based largely on the intuition that the ultra-violet structure must be unaffected by the Higgs mechanism, as is the case without γ_5 couplings.[3] But such a general argument needs rigorous verification.

To nail down my result, it appears necessary to complete a 6th order calculation to find whether or not the new anomaly somehow cancels. In partial calculations done so far, no sign of such a cancellation is seen, but such work is still incomplete. It is

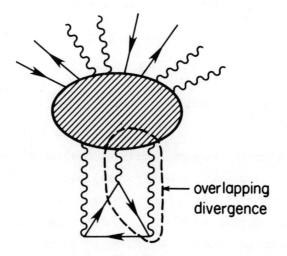

Fig. 1. General Feynman diagram containing triangle in overlapping divergence.

ANOMALIES, UNITARITY AND RENORMALIZATION

technically difficult.

Also, one should show that the result is independent of the dimensional regularization method. There exists a method using higher derivatives which preserves invariance[9]--this could show the $(4-n)$ expansion is not crucial in obtaining the result.

REFERENCES

1. P. H. Frampton, Ohio State Preprint COO-1545-248 (Nov. 1978).
2. P. H. Frampton, Ohio State Preprint COO-1545-249 (Dec. 1978).
3. G. 't Hooft, Nucl. Phys. $\underline{B35}$, 167 (1971).
4. For a recent review and complete set of references, see P. H. Frampton, Lectures on Gauge Field Theories, Part Three: Renormalization, Ohio State Preprint COO-1545-245 (1978).
5. S. L. Adler, Phys. Rev. $\underline{177}$, 2426 (1969); J. S. Bell and R. Jackiw, Nuovo Cimento $\underline{60A}$, 47 (1969).
6. C. Bouchiat, J. Iliopoulos and Ph. Meyer, Phys. Letters $\underline{38B}$, 519 (1972); D. J. Gross and R. Jackiw, Phys. Rev. D $\underline{6}$, 477 (1972); H. Georgi and S. L. Glashow, Phys. Rev. D $\underline{6}$, 429 (1972).
7. W. A. Bardeen, Proceedings of the Sixteenth International Conference on High Energy Physics, Fermilab (1972) Vol. 2, p. 295.
8. It is necessary but not sufficient for a prescription to give unique, finite answers satisfying all Ward identities. There is also unitarity.
9. M. Baker and C. Lee, Phys. Rev. D $\underline{15}$, 2201 (1977).

[Editors' note: We wish to point out that other workers have claimed to have refuted Prof. Frampton's conjecture; in particular, the interested reader is referred to K. Ishikawa and K.A. Milton, Non-contribution of $n \neq 4$ Axial-Vector Anomaly, U.C.L.A. preprint UCLA/79/TEP/4 February 1979.]

CHARM PARTICLE PRODUCTION BY NEUTRINOS

N. P. Samios

Brookhaven National Laboratory

Upton, New York 11973

Today I am going to discuss further evidence, indeed confirmation, of the existence of the Λ_c^+ (2260) and Σ_c^{++} (2424) charmed baryons, as well as the observation of the D^{*+}(2000) charmed meson, these states all being produced by neutrino interactions. In addition I will briefly review the status of ($\mu^- e^+$) di-lepton neutrino production in bubble chambers, especially as it affects charm production.

The specifics of the first experiment - in simple liquids, H_2 and D_2, at relatively low neutrino energies (0-20 Gev) - are shown in Fig. 1. The main point is that there is a sample of 5,000 charged-current neutrino events and 35 strange particle events and, as I will shortly show, 2 baryonic charm events. In Fig. 2, the fine resolution of this device and technique is demonstrated where W, the recoil hadron mass, is plotted and where the neutron and Δ^{++} peaks are clearly seen. The first example of charmed baryon production is shown in Fig. 3 and has been published in Phys. Rev. Lett. 34, 1125 (1975). I remind you that this was a clear example of $\Lambda S = -\Lambda Q$, the reaction being $\nu p \to \mu^- \Lambda^0 \pi^+ \pi^+ \pi^- \pi^-$. There is only one strange particle, transverse momentum is balanced, with alternate explanations being at the

---

*Research supported by U.S. Department of Energy Contract Number EY-76-C-02-0016.

BNL - 7' Chamber

    Cnops, Connolly, Kahn, Kirk, Murtagh, Palmer, Tanaka and Samios

$\nu(H_2,D_2)$ Volume = $9m^3$ (Fiducial $6m^3$)

                 Plates: 4 2" S.S. (2IM.Fp)

                 Kimematics - low energy

                           few missing particles.

                           (80% of all events

                              uniquely identified)

H_2 200 k pix   400 events ⎫ analyzed
D_2 550 k pix   2300 events ⎭
       450 k pix   1300 "   found
                     1000 "   expected
                     5,000

35 Strange particle events $\Lambda^0 \to p\pi^-$
                                       $K_S^0 \to \pi^+\pi^-$
                                       $\Sigma^+ \to \pi^+\nu$

     2/3 nothing missing
     i.e. $\nu n \to \mu^- \Lambda^0 K^+$ (8)
           $\nu p \to \mu^- \Lambda^0 K^+ \pi^+$ (6)

FIGURE 1

The characteristics of ν interaction experiment in H_2 and D_2 at low energies (0-20 GeV).

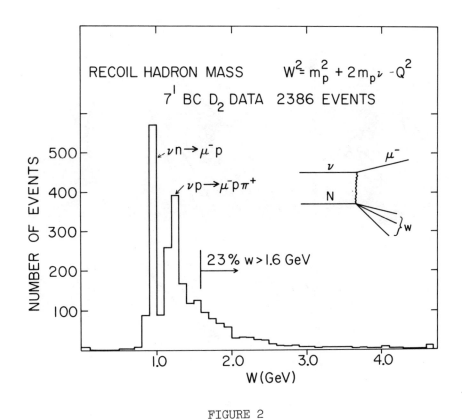

FIGURE 2

Recoil hadron mass distribution for deuterium events.

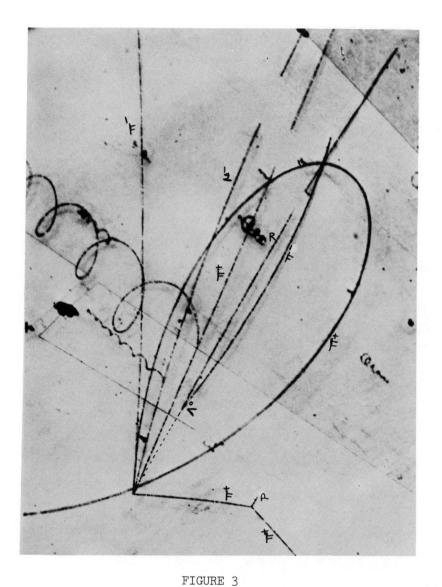

FIGURE 3

The first example of a charmed baryon (Phys. Rev. Lett. 34, 1125 (1975)).

10^{-4} level. The pertinent masses are Λ_c^+ ($\Lambda^0 \pi^+ \pi^+ \pi^-$) = 2260 MeV and the Σ_c^{++} ($\Lambda_c^+ \pi^+$) = 2426 MeV. The second event is shown in Fig. 4 and has been published in Phys. Rev. Lett. 42, 197 (1979). In this instance the reaction occurred on a neutron (in deuterium), the reaction fitting $\nu n \rightarrow \mu^- \bar{K}^0 \pi^- \pi^+ p$, all alternate hypotheses being less than 1% probable. The exciting feature is that the $K^0 \pi^-$ mass is consistent with $K^*(892)$, thereby fixing the K_s as a $\bar{K}^0$ i.e. S = -1, again an example of $\Delta S = -\Delta Q$. The hadronic mass $M(\bar{K}^0 \pi^- \pi^+ p)$ = 2254 ± 10 MeV is that of the Λ_c^+ in this case decaying via a K^0 mode. This second event clearly confirms the existence of the Λ_c^+ (2260).

The second experiment which I will discuss involves the study of neutrino interactions at higher energies, 10-100 Gev, but in a Ne/H_2 mixture. The particulars are shown in Fig. 5, the main data comparing about 200 di-lepton events and 3,000 charged particle events. The first results of the $\mu^- e^+$ study have been published in Phys. Rev. Lett. 39, 62 (1977). The present status is a yield of 204 $\mu^- e^+$ events from a sample of 83,000 charge current events. This corresponds to a corrected production rate of about .5%. These events are accompanied by 43 V^0s (a clear excess over the 12 events expected randomly). The corrected V^0 rate is $\mu^- e^+ V^0 / \mu^- e^+ \simeq .6$. If neutral and charged strange particles are produced at equal rates this would correspond to approximately 1.2 strange particles per event, in good agreement with charm production. A summary of a wide variety of experiments on this subject is shown in Table I. The main point is that the sum of all other experiments is similar to the BNL-Columbia results. If one examines the x, y distribution for the BNL-Columbia di-lepton events as shown in Fig. 6, one notes the excess at small x, corresponding to a 37% sea contribution and 1.4 strange particles per event, in agreement with the above rate expectation. In summary one would conclude that the di-lepton events have all the characteristics of charm production.

I will now turn to a discussion of the non-leptonic decay of charm particles. As noted earlier, about 3,000 strange particle

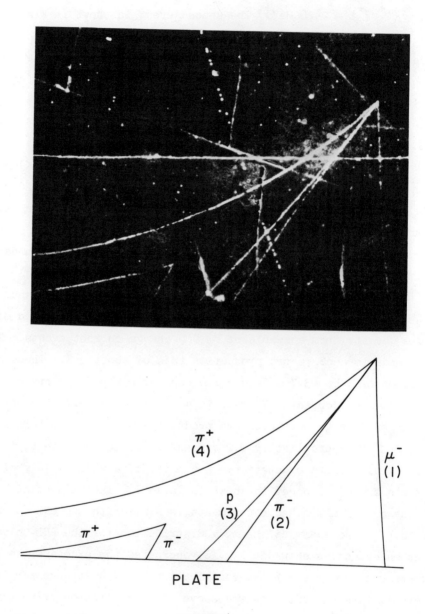

FIGURE 4

The second charmed baryon event (Phys. Rev. Lett. 42, 197 (1979)).

BNL - Columbia Experiment: 15' Chamber. FNAL

ν_μ Ne - H_2 60% mixture
  28 tons
  ℓ_R = 40 cm   9RL ϕ
  ℓm+ 120 cm   3IL ϕ

134 k pix=106,000   c.c. events
  $\nu_\mu N_e \to \mu^-$ + anything

Reaction	Measured	Fraction	
$\nu_\mu Ne \to \mu^- + ..$	2,000	2%	
$\nu_\mu Ne \to \mu^- v^0 + .$	3,000	50%	
$\nu_\mu Ne \to \mu^- e^+ - ..$	200	70%	
$\nu_\mu e^- \to \nu_\mu e^-$	11	100%	$sm^2\theta_{w\overline{3}} = .2^{+.16}_{-.08}$
$\nu_e N_e \to e^- ...$	180	25%	
$\nu_e N_e \to e^+$	30	25%	

BNL - Cnops, Connolly, Kahn, Kirk, Murtagh, Palmer, Tanaka and Samios

Columbia: Baltay, Caroumbalis, French, Hibbs, Hylton, Kalelkar, Shastri

FIGURE 5

The characteristics of ν interaction experiment in a Ne/H_2 mixture at higher energies (10-100 GeV).

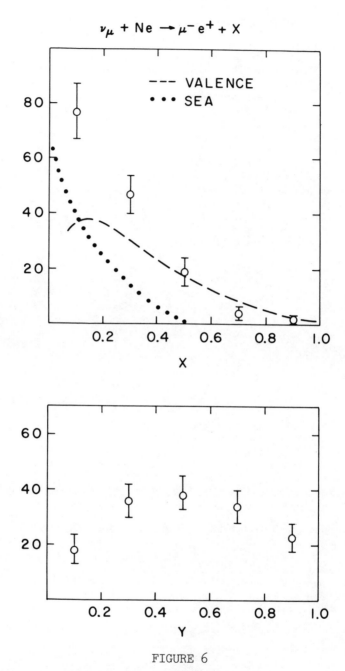

FIGURE 6

x and y distributions for the BNL-Columbia dilepton events.

TABLE I

$\mu^- e^+$ From Other Experiments

Experiment	$<E_\nu>$ BeV	Liquid	Events Observed	Vees* Observed	$\mu^- e^+/\mu^-$ Rate (%)
Gargamelle(11) CERN PS	1-8	Freon	14 $\mu^- e^+$	3	
Wisconsin-CERN-Hawaii-Berkeley(12) Fermilab 15 foot B.C., E28	~30	21% Ne	17 $\mu^- e^+$	11	0.8 ± 0.3
Columbia-Brookhaven Fermilab 15 foot B.C., E53	~30	64% Ne	164 $\mu^- e^+$	33	0.5 ± 0.15
Berkeley-Seattle-LBL-Hawaii(6) Fermilab 15 foot B.C., E172	~30	64% Ne	6 $\mu^- e^+$	1	0.34 + 0.23 − 0.13
Fermilab-LBL-Hawaii(13) Fermilab 15 foot B.C., E546	~30	50% Ne	40 $\mu^- \mu^+$	5	0.43 ± 0.16
BEBC Narrow band(14) CERN SPS	~75	60% Ne	11 $\mu^- \mu^+$ 5 $\mu^- e^+$	6 2	0.7 ± 0.3
BEBC Wide band(15) CERN SPS	~30	60% Ne	21 $\mu^- e^+$	6	0.5 ± 0.17
Gargamelle Wide Band(16) CERN SPS	~30	Freon Propane	46 $\mu^- \mu^+$	6	0.7 ± 0.2
			160 164	40 Others 33 BNL-Columbia	

*Vees stand for $K_S^0 \to \pi^+ + \pi^-$ or $\Lambda^0 \to p + \pi^-$ decays events have been accumulated, slightly more K^o's than Λ^o's. With two thirds of the data, the $D^o \to K^o \pi^+ \pi^-$ was evident (see Fig. 7, and Phys. Rev. Lett. 41, 73(1978)) inclusively produced via the reaction $\nu Ne \to \mu^- D^o$ + anything. A similar search for an inclusive production of Λ_c^+ with its subsequent decay into $\Lambda \pi^+$ showed an excess of 20 events, a 2σ effect at the expected mass (Fig. 8). It is clear then that inclusive studies at high energies and in complex liquids are not likely to be the most productive means of deciphering charm particle production; they work for the D^o and give some evidence for the Λ_c^+ but the latter is far from overwhelming. As such we have exploited the mass difference between two charm particles, one decaying into the other via a pion emission. There is the well known example of $D^{*+} \to D^o \pi^+$ which has a small Q = 5 MeV value. Therefore, we have taken the approximately 40 events in the D^o signal (Fig. 7) and combined it with a single π^+, at the same time performing the same procedure above and below the D^o mass as control regions. In Fig. 9a, one sees a cluster of 6 events at a mass difference of 146 MeV for the D^o region with an absence of such an excess in the control region (Fig. 9b). We attribute this effect to the well known D^{*+} to D^o decay.

Of greater interest is the application of this same procedure to the charmed baryon possibilities. The particular sequence exploited to date has been

$$\Sigma_c^{++} \to \Lambda_c^+ \pi^+$$
$$\hookrightarrow \Lambda^o \pi^+ \quad \text{(a)}$$
$$\hookrightarrow (\Lambda \pi^+)^* \pi^+ \pi^- \quad \text{(b)}$$

In essence we have observed 7 examples of (a) with an expected background of one event and 3 examples of (b) again with a background of one event. Summing this gives a total of ten events where two are expected, a quite significant result. The two dimensional distribution of $M(\Lambda \pi^+)$ versus $\Delta m\ [(\Lambda \pi^+ \pi^+) - (\Lambda \pi^+)]$ is shown in Fig. 10 for the observed events, the expected number of events as deduced from

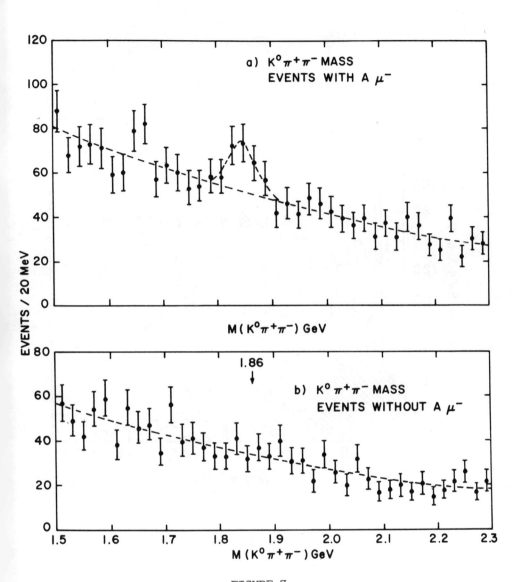

FIGURE 7

$K^0\pi^+\pi^-$ mass distributions with and without a μ^-.

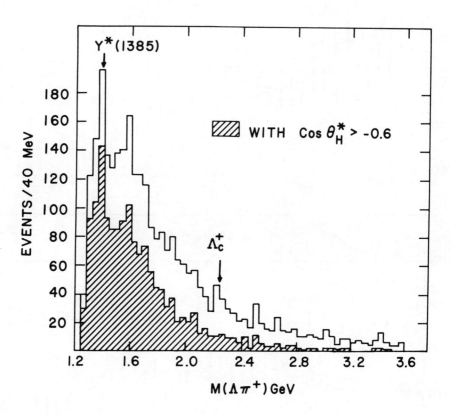

FIGURE 8

Mass distribution of $\Lambda^0\pi^+$ events.

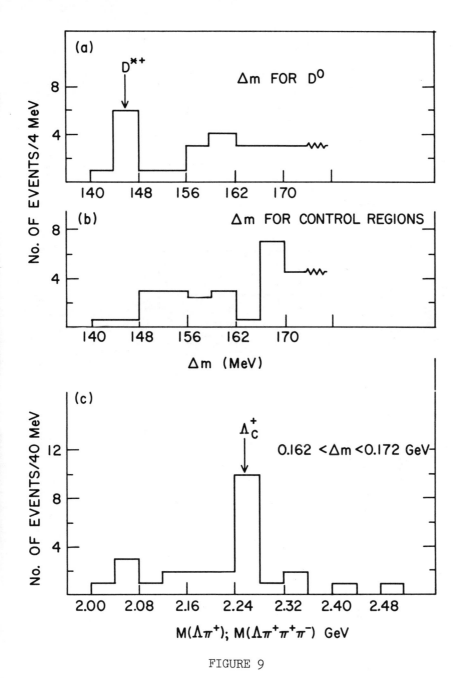

FIGURE 9

Mass plots for a) D^{*+} b) control region c) Λ_c^+

	OBSERVE								EXPECT						
M	.142–.152	.152–.162	.162–.172	.172–.182	.182–.192	.192–.202			51.7	62.7	73.7	84.7	95.7	106.7	No.
2.00	0	1	0	2	1	0		136	0.9	1.1	1.3	1.5	1.7	1.9	.017
2.04	0	2	3	0	1	0		119	0.8	1.0	1.2	1.4	1.6	1.7	.016
2.08	0	0	1	1	2	4		148	0.8	0.9	1.1	1.3	1.4	1.6	.015
2.12	1	0	0	0	1	0		110	0.7	0.9	1.0	1.2	1.3	1.5	.013
2.16	0	0	1	3	1	2		92	0.7	0.8	0.9	1.1	1.2	1.4	.012
2.20	0	0	1	0	4	4		124	0.6	0.7	0.9	1.0	1.1	1.2	.011
2.24	1	0	7	1	0	0		106	0.5	0.7	0.8	0.9	1.0	1.1	.010
2.28	0	0	1	1	0	0		47	0.5	0.6	0.7	0.8	0.9	1.0	.009
2.32	0	0	2	0	1	0		64	0.4	0.5	0.6	0.7	0.8	0.9	.008
2.36	0	1	0	1	0	0		58	0.4	0.4	0.5	0.6	0.7	0.7	.007
2.40	0	0	0	1	0	1									
				Δm								Δm			

FIGURE 10a

Two-dimensional distribution of $M(\Lambda\pi^+)$ vs Δm – observed and expected numbers of events.

POISSON PROBABILITIES

M	.142	.152	.162	.172	.182	.192	.202
2.00							
2.04	.41	.37	.27	.25	.31	.15	
2.08	.45	.18	.09	.25	.32	.18	
2.12	.45	.41	.37	.35	.24	.06	
2.16	.35	.41	.37	.30	.35	.22	
2.20	.50	.45	.37	.07	.36	.24	
2.24	.55	.50	.37	.37	.02	.03	
2.28	.30	.50	$.18 \times 10^{-4}$	.37	.37	.33	
2.32	.61	.55	.35	.36	.41	.37	
2.36	.67	.61	.10	.50	.36	.41	
2.40	.67	.27	.61	.33	.50	.35	

Δm

FIGURE 10b

The Poisson probability distribution for Figure 10a$_1$.

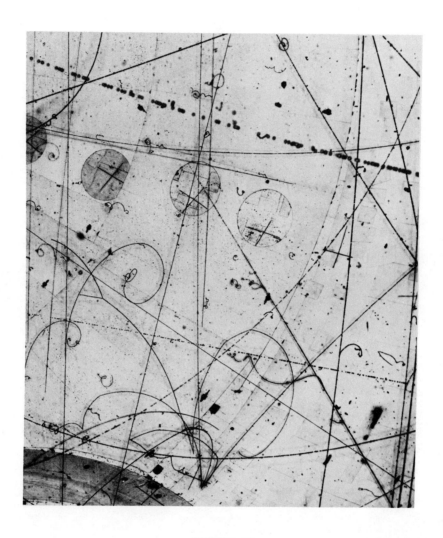

FIGURE 11

A further example of an elastic production of $\Lambda_c \to \Lambda \pi^* \pi^+ \pi^-$.

non-correlated M, m distributions and the resultant Poisson probabilities. The effect is striking, even for this one channel. Similar conclusions with less data are reached with channel (b), as noted earlier the signal here consisting of 3 events. The effective mass distribution for the Λ_c^+ in $\Lambda\pi^+$ and $\Lambda\pi^+\pi^+\pi^-$ for both these channels is shown in Fig. 9c for these events where the mass difference Δm is between .162 and .172 Gev. The excess of 10 events is clearly observable. In the sample consisting of the $\Lambda\pi^+\pi^+\pi^-$ final states, one of the events is elastic and very similar in its characteristics to the first BNL event. A picture of this event is shown in Fig. 11 where the fast μ^- is seen to traverse the chamber in the forward direction, the Λ^0 is observed to decay and the three positive and one negative pions are seen to interact in the chamber. The kimematics fit exceedingly well for the reaction $\nu p \to \mu^- \Sigma_c^{++}$; $\Sigma_c^{++} \to \Lambda_c^+ \pi^+$ and $\Lambda_c^+ \to \Lambda^0 \pi^+\pi^+\pi^-$. The derived masses, from the above events are $\Lambda_c^+ = 2257 \pm 10$ Mev and $\Sigma_c^{++} - \Lambda_c^+ = 168 \pm 5$ Mev. The decay modes observed to date for the Λ_c^+ are therefore $\Lambda\pi^+, \Lambda\pi^+\pi^+\pi^-$ and $\bar{K}^0 p\pi^+\pi^-$.

My final comment has to do with production rates. If one considers the (10-2) $\Sigma_c^{++} \to \Lambda_c^+$ events and corrects for: lifetime + Λ^0 neutral decay modes (factor of 3); other possible charged decay modes (x2); estimate of Λ_c^+/Σ_c^+ production (x3) and $\Lambda_c^+ \to$ decay modes with neutrals (x2), one obtains a rate of 8x2x3x2x2 events out of total of 75,000 charged current events or $\simeq 1/3$ % Λ_c^+ production - quite reasonable.

In summary I would say that the existences of the Λ_c^+, Σ_c^{++} are in good shape, with their respective masses being quite well measured, as well as clear indication of several decay modes for the Λ_c^+. The rates and properties of leptonic as well as non-leptonic modes are in good agreement with those expected of charmed particles. The remaining task is to now find the host of other expected states, especially the F, F* and Ξ_c's.

TECHNIQUES TO SEARCH FOR PROTON INSTABILITY TO 10^{34} YEARS

David B. Cline

Fermilab, Batavia, Illinois and

University of Wisconsin, Madison, Wisconsin

ABSTRACT

We discuss the experimental techniques to search for proton instability at very long lifetime. A large water detector and the photosensitive elements needed to observe the decay products are considered. For small detectors (~ 100 tons) ordinary photomultipliers are adequate - for much larger detectors ($\sim 10,000 - 100,000$ tons) new techniques of light collection are required. We also discuss the most common backgrounds from cosmic ray neutrinos and how to reject these at the level of 10^{34} years.

The subject of proton stability has been of interest since at least 1929.[1] Periodically through the years there has been interest in searching for proton decay at longer and longer lifetimes. Early experiments by Reines and collaborators have established a lifetime in excess of 10^{28} years for $p \to \mu^+ \ldots$ decays.[2] Recently there has been renewed interest in this subject for two reasons. (1) The net positive baryon number of the universe may be connected in some way to the stability of baryonic matter in the early universe.[3,4] (2) Gauge theories such as that due to Pati and Salam and Georgi and Glashow[5] incorporate transitions between leptons and quarks and therefore allow for proton disintegration.[6]

TABLE 1

Physical Properties of Pure Water

Density	1 Ton/m^3
Specific Energy Loss	203 MeV/m
Radiation Length	0.36 m
Critical Energy	73 MeV
Nuclear Collision C.	0.56 m
Transparency	
visible (480 nm)	> 25 m
(300nm)	4 m

TABLE 2

Conversion Efficiencies for Wavelength Shifter Approach

Absorption BBQ	.90
Emission BBQ	0.90
Transmission BBQ	0.35
Captured Light (n = 1.5 to 1.33)	0.05
Light Pipe	0.75
Boundary BBQ/H_2O	0.90
	$0.97 \cdot 10^{-2}$

We now turn to the present experimental limits on the proton lifetime. In the experiment of Reines et al. in which a 19 ton detector is used 6 candidates for $p \rightarrow \mu^+$ were observed and ~ 600 events with very small pulse height were also observed. Using the μ^+ events to put a limit on $p \rightarrow \mu^+$'s... an approximate limit of 10^{30} years is obtained; for the e^+ decays lacking a clear understanding of the origin of these events the limit would be about 100 times worse ($\sim 10^{28}$ years). We note that the acceptance of the detector for stopping μ^+ will be dependent on the specific decay mode considered. For example, $p \rightarrow \pi^+ + \nu$ and $p \rightarrow \pi^+ + 3\nu$ may have considerably different detection efficiency.

In order to observe a lifetime of 10^{34} years it is necessary to have approximately 10^{34} nucleons in a sensitive volume. We believe that water is a logical choice for this medium because it is extremely transparent but has a radiation length (.36m) and hadron absorption (.7m) length that are much smaller than the light absorption length ($\ell > 20$m). The properties of water are given in Table 1. The natural limiting volume of such a detector is $[2\ell]^3$ which contains approximately 64,000 tons of material and 4×10^{34} nucleons. We assume that the detector is placed under an earth shield either in a mine or a cave under a mountain such that there is approximately 1-3 KM of material around the detector of density $\sim 1-3$ gm/cm^3.

The Cerenkov radiation emitted by the decay products of nucleon disintegration can be detected in such a large water detector by collecting the light at the boundaries and using a wavelength shifter to convert and trap the light. However, as shown in Table 2 it is difficult to achieve more than a percent of active coverage with this technique.

The wavelength spectrum of light transmitted by water is shown in Fig. 2.[7] The spectrum is well matched to the S11 photocathode response. As will be shown later the number of photo-electrons that can be collected under ideal conditions from a proton decay

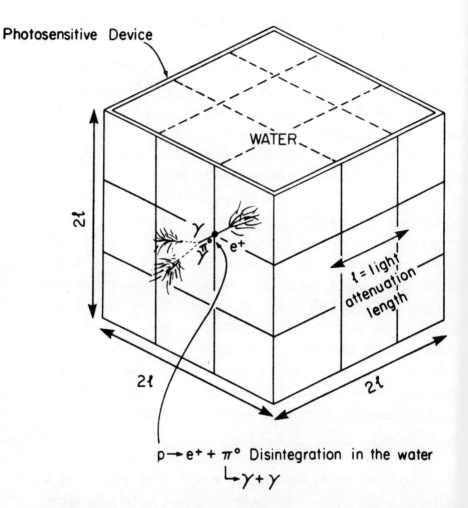

Figure 1. Schematic of a large water counter used to search for nuclear disintegration. The walls of the counter are lined with wavelength shifter material or other suitable detectors. Also shown is a schematic representation of $p \to e^+ + \pi^0$ decay in the counter.

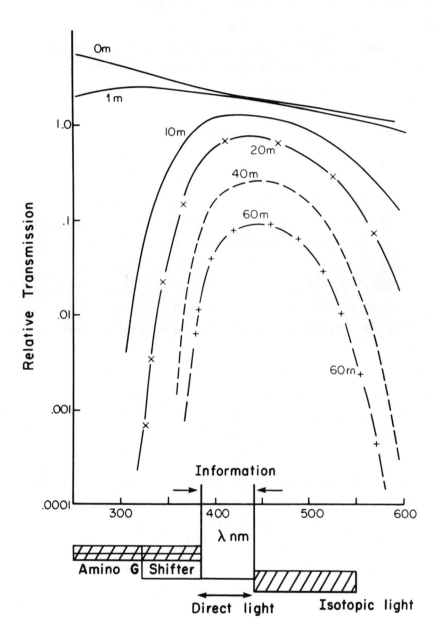

Figure 2. Transmission spectrum of light by water as function of wavelength at various depths of water.

is limited to a few hundred; thus it is extremely important to maximize the surface area of the light collecting device. Fig. 3 shows the number of 5" photomultipliers needed as a function of detector mass and of the surface area covered.[8] Clearly very large detectors require an unreasonable number of photomultipliers.

There are two possible solutions:

1. Use a wavelength shifter to collect light from the ultraviolet region - this in principle gives an increase of a factor of ~ 4. However in order to use the directionality of the light for background rejection some tricks are needed.[9] For example some wavelength shifters leave part of the spectrum unshifted and these photons can in principle be used to give directional information.[9]

2. Develop very large, inexpensive photosensitive devices (LSD = Light Sensitive Devices).[10]

Both of these options are being followed by a group that is designing a large detector to search for proton decay.

As an example we further consider two types of proton or neutron decay producers. Consider first the decay process

$$p \to e^+ + \pi^0 \atop \hookrightarrow \gamma + \gamma \qquad . \qquad (1)$$

The pulse height spectrum should show a peak structure as illustrated in Fig. 4a. We can estimate the number of Cerenkov photons arising from the electromagnetic shower of the e^+ and the conversion of the photons in the water to be $N \sim 1.5 \times 10^5$ in the frequency band from 350 nm to 550 nm. The transmission of light at different frequencies through pure water and the response of various photo cathodes is shown in Fig. 2.[7] Note that the attenuation length of water is approximately 20M.[8] Another property of process 1 is the directionality of the decay products. The resulting Cerenkov light will transmit this directionality to the collecting boundaries of the detector. If the Cerenkov cone can be

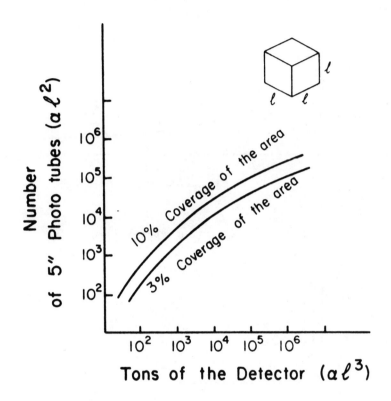

Figure 3. Number of 5" phototubes needed to cover 3% and 10% of the area as a function of detector mass.

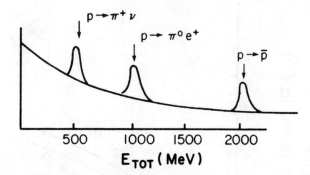

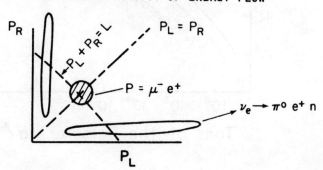

Figure 4. Schematic of the technique to suppress neutrino induced background by using the pulse height or directionality of the interaction and the nucleon decay products.

SEARCH FOR PROTON INSTABILITY TO 10^{34} YEARS

detected it may even be possible to identify the e^+ in this reaction. As an example we show schematically the pulse height in the left-right sides of the detector in Fig. 3b. This is the essential technique needed to reject the neutrino induced background.

A second type of proton or neutron decay could involve charged hadrons such as

$$p \to e^+ + \rho^0 \hookrightarrow \pi^+ + \pi^- \qquad (2)$$

$$n \to e^+ + \pi^- \qquad (3)$$

In this case the e^+ energy is measured through the electromagnetic cascade whereas the $\pi\pm$ energy is measured by the hadronic interaction and subsequent electromagnetic energy. We expect the intrinsic energy resolution will be poorer in this case. We conservatively estimate that greater than 6×10^4 photons in the 550 - 350 nm wavelength band will be produced in these decays. Again the decay products are very directional.

Processes 1, 2 and 3 are examples of nucleon decays in which all the energy appears in the final state in positrons or hadrons. There are two other types of decay modes, for example

$$p \to \mu^+ + \pi^0 \qquad (4)$$
$$\longrightarrow \mu^+ + \rho^0$$

and

$$p \to \nu + \pi^+ \qquad (5)$$

Process 4 will also be readily detectable but the muon energy will be underestimated in a Cerenkov detector. Process 5 will give a π^+ with an energy of 1/2 the proton rest mass, but of course the neutrino energy is not detectable.

We may estimate the number of collected photo electrons in processes 1 - 4 by the following procedure: Assume a photosensitive surface of 3% and that the detector is 40 M on a side and $(.5 - 1.5) \times 10^5$ photons are produced in the decay. On average the light will go one light absorption length and be reduced by e^{-1}. The collection efficiency of most photosensitive devices is $\sim 50\%$ and photo cathode efficiency is 15%. Thus the number of photoelectrons is

$$N_{p_e} = 3 \times 10^{-2} \times e^{-1} \times 1/2 \times 0.15 \times (0.5 - 1.5) \times 10^5$$
$$= (50 - 100) \text{ photoelectrons} .$$

We believe that this is barely adequate to detect proton decay. Of course the coverage of 3% of the area with photomultipliers is probably an unreasonable possibility.

We now turn to the various background processes that can mimic proton or neutron decay. First consider neutrino interactions in the wall of the cave or the penetration of atmospheric muons into the detector from above. We assume a veto layer of counters (which could be the wavelength shifter array, for example, or the outer part of the detector) such that this background is eliminated from consideration.

The neutrino induced backgrounds that must be considered are

$$\nu_\mu + N \rightarrow \mu + N \tag{6}$$
$$\rightarrow \mu + \pi^0 + N \tag{7}$$
$$\nu_\mu + N \rightarrow \nu_\mu + \pi^0 + N \tag{8}$$
$$\nu_e + N \rightarrow e + N \tag{9}$$
$$\rightarrow e + \pi^0 + N . \tag{10}$$

The calculated flux of cosmic ray neutrinos is shown in

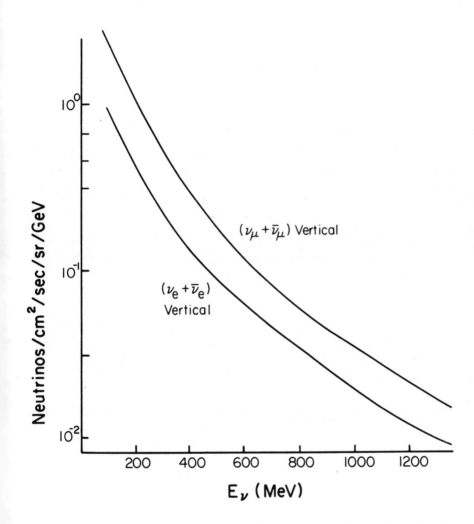

Figure 5. Electron and muon neutrino spectrum from atmospheric interaction of cosmic ray primaries. Only the vertical neutrino spectrum is shown.

Fig. 5.[11,12] It is convenient to compare the rate for these processes with the expected hypothetical rates for proton or neutron decay into specific channels for a given proton or neutron lifetime. The estimated rates for neutrino interactions and proton decays into various modes are shown in Fig. 6. We have considered an 8000 ton water detector for these calculations. Neutrino reactions 6 and 9 produce final states with an energetic lepton and low energy nucleon. Reactions 7, 8 and 10 produce final states with low energy nucleons and pions. None of these processes are similar to the signature of p or n decay described previously. However the rates of these processes occur at a level equivalent to a nucleon decay at $\sim 10^{32}$ years and therefore in order to reach lifetimes of $10^{33} - 10^{34}$ years a technique for discriminating against these backgrounds is essential.

A simple procedure to reject the cosmic ray neutrino background can be provided by using the directionality of the neutrino interactions as mentioned previously and the directional information available in the counter array. Because low energy neutrino interactions are dominately elastic the charged lepton carries the major fraction of the neutrino energy and the incident direction. The decay of a p or n will produce back to back or isotropic decay products since the nucleon decays at rest. Thus the angular distribution of the decay products can be used to discriminate against background. The power of such a detector in rejecting nearly all types of neutrino or cosmic ray muon background can be appreciated if the events are imagined to be reconstructed by the Cerenkov light as shown in Fig. 4.

In practice events occur over the entire fiducial volume leading to a slight smearing of the left - right separation; thus timing and vertex reconstruction are essential. We estimate that a rejection factor of at least 100 can be obtained by using the directional information in the p or n decay. This allows a search for lifetimes to the level of $\sim 10^{34}$ years.

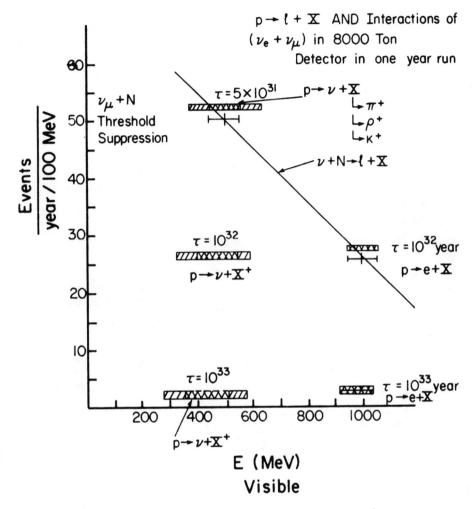

Figure 6. Event rate for a one year interval in an 8000 ton detector for various nucleon decay modes as a function of the visible energy and the particle lifetime of the decay mode. Also shown is the neutrino interaction rate as a function of visible energy. The estimated energy resolution of the counter for each decay mode also shown.

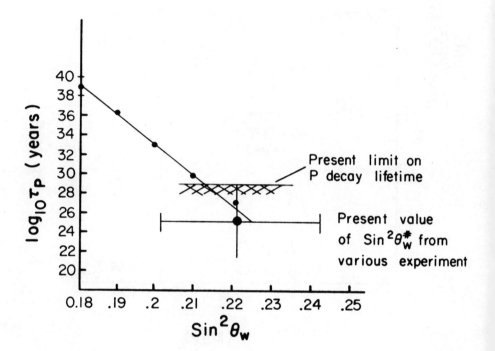

Figure 7. Proton Decay Lifetime as a function of the Weinberg angle $\sin^2\theta_w$ in the SU(5) model.

In summary, a natural detector to search for proton decays at very long lifetimes is a large water detector utilizing the Cerenkov radiation. It is essential to reject the large neutrino backgrounds and therefore the detector will be most sensitive to nucleon decays in which all the nucleon mass appears in the final state in detectable particles. The detection of final states with neutrinos is more problematic. However it may be possible to obtain sufficient background rejection by using the sequence

$$p \to \nu + \pi^+$$
$$\hookrightarrow \mu \to e \quad .$$

Reactions (6) arising from $\bar{\nu}_\mu$ interactions give similar signature. Since we estimate these rates to be $\sim 1/9$ smaller than the corresponding neutrino reaction it should be possible to search for these neutrino decay modes out to a partial lifetime of $\sim 10^{32}$ years. We note that the Pati - Salam model predicts proton lifetimes that are considerably shorter but with multiple neutrino final states.[5]

After we gave this talk we received a preprint on the relation of the proton lifetime to the Weinberg angle $\sin^2\theta_w$. This relationship is plotted in Fig. 7, indicating that any lifetime from $10^{30} - 10^{35}$ is compatible with the present experimental data. Such a broad range of possibilities clearly requires a new, extremely large detector to search for proton decay at every level.

REFERENCES

1. A. Pais, "The Early History of the Theory of the Electron", 1897-1947, Chapter 5, in Aspects of Quantum Theory, edited by A. Salam and E.P. Wigner, p. 79 (Cambridge University Press, 1972); E.C.G. Stueckelberg, Helv. Phys. Acta. 11, 299 (1939); E.P. Wigner, Proc. Am. Philos. Soc. 93, 521 (1949) [see also Proc. Nat. Acad. Sci. 38, 449 (1952)]; M. Goldhaber, "Status

of Conserved Quantum Numbers", talk given at the Ben Lee Memorial Conference (edited by D. Cline and F. Mills).

2. F. Reines, C.L. Cowan, Jr., and M. Goldhaber, Phys. Rev. 96, 1157 (1954); F. Reines and M.F. Crouch, Phys. Rev. Lett. 32, 493 (1974) and subsequent analysis (1977).

3. For one example, see S. Weinberg, in Lectures on Particles and Field Theory, edited by S. Deser and K. Ford (Prentice-Hall, Inc., Englewood Cliffs, N.J., 1964), p. 482. The subject has been considered in recent papers by M. Yoshimura, Phys. Rev. Lett. 41, 381 (1978); S. Dimopoulos and L. Susskind, SLAC-PUB-2126, to be published in Phys. Rev.; B. Toussaint, S.B. Treiman, F. Wilczek, A. Zee (to be published); A.Yu. Ignatiev, N.V. Krosnikov, V.A. Kuzmin, and A.N. Tavkhelidze, Phys. Lett. 76B, 436 (1978).

4. S. Weinberg, Cosmological Production of Baryons", Harvard AUTP-781 A040; private communication, L. Okun, G. Goebel, and G. Steigman.

5. J.C. Pati and A. Salam, Phys. Rev. D8, 1240 (1973) and Phys. Rev. D10, 275 (1974); H. Georgi and S.L. Glashow, Phys. Rev. Lett. 32, 438 (1974); F. Gursey and P. Sikivie, Phys. Rev. Lett. 36, 775 (1975).

6. A. Buras, J. Ellis, M.K. Gaillard and D.V. Nanopoulos, Nuclear Physics B135, 66 (1978); H. Fritzsch and P. Minkowski, Ann. Phys. (N.Y.) 93, 193 (1975); M. Gell-Mann, P. Ramond and R. Slansky, to be published.

7. Dumand Workshop proceedings, Fermilab report T.M.

8. D. Cline and C. Rubbia, Comments on the Search for Proton Decay at Very Long Lifetime, published in the Proceedings of the Madison Seminar on Baryon Instability, D. Cline, Editor, University of Wisconsin, 1979.

9. D. Cline, unpublished.

10. C. Rubbia and D. Winn, unpublished.

11. E.C.M. Young, Chapter 6 in "Cosmic Rays at Ground Level", edited by A.W. Wolfendale (Institute of Physics, London, 1973); J.L. Osborne, Chapter 5 in "Cosmic Rays at Ground Level" edited by A.W. Wolfendale (Institute of Physics, London, 1973); J.L. Osborne, S.S. Said, and A.W. Wolfendale, Proc. Phys. Soc., <u>86</u>, 93 (1965).

12. F. Reines, W.R. Kropp, H.W. Sobel, H.S. Gurr, J. Lathrop, M.F. Crouch, J.P.F. Sellschop, and B.S. Meyer, Phys. Rev. <u>D4</u>, 80 (1971); M.R. Krishnaswamy, M.G.K. Menon, V.S. Narasimham, K. Hinotani, N. Ito, S. Miyake, J.L. Osborne, A.J. Parsons and A.W. Wolfendale, Proc. Roy. Soc. <u>A323</u>, 489 (1971).

13. "Search for Proton Decay Using a Large Pb Shower Counter Array", D. Cline and C. Rubbia, unpublished.

14. "The Weak Mixing Angle and Grand Unified Gauge Theories, W. Marciano, Rockefeller Preprint, 1979.

CHARGED AND NEUTRAL-CURRENT INTERFERENCE: THE NEXT HURDLE FOR WEINBERG-SALAM

S. P. Rosen

Purdue University

West Lafayette, Indiana 47907

ABSTRACT

I discuss the interference between charged and neutral currents in ν_e-e and $\bar{\nu}_e$-e scattering as a further test of the Weinberg-Salam model. For $\sin^2\theta_{GW} \simeq 1/4$, the model predicts <u>destructive</u> interference, but present data does not establish either that the interference is destructive, or that it is even present in the first place. With this in mind, I construct a model with two neutral bosons which reproduces recent fits to the lepton-quark data, but predicts <u>constructive</u> interference in ν_e-e and $\bar{\nu}_e$-e scattering. This model has observable consequences in $e^+e^- \to \mu^+\mu^-$, in particular that the forward-backward asymmetry must be several times larger than the Weinberg-Salam prediction if μ-e universality holds. Problems of constructing a gauge theory are briefly discussed.

The impressive success of the Weinberg-Salam model[1] in fitting the data on neutral-current interactions[2] in neutrino-quark[3] and electron-quark[4] scattering makes it very tempting for one to believe that, as far as the low energy phenomenology of neutral currents are concerned, the game is over: Weinberg-Salam is <u>the</u> model of the electro-weak interaction! In my talk today I would like to express a few words of caution about yielding to this temptation too easily. The spirit in which I offer them is a positive one: the model has achieved remarkable success in overcoming previous hurdles in lepton-quark scattering, and it is certainly to be hoped that it will overcome the next hurdle, one which occurs in lepton-lepton scattering. My collaborators in this effort are Boris Kayser, Ephraim Fischbach, and Harvey Spivack[5]; they deserve credit for all the good thoughts that are about to be expressed, and I alone am responsible for any thought which may be construed as bad.

It is generally believed that, whereas the elastic scattering of muon-type neutrinos by electrons is engendered by neutral-current alone, the elastic scattering of electron-type neutrinos is brought about by a combination of both neutral and charged currents. Under the right circumstances, these two interactions will interfere co-herently with one another,[6,7] and the sign of the interference, that is whether it is constructive or destructive, will depend upon the sign of the coupling of the electron in the neutral current. Since the Weinberg-Salam model makes a specific prediction for this sign, the detection of coherent interference in elastic ν_e-e scattering presents the model with yet another hurdle to overcome. It is, moreover, a unique hurdle because there is no other way of observing the absolute sign of the neutral current couplings of electrons and neutrinos in a model-independent fashion.

To ensure coherent interference, I shall assume that the scattered neutrino in neutral current reactions is of the same type[7] as the incident one. For reasons which will emerge shortly, I shall also consider the possibility that the neutral current is engendered by more than one neutral boson[8] Z_i^0 (i = 1,2, ... n). The diagrams contributing to the total cross-sections for $(\nu_\mu e)$ and $(\nu_e e)$ elastic scattering are illustrated in Fig. 1. In general the amplitude arising from the sum over Z_i^0 diagrams is an admixture of (V - A) and (V + A) components:

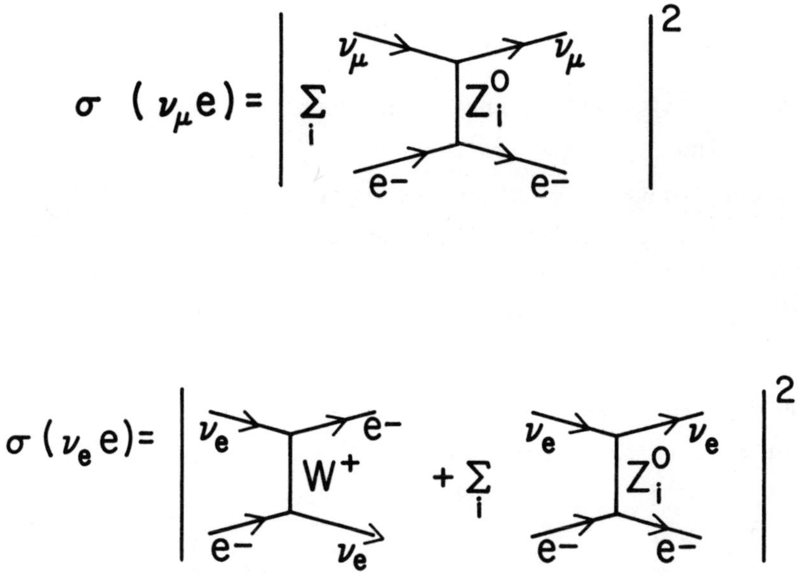

$$\sigma(\nu_\mu e) = \left| \sum_i \begin{array}{c} \nu_\mu \quad \nu_\mu \\ Z_i^0 \\ e^- \quad e^- \end{array} \right|^2$$

$$\sigma(\nu_e e) = \left| \begin{array}{c} \nu_e \quad e^- \\ W^+ \\ e^- \quad \nu_e \end{array} + \sum_i \begin{array}{c} \nu_e \quad \nu_e \\ Z_i^0 \\ e^- \quad e^- \end{array} \right|^2$$

Fig. 1: The cross-sections for ν_μ-e and ν_e-e scattering as sums over neutral and charged current diagrams. An arbitrary number of weak neutral bosons is assumed.

$$\text{Amp}(\Sigma Z_i^0) = \frac{G}{\sqrt{2}} (\bar{\nu}_\ell \gamma_\lambda (1 + \gamma_5) \nu_\ell)(\bar{e} \gamma_\lambda [g_+ (1 + \gamma_5) + g_- (1 - \gamma_5)] e),$$

where

$$g_\pm = \frac{1}{2} (g_V \pm g_A) \quad \text{and} \quad \ell \equiv e, \mu. \tag{1}$$

The effective V and A coupling constants can be expressed in terms of lepton-boson couplings:

$$\frac{G g_x}{\sqrt{2}} = \Sigma \frac{c_{Li}^\nu c_{xi}^e}{(M_i)^2}, \quad \text{where} \quad x \equiv V, A. \tag{2}$$

I assume μ-e universality, and so the coupling constants of the left-handed neutrinos ν_e and ν_μ to the boson Z_i^0 are both given by c_{Li}^ν; the V,A coupling constants of the electron to Z_i^0 are c_{Vi}^e and c_{Ai}^e respectively, and from μ-e universality the corresponding muon constants are

$$c_{xi}^\mu = c_{xi}^e \quad \text{with} \quad x \equiv V, A. \tag{3}$$

The charged current diagram for ν_e-e scattering is pure V - A, and after a Fierz transformation it takes the usual form

$$\text{Amp}(W^+) \equiv \frac{G}{\sqrt{2}} (\bar{\nu}_e \gamma_\lambda (1 + \gamma_5) \nu_e)(\bar{e} \gamma_\lambda (1 + \gamma_5) e). \tag{4}$$

The total amplitude for ν_e-e scattering is, from Eqs.(3) and (4):

$$\text{Amp}(\nu_e e) = \frac{G}{\sqrt{2}} (\bar{\nu}_e \gamma_\lambda (1 + \gamma_5) \nu_e)(\bar{e} \gamma_\lambda [(1 + g_+)(1 + \gamma_5) + g_-(1 - \gamma_5)] e) \tag{5}$$

In the limit of zero electron mass, V - A and V + A amplitudes do not interfere with one another, and so it follows from Eq.(5) that the cross-section for elastic ν_e-e scattering $\sigma(\nu_e e)$ is sensitive to the sign of $(g_V + g_A)$ but not to the sign of $g_V - g_A$.

In the Weinberg-Salam model with a single Z^0, the neutral current coupling constants of the electron are[1]

$$g_V = -\frac{1}{2} + 2\sin^2\theta_{GW}$$
$$g_A = -\frac{1}{2}$$
(6)

where θ_{GW} is the Glashow-Weinberg mixing angle. Thus, for ν_e-e scattering the V - A coefficient is

$$1 + g_+ = \frac{1}{2}(2 + g_V + g_A) = \frac{1}{2}(1 + 2\sin^2\theta_{GW}) \,. \quad (7)$$

Now, all the latest fits to neutrino-quark and electron-quark neutral current data indicate that[3,4]

$$\sin^2\theta_{GW} \simeq 1/4 \quad (8)$$

and hence that

$$1 + g_+ \simeq 3/4 < 1. \quad (9)$$

Thus the Weinberg-Salam model predicts <u>destructive interference</u> in ν_e-e scattering.

Two types of experiments to measure the scattering of electron-type neutrinos by electrons have been discussed up to now. One is a reactor experiment in which one uses an atomic reactor as an intense source of low-energy ($0 \le E_\nu \le 5$ Mev) anti-neutrinos, $\bar{\nu}_e$.[9] The other is a beam dump experiment in which the decay sequence

$$\pi^+ \to \mu^+ + \nu_\mu$$
$$\hookrightarrow e^+ + \nu_e + \bar{\nu}_\mu$$
(10)

for pions and muons essentially at rest provides one with a known spectrum of neutrinos in the range $0 \le E_\nu \le 54$ Mev.[10] As far as I know, there are no experiments being planned with higher energy-type neutrinos.

Presently available data are rather sparse. The only published measurements come from a reactor experiment by Reines, Gurr, and Sobel,[9] who report the following cross-sections:

$$\sigma_{expt}(\bar{\nu}_e e) = (0.87 \pm 0.25)\sigma_{V-A} \quad 1.5 \le E_\nu \le 3.0 \text{ Mev}$$

$$= (1.7 \pm 0.44)\sigma_{V-A}. \quad 3.0 \le E_\nu \le 4.5 \text{ Mev.} \quad (11)$$

They express their results in terms of σ_{V-A}, the cross-section for a pure charged-current interaction, for historical reasons: when the experiment was originally planned, its purpose was to search for neutrino-electron scattering as evidence for the existence of an intermediate charged vector boson W^+! That we now want to use it as a means of seeing the interference between charged and neutral currents indicates how long a time scale the experiment requires!

To interpret the results of Eq.(11), my colleagues and I[5] have calculated the expected cross-section in three separate cases using the spectrum of Avignone and Greenwood.[11] The three cases are: (i) the Weinberg-Salam model with destructive interference and $\sin^2\theta_W = 1/4$; (ii) exactly as in (i) except that the sign of the interference has been changed from destructive to constructive; and (iii) exactly as in (i) except that there is no coherent interference between charged and neutral currents. Our reasons for considering case (ii) will be discussed a little later in the talk; case (iii) applies if the outgoing neutrino in the neutral-current diagram is different from the incident one, and numerically it is the mean of cases (i) and (ii). Our results, expressed as multiples of σ_{V-A}, are presented in Table I.

A perusal of Table 1 reveals that in the lower energy bin, the experimental cross-section is in good agreement with case (i), destructive interference, but the errors are sufficiently large that it is within three standard deviations of case (iii), incoherent interference. By contrast, the data in the higher energy bin is in good agreement with case (iii), but it is also within 1 1/2 standard deviations of case (i) and within 2 of case (ii). Thus, we may con-

TABLE I: Reactor cross-sections for $\bar{\nu}_e e$ scattering in units of σ_{V-A}, the pure charged-current cross-section.

$E_{\bar{\nu}}$ (Mev)	THEORETICAL			EXPERIMENTAL
	Case (i) Destructive Interference $\sin^2\theta_{GW} = 1/4$	Case (ii) Constructive Interference	Case (iii) Incoherent Interference	
1.6 - 3.0	0.85	2.2	1.5	0.87 ± 0.25
3.0 - 4.6	1.1	2.7	1.9	1.7 ± 0.44

clude that:

(a) there is weak evidence for destructive interference;

(b) it is not easy to exclude the possibility of constructive interference; and

(c) it is not possible to establish that coherent interference between charged and neutral currents really does occur.

In connection with point (c) it is worth noting that the observation of a cross-section which is definitely smaller than σ_{V-A} would automatically imply the presence of coherent, destructive interference; on the other hand a cross-section larger than σ_{V-A} could arise either from coherent, constructive interference, or from the incoherent case. Obviously, at least one more reactor experiment is absolutely necessary to decide these issues.

No beam dump experiment has yet been performed, but one is being developed at LAMPF.[10] My colleagues and I[5] have computed the theoretical event rates for the same three cases as were considered in the reactor case. We use the known neutrino spectrum from π- and μ- decay at rest, and determine the number of events in which the scattered electron has an energy lying between some lower cut E_c and the maximum energy of about 54 Mev. Our results, in arbitrary units are displayed in Table II; $\sin^2\theta_{GW}$ is again taken to be 1/4.

TABLE II: Event rates, in arbitrary units, for beam dump experiment with $E_c < E_e < E_{max} \simeq 54$ Mev.

E_c/E_{max}	3/8	1/2	3/4
Case (i) destructive interference	20	12	2
Case (ii) constructive interference	50	30	5
Case (iii) incoherent interference	35	21	3.5

It can be seen from Table II that, independent of the lower energy cut-off, the predictions of cases (i), (ii), and (iii) are approximately in the ratios

$$(i) : (ii) : (iii) = 1 : 2.4 : 1.75 \qquad (12)$$

If we are willing to assume that there is a coherent interference between charged and neutral currents, we can conclude from Eq.(12) that an experiment of 25% accuracy would be sufficient to distinguish the destructive case (i) from the constructive case (ii). On the other hand, if we want to prove that coherent interference occurs, we would require a more precise experiment, at the level of 10 - 15% accuracy.[5]

Since there is no strong evidence for destructive interference at the present time, it is useful to consider what options may be open to us. Any alternatives to the Weinberg-Salam model which we might put forward must be consistent with the constraints that:

(i) the neutrino-quark data are very well described by the

Weinberg-Salam model with $\sin^2\theta_{GW} \simeq 1/4$;[2,3,4]

(ii) the electron-quark data (from polarized electron-deuteron scattering[2]) fall within one standard deviation of the model[4]; and

(iii) within large experimental uncertainties, $\sigma(\nu_\mu e)$ and $\sigma(\bar{\nu}_\mu e)$ are also in agreement with the model.[2,4,12]

Undoubtedly, the area with the greatest room for maneuvering is that of lepton-lepton scattering, where we could easily change the magnitudes of the coupling constants, as compared with the Weinberg-Salam model, and also their signs. In this talk I want to concentrate on signs.

Accordingly, I shall consider a somewhat idealized situation in which the lepton-quark data and the cross-sections for ν_μ-e and $\bar{\nu}_\mu$-e scattering are all assumed to agree with the Weinberg-Salam model. This means that the lepton-quark neutral current interactions must be exactly the same as in the Weinberg-Salam model, whereas the neutrino-electron neutral current interactions must have the same magnitude, but not necessarily the same signs. Opposite signs will, of course, give rise to constructive interference in ν_e-e and $\bar{\nu}_e$-e scattering.

The question now arises as to whether it is possible to construct models which yield different signs for neutrino-electron couplings without affecting the lepton-quark couplings. The answer depends upon the number of neutral bosons: if there is only one boson, it is "No"; but if there are several bosons, it may be "Yes". With only one Z^o we are essentially locked into all the signs of the original model by the constraints imposed on the lepton-quark sector; but with two or more bosons, we have the freedom to arrange things differently.

To illustrate the point, let us examine a model with two bosons, Z_1^o and Z_2^o.[5] Let us assume that:

(i) Z_1^o couples to leptons and to quarks <u>exactly as in the Weinberg-Salam model</u>; and

(ii) Z_2^o couples only to leptons.

As a result, the model automatically fits the lepton-quark data as described above. For neutrino-electron scattering it gives rise to phenomenological coupling constants (see Eq.(1)) of the form:

$$g_x = g_x^1 + g_x^2$$
$$g_x^1 = g_x(W-S) \qquad (x = V, A) \tag{13}$$

where the $g_x^{(i)}$ are associated with the boson Z_i^0 ($i = 1,2$), and the $g_x(W-S)$ have the same values as in Eq.(6):

$$g_V(W-S) = -\frac{1}{2} + 2\sin^2\theta_{GW}$$
$$g_A(W-S) = -\frac{1}{2} \tag{14}$$

The term whose sign we wish to change is $g_+ \equiv (g_V + g_A)$; the other combination, $g_- \equiv (g_V - g_A)$, always enters cross-sections through its squared modulus, and so we are free either to change its sign, or to leave it unchanged. Thus one possibility is to choose

$$g_x^2 = -2g_x^1 = -2g_x(W-S) \tag{15a}$$

in which case:

$$g_+ = -[g_V(W-S) + g_A(W-S)] = -g_+(W-S)$$
$$g_- = -[g_V(W-S) - g_A(W-S)] = -g_-(W-S) \tag{16a}$$

and the other possibility is to choose

$$g_V^2 = g_A^2 = -[g_V(W-S) + g_A(W-S)] \tag{15b}$$

in which case

$$g_+ = -g_+(W-S)$$
$$g_- = +g_-(W-S) \tag{16b}$$

Since this model is designed to give constructive interference in ν_e-e and $\bar{\nu}_e$-e scattering, we can ask in what other phenomena it may differ from the Weinberg-Salam model. One obvious place is e^+e^- annihilation into muon pairs, which will now take place through both Z_1^0 and Z_2^0 intermediate states, as well as through the photon[13] (see Fig. 2). Interference between weak and electromagnetic interactions will now involve two contributions; thus the forward-backward asymmetry, which comes from the interference of the weak axial-vector current and the photon, will now be proportional to

$$A(F-B) = \frac{G}{\sqrt{2}} \left[\frac{c_{A1}^e c_{A1}^\mu}{(M_1)^2} + \frac{c_{A2}^e c_{A2}^\mu}{(M_1)^2} \right] \qquad (17)$$

and parity violating effects will be proportional to

$$A(P-V) = \frac{G}{\sqrt{2}} \sum_{i=1,2} \left[\frac{c_{Ai}^e c_{Vi}^\mu}{(M_i)^2} + \frac{c_{Vi}^e c_{Ai}^\mu}{(M_i)^2} \right] . \qquad (18)$$

The lepton-boson coupling constants and boson masses in Eqs. (17) and (18) are the same as those in Eqs. (2) and (3). If μ-e universality is assumed (Eq. (3)), then

$$A(F-B) = \frac{G}{\sqrt{2}} \left[\frac{(c_{A1}^e)^2}{M_1^2} + \frac{(c_{A2}^e)^2}{M_2^2} \right]$$

$$A(P-V) = \frac{G}{\sqrt{2}} \sum_{i=1,2} \left[\frac{2 c_{Ai}^e c_{Vi}^e}{M_i^2} \right] . \qquad (19)$$

It follows immediately from Eq. (19) that, since the contribution of the boson Z_1^0 to the forward-backward symmetry is equal to that of Weinberg-Salam model, the 2-boson model must lead to a forward backward asymmetry no smaller than that of Weinberg-Salam:

$$A(F-B, 2Z_0) \geq A(F-B, W-S) \qquad . \qquad (20)$$

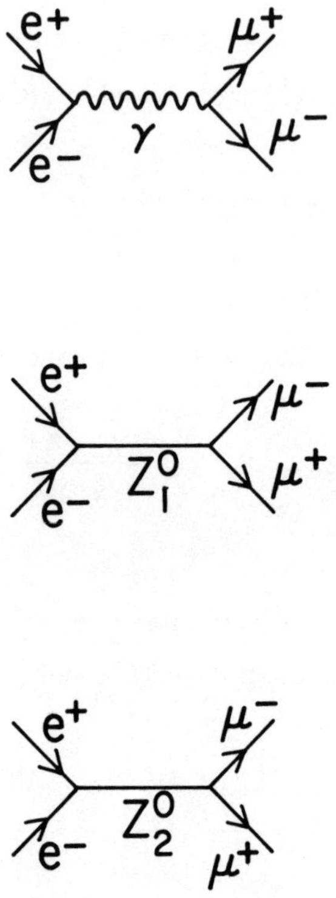

Fig. 2: Electromagnetic and weak neutral current diagrams for $e^+e^- \to \mu^+\mu^-$: Two weak neutral bosons are assumed.

CHARGED AND NEUTRAL-CURRENT INTERFERENCE

To estimate by how much the forward-backward asymmetry may increase, we assume that the couplings of all leptons to Z_2^o are comparable: For example

$$c_{L2}^\nu = c_{A2}^e \quad . \tag{21}$$

For the choice in Eq.(15a) and $\sin^2\theta_{GW} = 1/4$, we find that

$$A(F\text{-}B, 2Z_o) = 3\, A(F\text{-}B, W\text{-}S) \tag{22a}$$

and for the choice in Eq.(15b), we find that

$$A(F\text{-}B, 2Z_o) = 2\, A(F\text{-}B, W\text{-}S) \quad . \tag{22b}$$

These estimates will increase or decrease, to the extent that c_{A2}^e in Eq.(21) is greater than, or less than c_{L2}^ν.

There is always the possibility that μ-e universality either breaks down for Z_2^o, or has to be applied in a way more sophisticated than simple substitution.[14,15] In this case it is conceivable that the Z_2^o contribution in Eq.(17) might tend to cancel the Z_1^o contribution, and hence lead to a forward-backward asymmetry which is much smaller[15] than the prediction of the Weinberg-Salam model.

As far as parity-violating effects are concerned, the difference between the choices in Eq.(15a) and (15b) is much more striking than that for the forward-backward asymmetry. When $\sin^2\theta_{GW} \approx 1/4$, the vector coupling constant g_V of Eq.(6) vanishes, and hence the electron-boson coupling constant c_{V1}^e must also vanish. It follows that, for the choice of Z_2^o couplings in Eq.(15a), the constant c_{V2}^e also vanishes, and with it the parity-violating coefficient of Eq.(19):

$$A(P\text{-}V) = 0 \quad . \tag{23a}$$

In the case of Eq.(15b) the coefficient c_{V1}^e still vanishes, but c_{V2}^e no longer does. As a result, parity violating effects will occur, and the relevant coefficient of Eq.(19) will be "large":

$$A(P\text{-}V) = 1 \quad . \tag{23b}$$

It is natural to consider whether this model can be made into a gauge theory. Since we have introduced one extra neutral boson, the simplest possible group beyond Weinberg-Salam is SU(2) × U(1) × U(1) in which the second U(1) is gauged by Z_2^o. For $\sin^2\theta_{GW} \simeq 1/4$, the choice of constants in Eq.(15a) yields a pure axial-vector current of leptons coupled to Z_2^o:

$$J_\lambda Z_{2\lambda}^o = \sum_\ell c_{A2}^\ell (\bar{\ell}\gamma_\lambda \gamma_5 \ell) Z_{2\lambda}^o \quad . \tag{24a}$$

Independent of $\sin^2\theta_{GW}$, the choice of Eq.(15b) always yields a V-A leptonic current:

$$J_\lambda Z_{2\lambda}^o = \sum_\ell c_{L2}^\ell (\bar{\ell}\gamma_\lambda (1+\gamma_5)\ell) Z_{2\lambda}^o \tag{24b}$$

In a way it is unfortunate that neither of those choices yields a pure vector current, because the Z_2^o sector of the theory would then automatically be free of anomalies;[16] one could then adjust the constants c_2^ℓ (which is always possible in a U(1) gauge), or even add new leptons in order to make the Z_1^o - Z_2^o sector anomaly-free. As things stand, one probably cannot avoid introducing new particles and making the τ-lepton couplings different from those of the muon and electron (which are equal by virtue of μ-e universality) in order to remove all possible anomalies.

To generate lepton and bosons masses, it would probably be necessary to introduce at least one new Higgs scalar $\Phi(Z_2^o)$ which behaves as a doublet with respect to the original SU(2) × U(1) and has a non zero charge with respect to the new U(1). Without this latter requirement, it would not be possible, especially with the choice of Eq.(15b), to construct a gauge invariant lepton-Higgs interaction. An alternative to this might be to use the device employed by Ma, Pramudita, and Tuan[17] of making the new Higgs field a scalar with respect to SU(2) × U(1); particle masses would then be generated by terms which are bilinear in Higgs fields rather than linear. These and other questions remain to be explored should the need arise.

I began this talk by discussing the interference between charged and neutral currents in neutrino-electron scattering, and ended with weak effects in electron-positron annihilation into muon pairs. From an experimental point of view, this is probably the wrong way 'round because weak effects in $e^+e^- \to \mu^+\mu^-$, in particular the forward-backward asymmetry, are likely to be measured before the cross-sections for ν_e-e and $\bar{\nu}_e$-e scattering. This being so, I would like to conclude by observing that a measurement of the forward-backward asymmetry either in excess of the Weinberg-Salam prediction or well below it would strongly suggest the existence of two or more weak neutral vector bosons; on the other hand, a measurement in agreement with the prediction would mean that Weinberg and Salam would have jumped yet another hurdle. Whichever the case, it will still be very important to seek confirmation in neutrino-electron scattering.

The speaker is indebted to his colleagues, especially B. Kayser, for many provocative discussions. His research was supported in part by the U.S. Department of Energy.

REFERENCES

1. S. Weinberg, Phys. Rev. Letters 19, 1264 (1967); A. Salam in Elementary Particle Physics: Relativistic Groups and Analyticity, edited by N. Svartholm (Almquist and Wiksell, Stockholm, 1968) P. 367; S. Glashow, Nucl. Physics 22, 579 (1961).
2. For a review of the latest neutrino data see C. Baltay, Proceedings of XIXth International Conference on High Energy Physics, Tokyo, Japan 1978 (to be published).
3. P. Hung and J. J. Sakurai, Phys. Lett. 72B, 208 (1978); G. Ecker, Phys. Lett. 72B, 450 (1978); P. Langacker and D. Sidhu, Phys. Lett. 74B, 233 (1978); L. F. Abbott and R. M. Barnett, Phys. Rev. D18, 3214 (1978); E. Monsay, Phys. Rev. D18, 2277 (1978). For a review see L. M. Sehgal, Neutrinos - 78 edited by E. Fowler (Purdue University, W. Lafayette, Indiana, 1978) p. 253.

4. The experimental data on e-q scattering is given by C. Y. Prescott et al., Phys. Lett. 77B, 347 (1978); for a phenomenological analysis see L. F. Abbott and R. M. Barnett, SLAC-PUB-2227 (November 1978).

5. B. Kayser, E. Fischbach, S. P. Rosen and H. Spivack (to be published); B. Kayser, Neutrinos - 78 edited by E. C. Fowler (Purdue University, West Lafayette, Indiana, 1978), p. 979.

6. B. Kayser, G. Garvey, E. Fischbach, and S. P. Rosen, Phys. Lett 52B, 385 (1974).

7. L. M. Sehgal, Phys. Lett. 55B, 205 (1975).

8. L. M. Sehgal, Phys. Lett. 48B, 60 (1974), and Nucl. Phys. B70, 61 (1974).

9. F. Reines, H. Gurr, and H. Sobel, Phys. Rev. Lett. 37, 315 (1976).

10. H. H. Chen, F. Reines, R. Burman et al., LAMPF experiment N⁰ 225 (1975); G. A. Brooks, H. H. Chen, J. F. Lathrop, Proceedings of Neutrino - '77 (Nauka, Moscow 1978) Volume 2, p. 376.

11. F. T. Avignone III and Z. D. Greenwood, Phys. Rev. D16, 2383(1978)

12. For a review of the data on ν_μ-e scattering see M. Baldo-Ceoli Neutrino - 78 edited by E. C. Fowler (Purdue University, W. Lafayette, Indiana 1978) p. 387; C. Pascaud, ibid. p. 399; and C. Baltay et al., ibid, p. 413.

13. See, for example, B. Kayser, E. Fischbach, and S. P. Rosen, Phys. Rev. D11, 2547 (1975).

14. See, for example, S. Meshkov and S. P. Rosen, Phys. Rev. Lett. 29, 1764 (1972).

15. K. Higashijima, and R. Sasaki, Prog. Theoret. Phys. (Kyoto) 56 1939 (1976).

16. H. Georgi and S. L. Glashow, Phys. Rev. D6, 429 (1972).

17. E. Ma, A. Pramudita, and S. F. Tuan, Phys. Lett. 80B, 79 (1978)

THE QUARK MODEL PION AND THE PCAC PION*

K. Johnson

Massachusetts Institute of Technology

Cambridge, Massachusetts 02139

I. INTRODUCTION

An important problem in the theory of strong interactions is furnished by the least massive hadrons, the pseudoscalar meson octet. The difficulty is that there are two such octets.

The first is the so-called PCAC octet.[1] According to current opinion, the basis for the theory of strong interactions is QCD. In this theory the quark masses are arbitrary parameters. In particular the masses of the light quarks, u, d and s are small. In the limit where these masses are zero the equations of QCD are chirally invariant. However, the belief is that the ground state spontaneously breaks this symmetry. As a consequence there are zero mass collective states or Goldstone bosons provided by the pseudoscalar octet, the pi's and K's. When quark masses are included, the chiral symmetry is broken, and the pseudoscalar mesons acquire their observed masses. The associated conservation laws are approximate (PCAC). Because the up and down quark masses are considerably smaller than the strange quark mass, the partial symmetry associated with this doublet

---

*This work is supported in part through funds provided by the U.S. Department of Energy (DOE) under contract EY-76C-02-3069.

(SU(2)×SU(2)) is more accurate than that associated with the triplet (SU(3)×SU(3)). The belief that this description is true is not based upon a demonstration that the QCD vacuum spontaneously breaks chiral symmetry, but upon the phenomenological success of PCAC.[1] The dynamical basis for the spontaneous symmetry breaking is at present unknown.[2]

The other pseudoscalar octet is that associated with the quark model. In the quark model the mesons are quark-antiquark composite hadrons with total quark spin angular momentum zero and are closely related to vector mesons with total quark spin one. In the SU(6) approximation, the singlet and triplet meson states have equal masses. In the quark model, the mass of singlet mesons is not zero. However the singlet state is lighter than the triplet as a consequence of the spin dependent quark-quark interaction provided by the colored gluon magnetic exchange interaction.[3,4] Within this model there is no particular reason for the pseudoscalar state to have zero mass when the quark masses are zero. Further, no reason is provided for the validity of the partial symmetries SU(2)×SU(2) or SU(3)×SU(3).

Further in the quark model, the mesons are the L=0 members of rotational bands. The triplet states belong to a rotational family whose masses accurately follow the relation $J=\alpha'M^2+\alpha_o$. However, singlet states do not belong to a family whose masses are observed to follow such a simple relation, even though the other states are present in the data. Hence, even in the quark there seems to be a significant difference between the character of the quark spin singlet and triplet states.[5] In summary, in the quark model the pseudoscalar octet is not qualitatively different from other states and indeed it shares family ties with other hadrons, none of which reveal any obvious indication of a deep connection to the collective aspect of a spontaneously broken symmetry.

Here we wish to give a preliminary report on an attempt to provide a resolution of this dichotomy. We shall show that in the MIT bag model the pseudoscalar quark model mesons can be made to satisfy some necessary conditions to do this. Further, it will be possible to achieve this with the same parameterization which is used to obtain the mass spectrum of all the other hadrons.

For light quarks, the bag interior is defined by the non-chirally invariant condition $q\bar{q}>0$. Attempts have been made to construct a phenomenological bag model with a spontaneously broken chiral symmetry by employing the same mechanism as employed in the well-known σ model.[6,7,8] The difficulty that this approach faces is double counting, since the quark model pion is also present. We shall not introduce a pion field. We shall try to enforce in an ad hoc way on the quark model meson[9] some of the necessary conditions that it must fulfill if it is to properly approximate the one which comes from a chirally invariant microscopic theory with a ground state which spontaneously breaks the symmetry. The principal requirement is that its mass be zero, in the limit where the Lagrangian is symmetric.

For simplicity, here we shall restrict our attention to the most accurate chiral symmetry, $SU(2) \times SU(2)$, which is associated with the up and down quarks. By requiring that the mass of the pi be zero when the quark masses are zero, we shall show that when masses are added to the quarks the mass needed to shift the pi meson to its observed value is about 50 MeV. We shall obtain a size for the pi. The radius is $\sim .6$ fermi, about half the radius of the proton. We shall also estimate the value of the pion decay constant and find $F_\pi \sim 140$ MeV (in comparison to $F_\pi^{observed}=95$ MeV).

II. BRIEF REVIEW OF THE MIT BAG MODEL WITH A SLIGHTLY IMPROVED STATIC BAG APPROXIMATION

Let us adopt the position that the MIT bag[10] model is a phenomenological version of QCD. The space in a bag is a bubble of free field vacuum embedded in the exact QCD vacuum. The free field phase has a higher energy than the QCD vacuum but an amount B per unit volume. B is the MIT bag constant which is a phenomenological parameter with dimension E^4. The interior phase or bag can be obtained in either of two alternative ways. If the color singlet scalar quark density $\bar{q}(x)q(x)$ is positive, then the phase is a free one. Alternatively if $-\frac{1}{4}\text{Tr}[F^{\mu\nu}F_{\mu\nu}] > B$ where $F^{\mu\nu}$ is the color field strength, we also have the free vacuum phase. The Lagrangian which describes this system is [11]

$$L = L_{QCD} - B\left[\theta(\bar{q}q) + \theta(-\tfrac{1}{4}\text{Tr}[FF]-B) - \theta_1 \cdot \theta_2\right] \qquad (2.1)$$

where L_{QCD} is the conventional QCD Lagrangian and $\theta(x) = \begin{cases} 1 & x>0 \\ 0 & x\leq 0 \end{cases}$.

We note that this Lagrangian is <u>not</u> chirally invariant even when the quark mass terms which are present in the conventional QCD Lagrangian are zero. This is because of the presence of the non-chirally invariant operator, $\bar{q}q$, in (2.1). In our view this lack of invariance is a consequence of the spontaneous symmetry breaking to be associated with the exterior, QCD vacuum. The interior, free vacuum is chirally invariant. The boundary between the phases, when that is given by quarks, is the surface where $\bar{q}q=0$, with $\bar{q}q>0$ in the bag. Thus, $\bar{q}q$ defines the separation between an interior, chirally invariant vacuum, and an exterior, non-chirally invariant vacuum. For massive quarks, $\bar{q}q$ is approximated by $\sum_i \rho(x-x_i)$ where $\vec{x}_i$ are the positions of the heavy quarks, and $\rho(\vec{\xi})$ is spread over a region with spatial extent of the order 1/M, where M is the mass of the quark. Hence, $\bar{q}q$ by itself would define only small bags in the neighborhood of the quarks. However, when the quarks are separated by a distance large compared to their Compton size they are connected to each other by large electric colored fields. Thus, the bag extends beyond the quarks since $-\tfrac{1}{4}\text{Tr}[FF]>>B$. The bag is in this case "inflated" by the colored fields carried by the quarks.[12]

In contrast, in light-quark hadrons $\bar{q}q>0$ over the entire domain occupied by the quarks, and because the quarks are in a color singlet combination no large colored fields are present unless the quarks are constrained to be spatially separated as for example by an angular momentum barrier. As a consequence the bag and its boundary are defined by the quarks. This is the circumstance which corresponds to the light quark hadrons with lowest total angular momentum.

To describe the masses and properties of the light hadrons the "static bag" approximation has proved to have remarkable success.[4,13] In this case it is assumed that the hadron is a collection of quarks which in lowest approximation move freely in a sphere with a given radius. The radius is determined by minimizing the total energy

$$E_{STATIC} = E_{QUARKS} + \frac{4\pi}{3} B R^3 \quad . \qquad (2.2)$$

In our earlier work, the mass of the corresponding hadron was taken as the value of (2.2) at the minimum. The wave functions and radius so determined are solutions of the classical bag equations obtained from (2.1). Included in the energy of the quarks is the kinetic energy of the quarks and the lowest order QCD interaction energy calculated perturbatively using the zeroth order quark wavefunctions.

In our earlier work we also included in E_{quark} a term $-z/R$, where R is the radius of the spherical bag and z a dimensionless constant which was taken as a parameter. The term $-z/R$ was added on the expectation that an improvement in the quantum treatment would likely result in yielding a term in the energy of this form, since most of the hadronic constituents were of low mass so the only relevant mass scale would be set by the size of the bag, whence the form $-z/R$ with z a numerical constant. Further, since this addition was expected to be a quantum effect associated with fluctuations within the bag, it should be the same for all hadrons.

The static bag approximation clearly violates translation invariance, that is, the state is not an eigenstate of the total momentum. We can attempt to include a center of mass correction in the following way. If we regard the static bag approximation as a variational guess to a complete quantum treatment,[14] the static bag wave function is a trial wave function which is not an eigenstate of the total momentum, so the static energy is

$$\langle S|\{M^2 + \vec{P}^2\}^{\frac{1}{2}}|S\rangle = E_{STATIC} \qquad (2.3)$$

where M^2 is the mass operator, and $\vec{P}$ is the total momentum. In our wave function, $\langle S|\vec{P}|S\rangle = 0$ but S is not an eigenstate of $\vec{P}$. When $M^2 \gg P^2$, our identification of E_{static} with M becomes exact. The first correction to it would have the form[15]

$$M = E_{STATIC} - \frac{1}{2E_{STATIC}} \langle \vec{P}^2 \rangle \qquad (2.4)$$

For a state with n quarks free in the same mode, $\langle S|\vec{P}^2|S\rangle = np^2$ where p is the momentum of each quark. For massless quarks $p \simeq 2.04/R$. In the lowest approximation,

$$E_{STATIC} = \frac{4}{3} np \qquad (2.5)$$

so

$$M = E_{STATIC} - \frac{3}{8} p = E_{STATIC} - \frac{.77}{R} \, . \qquad (2.6)$$

That is, this correction has the same form as our phenomenological guess $-z/R$. Since we found[4] $z \simeq 1.8$, however, not all of the z needed to give a reasonable fit to the mass spectrum can be associated with this center of mass correction. Although, the center of mass correction has the same form as our guess $-z/R$ for the massive states where $M^2 \gg P^2$, for the lightest states, we would expect that the static bag approximation is the worst, since the correction will no longer have this form. Thus, we come to the pi meson.

III. SHOEHORNING THE QUARK MODEL PI MESON INTO THE PCAC ROLE

Let us suppose that in the case of the pi, the mass comes out very small, that is, that most of E_{static} corresponds to the momentum fluctuation term $\vec{P}^2$. Then let us evaluate $<S|\{M^2+\vec{P}^2\}^{\frac{1}{2}}|S>$ for a quark-antiquark state where $\vec{P}^2 = (\vec{p}_1+\vec{p}_2)^2$ (the bag carries no momentum). Here $|S>$ is a state[14] where $\vec{p}_1^2 = \vec{p}_2^2 = p^2$ where p is single quark momentum. For a massless quark $p \simeq \frac{2.04}{R}$. In this case

$$<S|[M^2+(\vec{p}_1+\vec{p}_2)^2]^{\frac{1}{2}}|S> = \frac{1}{2}\int_{-1}^{+1} dz <S|[M^2+2p^2+2p^2 z]^{\frac{1}{2}}|S> \quad (3.1)$$

$$= \frac{1}{6p^2}[(M^2+4p^2)^{\frac{3}{2}} - M^3] \quad . \quad (3.2)$$

When $4p^2 >> M^2$ we find

$$= \frac{4}{3}p + \frac{M^2}{2p} + \ldots \quad (3.3)$$

Therefore, the mass formula becomes

$$M^2 = 2p\{E_{STATIC} - \frac{4}{3}p\} \quad . \quad (3.4)$$

In a quark-antiquark state

$$E_{STATIC} = 2p - \frac{z}{R} + E_{INTERACTION} + \frac{4\pi}{3}BR^3$$

so

$$M^2 = 2p\{\frac{2}{3}p - \frac{z}{R} + E_{INTERACTION} + \frac{4\pi}{3}BR^3\} \quad . \quad (3.5)$$

We recall that in our original fit $z \simeq 1.8$ and that of this for the massive states approximately 3/4 could be attributed to the center of mass correction. Therefore in (3.5) which now includes the center of mass effect in the form also suitable for low mass states, z is approximately +1.

For the (π/ρ) multiplet, the color spin interaction energy is

$\alpha_s \frac{4}{3} \frac{1}{R} \sigma_1 \cdot \sigma_2 (.175),^4$ which for the π $(\sigma_1 \cdot \sigma_2 = -3)$ is equal to $-\alpha_s (.70) \frac{1}{R}$. Hence we have a total quark energy

$$\frac{2}{3} p - \frac{z}{R} + E_{INTERACTION} \simeq (\frac{4}{3} - 1 - .70 \alpha_s) \frac{1}{R} \quad (3.6)$$

If we used the value of α_s (≈ 2.2)[16] determined to fit the heavier states,[4] the total static bag energy wouldn't even be positive. However, as α_s increases from 0, the mass of the particle which results from minimizing (3.6) would have a smaller and smaller radius. Hence, one would expect in calculations which include the higher order effects of asymptotic freedom that the "effective" value of α_s which would be relevant to a low mass, small extended hadron would be smaller than that relevant for the higher mass, larger hadrons.

To estimate this effect crudely let us use for α_s in (3.6) an R-dependent coupling parameter,

$$\alpha_s(R) = \frac{1}{\frac{9}{2\pi} \log(\frac{R_0}{R})} \quad (3.7)$$

which takes into account the asymptotic weakening of α which includes the effects of up, down and strange quark pairs, since the relevant hadron size is small in comparison to the corresponding Compton size (except for the strange quarks which do not make a large contribution in $\alpha_s(R)$). We choose R_0 so $\alpha_s(R) \approx 2.2$ when $\frac{1}{R} \approx \frac{1}{5}$ GeV, which is the characteristic size[4] of the heavier hadrons. This gives $\frac{1}{R_0} = .145$ GeV. As a check we find that when $\frac{1}{R} \sim 3$ GeV, $\alpha_s \cong .25$, which corresponds well with the effective coupling constant used at this distance in applications of perturbative QCD. We note from (3.5), as a consequence of the asymptotic weakening of the coupling constant α_s, that it becomes possible to choose z so that $M^2 = 0$ at the radius where $\partial M^2/\partial R = 0$. This is <u>not possible</u> with a constant α_s. Indeed, if z is so chosen, we find from the equation, $\partial M^2/\partial R = 0$, $M^2 = 0$;

$$23.9 \, BR^4 = R \frac{\partial \alpha_s(R)}{\partial R} \qquad (3.8)$$

so if α_s=constant, R=0.

A zero radius state would represent an instability which would not be consistent with the standard perturbative treatment of chiral symmetry breaking in field theory. We shall see in a moment that if we are consistently to fit our quark model pion into the PCAC framework, a necessary condition will be that when the quark masses vanish, the pion should have a finite radius. If we calculate the radius gotten from (3.8), using (3.7) for $\alpha_s(R)$, and $\frac{1}{R_0} = .145$, we find,

$$R_\pi = 3.4 \, \text{GeV}^{-1} \, . \qquad (3.9)$$

By choosing other forms for $\alpha_s(R)$ with the same asymptotic behavior when R→0, we find that $\frac{1}{R_\pi^2}$ varies at most by 20%. The value of z required to make M^2=0 may then be calculated from (3.6). We find z=.9, very close to the estimated value obtained from the phenomenology of the massive hadrons.

Let us now <u>assume</u> that if we had a description which was sufficiently accurate to describe the relation between the inside and outside vacua which was consistent with the quark-antiquark pion being also the Goldstone boson associated with spontaneously broken chiral symmetry, it would fix z so this state would have zero mass when the up and down quark masses are <u>exactly</u> zero. We can then determine how E_{quark} shifts when a non-vanishing mass is given to the up and down quarks. Only the average $\frac{1}{2}(m_d + m_u) = m$ is relevant since the pi contains just one quark of each flavor (for the π^0, on the average). When the quark mass is small, the only change required in (3.5) is to use the momentum relevant for a quark with a small mass,

$$p = \frac{x}{R} + \left.\frac{\partial p}{\partial m}\right|_{m=0} m + \ldots \qquad (3.10)$$

where $x = 2.04...$. (The quark energy is $\sqrt{p^2+m^2} = p + \frac{m^2}{2p} + ...$ but $m^2/2p$ will be negligible in comparison to m.) $\partial p/\partial m$ can be obtained from the eigenvalue equation associated with the Dirac wave function, or also

$$\left.\frac{\partial p}{\partial m}\right|_{m=0} = \int d^3x\, \bar{\psi}_o \psi_o = \frac{1}{2(x-1)} \quad (3.11)$$

where ψ_o is the zero mass bag wavefunction.

Thus we find, using (3.5) with z chosen so $M_\pi^2=0$ when m=0, and for a radius close to R_π,

$$M^2 = \frac{1}{2} a\, (R-R_\pi)^2 + \frac{2x}{3(x-1)} \frac{m}{R} . \quad (3.12)$$

Consequently, we find that to first order in m,

$$M_\pi^2 = \frac{2x}{3(x-1)} \frac{m}{R\pi} \quad (3.13)$$

If we evaluate (3.13) using our previously determined radius $R=3.4$ GeV^{-1} and $M=139$ MeV, we find $m=50$ MeV.

To check the consistency of this model of the pion, we may also compute the electromagnetic mass difference between the π^+ and π^o. The electromagnetic masses of the light-quark hadrons were calculated previously[17] using the linear formula $M \sim E_{BAG}$ valid for the heavier states. Generally satisfactory results were obtained with the single exception of the pi meson. Let us now transcribe our earlier work into the form it should take with a pion which is massless when symmetry breaking effects are absent. If we use formula (3.5), we find

$$M_+^2 - M_o^2 = \frac{2x}{R} \cdot \{E_+ - E_o\} \quad (3.14)$$

where E_+ and E_o are the <u>static</u> bag quark energies of the π^+ and π^o

which include the lowest order electromagnetic terms. Then,

$$M_+ - M_0 = \frac{x}{M_\pi R_\pi}(E_+ - E_0) \qquad (3.15)$$

In our earlier calculation[17] we found

$$E_+ - E_0 = 1.6 \text{ MeV} \qquad (3.16)$$

with our old pion radius $R'_\pi = 3.34$. Since E scales with R we now should put $E_+ - E_0 = 1.61 \times \frac{3.34}{3.40} = 1.58$ MeV. We then find

$$M_+ - M_0 = \frac{2.04}{(.139)(3.4)}(1.58) = 6.8 \text{ MeV} \qquad (3.17)$$

which is considerably more accurate $[(M_+ - M_0)_{observed} = 4.6 \text{ MeV}]$. Since $(M_+ - M_-)$ is proportional to $1/R_\pi^2$, we might alternatively use $M_+ - M_-$ to determine R_π. We then find

$$R = 4.1 \text{ GeV}^{-1} \qquad (3.18)$$

This would then give us using (3.13), a quark mass m=58 MeV. We could then compare the "observed" radius of R = 4.1 GeV to the calculated on R = 3.4 GeV. The difference is within that expected on the basis of our crude estimation of the effects of asymptotic freedom.

IV. THE PION DECAY CONSTANT

We shall finally estimate the magnitude of the pion decay constant, F_π. A method for calculating such an amplitude in the static bag model has been proposed earlier,[18] but it is only applicable for states where the mass is large in comparison to the energy contained in momentum fluctuations. For the π we have the opposite circumstance.

Let us start from the field theory formula for the π^-,

$$<0|\bar{u}(x)\gamma^\mu\gamma_5 d(x)|p> = i\sqrt{2}\, F_\pi\, p^\mu e^{ipx} \quad . \tag{4.1}$$

We first construct a wave packet localized at the origin in space, which is where we shall put the bag model pion. Then

$$|Bag> = \int d^3p\, \frac{\phi(p)}{\omega_p} |p> \tag{4.2}$$

will be such a state, with $\phi(p)$ describing the wave packet. The bag state should be normalized to unity so

$$1 = \int d^3p\, \frac{\phi^2(p)}{\omega_p^2}\, 2\omega_p\, (2\pi)^3 \tag{4.3}$$

since in (4.1) we have a covariantly normalized state, $<p|p'> = (2\pi)^3 2\omega_p \delta^{(3)}(p-p')$. We then have,

$$<0|\bar{u}(x)\gamma^0\gamma_5 d(x)|BAG> = i\sqrt{2}\, F_\pi \int d^3p\, \phi(p) e^{ipx} \quad . \tag{4.4}$$

We now <u>assume</u> that the static bag model wave functions may be used to compute the left hand side of (4.4). On combining (4.3) with (4.4) we obtain an expression for F_π. In making this approximation, we have taken as one the overlap between the "empty" bag and lump of free, unstable vacuum which remains in the true vacuum after the quarks have disappeared. Therefore we should expect to find that we obtain a larger value for F than the observed F_π. The evaluation of F_π is now straightforward and follows closely the standard static bag model computations.[4,13] We find

$$F_\pi = \frac{.501}{R_\pi} \tag{4.5}$$

(Again, we see why R_π must remain finite as $m_\pi \to 0$.) Using the radius estimated from the asymptotic freedom argument ($R_\pi = 3.4$ GeV^{-1}), we get

$$F_\pi = 150 \text{ MeV} \qquad (4.6)$$

to be compared with $F_\pi^{observed} = 95$ MeV. If we use the radius which fits the electromagnetic mass difference, $R_\pi = 4.1$ GeV^{-1}, we get $F_\pi = 120$ MeV.

IV. CONCLUSION

We have defined a static bag model with an estimated correction for the violation of translation invariance. We have shown that the lightest quark model state, the pi, has a mass which is quite small when parameters determined solely from the spectroscopy of massive hadrons are used. If we assume that a more accurate treatment would make this mass zero, when the quark masses are zero, we find that its mass shifts to its observed position with a quark mass ~ 50 MeV.[19] Further, we have shown that the radius also determined in this way yields an electromagnetic splitting, $\pi^+ - \pi^0$, in good agreement with the observed value. We have also obtained reasonable estimates for the decay constant F_π, and the pion radius, R_π.

Although the quark mass $\frac{1}{2}(m_u + m_d) \sim 50$ MeV we have obtained does not correspond closely to the value gotten in the context of current algebra,[20,21] it shares with those masses the qualitative feature that

$$\frac{1}{2} \frac{m_d - m_u}{(m_d + m_u)} \sim \frac{1}{10}$$

is larger than electromagnetic order. ($m_d - m_u$ is obtained from mass differences in the hadron spectrum and is ~ 3 to 4 MeV.)

V. ACKNOWLEDGMENT

I would like to thank my colleagues in the Center for Theoretical Physics for numerous discussions about these matters. In particular, I would like to acknowledge many helpful conversations with Dr. John Donoghue and Professor R.L. Jaffe. I would also like to thank Dr. M. Peshkin for pointing out an error in an earlier version of this work.

REFERENCES

1. H. Pagels, Physics Letters C, Vol. 16, 219 (1975).
2. C. G. Callan, R. Dashen and D. J. Gross, Phys. Rev. $\underline{D17}$, 2717 (1978); Princeton preprint, June 1978; R. Carlitz and D. Creamer Pittsburgh preprint; W. Bardeen and C. Lee, Fermilab-pub-79/18th
3. A. De Rújula et al., Phys. Rev. $\underline{D12}$, 147 (1975).
4. T. DeGrand et al., Phys. Rev. $\underline{D12}$, 2060 (1975).
5. C. Nohl and K. Johnson, Phys. Rev. $\underline{D19}$, 291 (1979).
6. A. Chodos and C. B. Thorn, Phys. Rev. $\underline{D12}$, 2733 (1975).
7. C. Callan et al., loc. cit.
8. G. Brown, M. Rho and V. Vento, Stony Brook preprint, 1979.
9. A similar idea has been explored by R. Friedberg and T. D. Lee, Phys. Rev. $\underline{D18}$, 2623 (1978).
10. A. Chodos, et al., Phys. Rev. $\underline{D9}$, 3471 (1974).
11. K. Johnson, Phys. Letters $\underline{78B}$, 259 (1978). Here we have a Lagrangian which is a combination of the two given in this paper
12. P. Gnädig, P. Hasenfratz, J. Kuti and A. S. Szalay, Phys. Letter $\underline{64B}$, 62 (1976); K. Johnson, AIP Conf. Proc. $\underline{48}$, 112 (1978).
13. A. Chodos et al., Phys. Rev. $\underline{D10}$, 2599 (1974).
14. To estimate the mean value of functions of P_{TOT}^2 we have assumed that the quarks move independently in the static bag with a magnitude of p^2 given by E^2-m^2, where E is the static bag energy of one quark.
15. This is equivalent to the form of the correction obtained in a complete quantum treatment valid for small amplitude bag deformations by C. Rebbi, Phys. Rev. $\underline{D12}$, 2407 (1975).
16. R. L. Jaffe, F. Low, private communication.
17. N. G. Deshpande et al., Phys. Rev. $\underline{D15}$, 1885 (1977).

18. P. Hays, M. V. K. Ulehla, Phys. Rev. D13, 1339 (1976); P. Hays, Erratum, Phys. Rev. D15, 931 (1977); the final expression for F_π differs from that in this paper in that
$$\int \frac{d^3p}{(2\pi)^3} \frac{1}{\omega_p} e^{i\vec{p}\cdot(\vec{x}-\vec{y})} = \frac{1}{2\pi^2} \frac{1}{(\vec{x}-\vec{y})^2}$$
replaces $\frac{1}{M_\pi} \delta^{(3)}(\vec{x}-\vec{y})$.

19. J. F. Donoghue, et al., Phys. Rev. D12, 2875 (1975).
20. S. Weinberg, Harvard preprint, HUTP-77/A057.
21. See however, J. Gunion et al., Nucl. Phys. B23, 445 (1977).

QUARK MODEL EIGENSTATES AND LOW ENERGY SCATTERING[†]

F.E. Low

Center for Theoretical Physics and Laboratory
for Nuclear Science and Department of Physics
Massachusetts Institute of Technology
Cambridge, Massachusetts 02139

In this talk, based on work with R. Jaffe, I wish to discuss the connection between the discrete mass eigenstates which are calculated in an approximate quark-gluon model and the observed low energy hadron-hadron scattering in the same mass region. The problem we claim to have solved can be stated crudely: where are the exotics? The quark model predicts exotic states, for example, in the I=2 ππ system at 1.15 GeV; the I=2 ππ S-wave phase shift, on the other hand, shows no sign of interesting behavior up to an energy of about 1.5 GeV, by which time it is approximately -.5 radians.

The problem is the following: If the mass eigenvalue is below the threshold for two particle emission, we have a true discrete state, a strongly stable object which can be directly observed, or indirectly observed through its possible weak or electromagnetic

---

[†]This work is supported in part through funds provided by the U.S. Department of Energy (DOE) under contract EY-76-C-02-3069.

decay. If the mass eigenvalue is above the two particle threshold, the situation is more complicated. In this case, we call the discrete eigenstate a "primitive". Its very existence depends on having made an approximation, that is, on having in some way decoupled it from its decay channels. Now, if the primitive is in fact sufficiently weakly coupled to its decay channels, we know that a narrow resonance will appear in the scattering and reaction amplitudes, and that the width of the resonance will go to zero with the coupling. The resonant energy will lie close to the primitive eigenvalue, as will the real part of the pole in the complex energy plane An example of this well understood situation is provided by the lowest vector meson nonet - the ρ, ω, k^* and ϕ. For these states, the approximation that creates the primitives might be the restriction to the $q\bar{q}$ sector of the Hilbert space for the vector mesons as well as for each of the pseudoscalar mesons into which they decay. The decays can proceed only via the two quark-four quark coupling, and are in addition inhibited by the $\ell=1$ angular momentum barrier of the final state.

On the other hand, if the primitive is strongly coupled to its decay channels, then understanding its reflection in the observed amplitudes requires an analysis that may depend in detail on the approximation that created the primitive in the first place. Of course, if we had a full understanding of the system, we could simpl calculate the observed amplitudes themselves. Lacking that, we exploit our (hoped for) understanding of the short range interaction of quarks and gluons to determine the properties of the primitives, and then analyze the low energy amplitudes to measure the masses and coupling of these primitives. Our technique is a modification of tl Wigner-Eisenbud[1] formalism suited to the case of primitives created by confining boundary conditions.

The kind of primitive we have in mind here would consist of four, five, six or more quarks and anti-quarks in SU(3) singlet states, i.e. $q^2\bar{q}^2$, $q^4\bar{q}$, q^6, etc. The approximation that creates th discrete primitives consists in confining the quarks and gluons to a spherical volume of radius R with bag boundary conditions at that radius.

In the following we shall consider in particular the primitives created by four quarks in even parity, $j=1/2$ states of the bag model with total angular momentum zero. These are then coupled to the S states of the two pseudoscalar meson system.

We digress now to give a simple example of how a confining boundary condition creates primitives, and how these primitives can be studied in the scattering. Consider the S-wave scattering of a nonrelativistic particle by a weak attractive square well of radius b and depth $V_o = U\hbar^2/2m$. It is clear that no resonant or bound states are created by this potential. However, if one imposes a boundary condition requiring the wave function to vanish at b, one creates an infinite set of primitive internal states at momenta k_n, where $b\sqrt{k_n^2+U}=n\pi$ $n=1,2,\ldots$. Now, although the phase shift does nothing spectacular as k varies from zero to infinity, we can precisely identify the primitive by looking for the poles of the quantity

$$P = k\omega + (kb+\delta(k))$$

which occur precisely at the k_n's of the primitives. This is because the external wave function $\sin(kb+\delta)$ clearly vanishes (by continuity) when the internal wave function satisfies the vanishing boundary condition at r=b. Therefore the study of the P matrix can reveal to us the primitives associated with a boundary condition requiring the wave-function to vanish at a radius b.

We make contact with the quark-gluon bag model calculations by noting that the two body relative probability density

$$\rho_Q(\vec{r}) = \int (\frac{\vec{r}_1+\vec{r}_2}{2} - \frac{\vec{r}_3+\vec{r}_4}{2} - \vec{r})\rho(\vec{r}_1)\rho(\vec{r}_2)\rho(\vec{r}_3)\rho(\vec{r}_4)d\vec{r}_1 d\vec{r}_2 d\vec{r}_3 d\vec{r}_4$$

vanishes very strongly at r=2R. In fact, it is almost indistinguishable from the density

$$\rho_M = |\phi(r)|^2 \quad,$$

where

$$\phi = \frac{1}{\sqrt{2\pi b}} \frac{\sin \pi r/b}{r} \quad,$$

with b adjusted to give the same mean-square radius as ρ_Q. The resulting relationship is b~1.4R. ρ_Q and ρ_M are shown in Fig. 1. We therefore interpret bag model calculations of primitives as corresponding to a vanishing wave-function at r=b; hence we study, for elastic scattering,

$$P = k\omega + (kb+\delta) \quad,$$

and for multi-channel reactions the matrix analogue. We take R (and hence b) from the calculations of Jaffe;[2] R is related to the expected mass of the primitive by the virial theorem

$$R = 5M^{\frac{1}{3}} \text{ GeV}^{-1} \quad,$$

and hence

$$b = 7M^{\frac{1}{3}} \text{ GeV}^{-1} \quad.$$

Before we look at data, we observe that a primitive which occurs at a wave-number k_c such that $k_c b=\pi$ requires a zero phase-shift at $k=k_c$, since if $P = \frac{k \cos(kb+\delta)}{\sin(kb+\delta)}$ has a pole at $k_c b=\pi$, $\delta(k_c)$ must be zero. We call this phenomenon compensation and refer to $M(k_c)=M_c$ as the compensation mass. Now, if $M_n<M_c$, then $\delta(k_n)>0$; if $M_n>M_c$, then $\delta(k_n)<0$. Thus a positive phase shift signals a primitive below the compensation energy, a negative phase shift signals a primitive above the compensation energy. Using $k=\pi/b(M)$ and $b(M)=7M^{\frac{1}{3}}$ we find compensation masses of .95 for the $\pi\pi$ and 1.08 for the πK systems.

Turning to the data, we consider first the exotic channels. Fig. 2(a) shows the I=2 S-wave $\pi\pi$ phase shift and Fig. 2(b) the

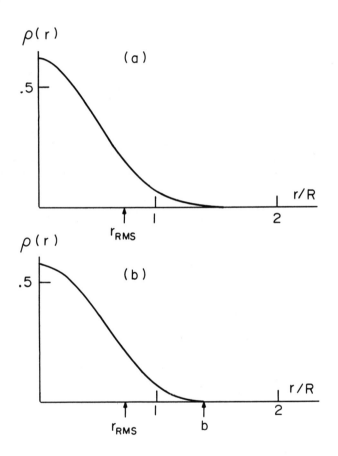

Figure 1

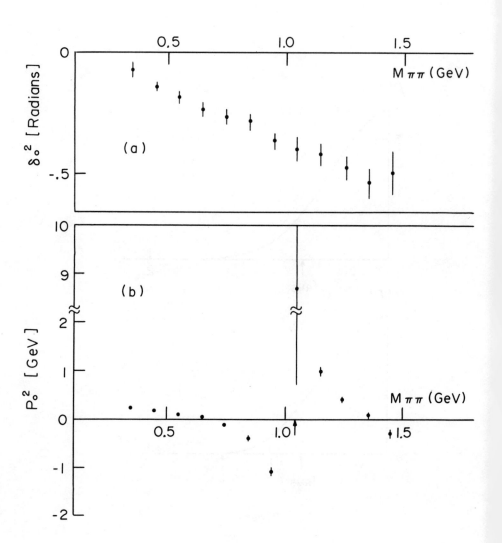

Figure 2

QUARK MODEL EIGENSTATES AND LOW ENERGY SCATTERING

corresponding P matrix with a pole at 1.04. The Jaffe prediction is 1.15.

Fig. 3(a) shows the I=3/2 S-wave $K\pi$ phase shift and Fig. 3(b) the corresponding P matrix with a pole at 1.19. The Jaffe prediction is 1.35. Note the repulsive phase shifts corresponding to primitives above the compensation energy.

We consider next the non-exotic states. Fig. 4(a) shows the $\pi\pi$ I=0 S-wave phase shift below $\overline{K}K$ threshold and Fig. 4(b) the P matrix element for the same region. There is a pole at .69; the Jaffe four quark prediction is .65. Going up in energy, there is a second state at .98; the Jaffe prediction is 1.10 - it is presumed to be a state with hidden strangeness - an $s\bar{s}$ pair - and the remaining $u\bar{u}$ and $d\bar{d}$ in an I=0 combination. Above $\overline{K}K$ threshold, we write

$$P_{ij} = N_{ij}/D$$

and

$$D = \cos(\theta_1+\theta_2) - \eta \cos(\theta_1-\theta_2) \quad ,$$

where

$$\theta_i = k_i b + \delta_i \quad ,$$

and

$$S = \begin{pmatrix} \eta e^{2i\delta_1} & i\sqrt{1-\eta^2}\, e^{i(\delta_1+\delta_2)} \\ i\sqrt{1-\eta^2}\, e^{i(\delta_1+\delta_2)} & \eta e^{2i\delta_2} \end{pmatrix} ,$$

and obtain η, δ_1, and δ_2 from the data. Fig. 5 shows D, N_{11}, N_{12} and N_{22} below 1.01 and 1.11 GeV. The pole of P (zero of D) is now at 1.04, 60 MeV higher than the displaced pole found earlier in the elastic data. The two numbers appear consistent, but the data are

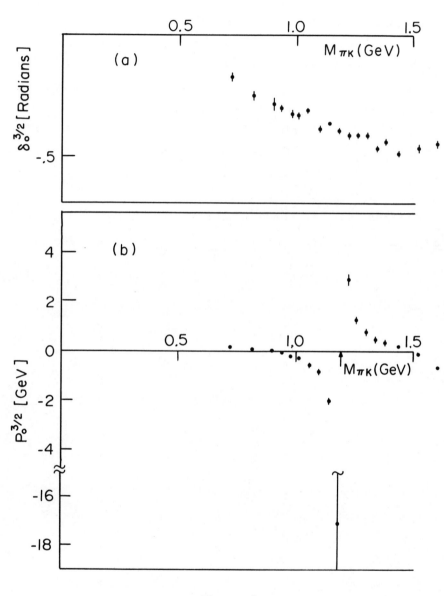

Figure 3

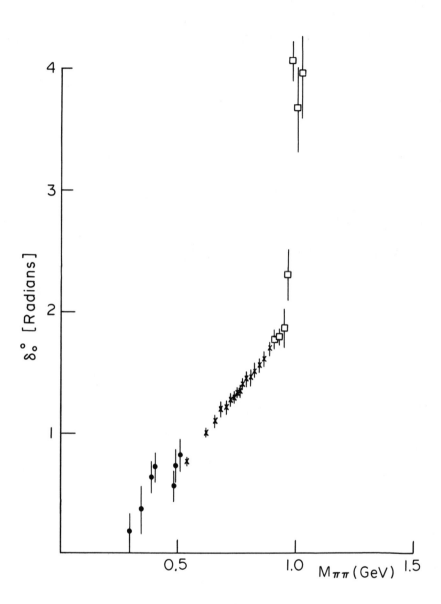

Figure 4a

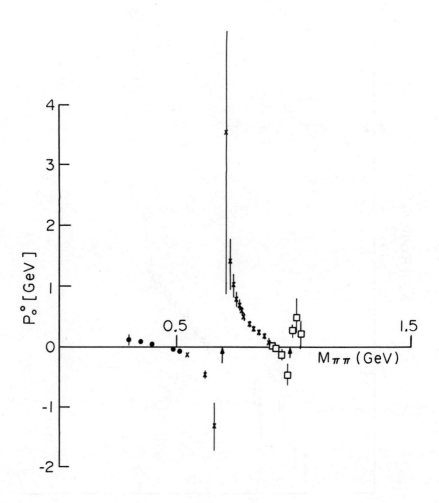

Figure 4b

QUARK MODEL EIGENSTATES AND LOW ENERGY SCATTERING

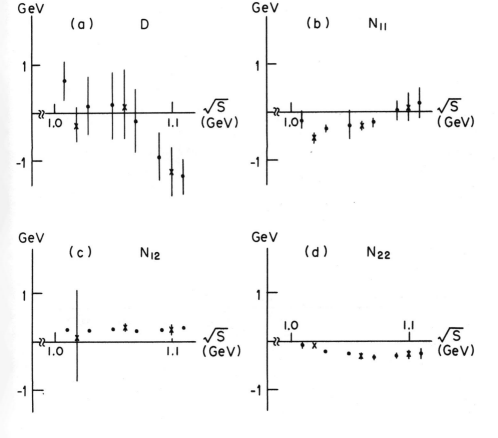

Figure 5

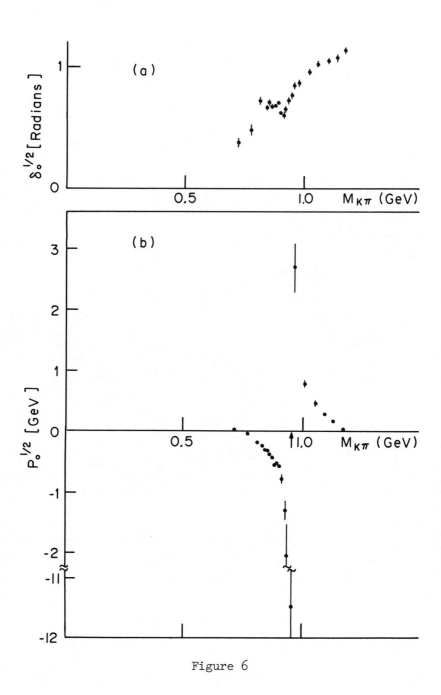

Figure 6

not sufficiently accurate to carry out the necessary extrapolation into the closed channel.

We finally have the I=1/2 Kπ system. The S-wave phase shift is shown in Fig. 6(a), the P matrix element in Fig. 6(b). The pole is at .96, the Jaffe prediction at .90.

In summary, the Jaffe flavor nonet (except for the experimentally inaccessible I=1) and the low lying members of his 36 are in the data, at approximately expected energies.

REFERENCES

1. E.P. Wigner and L. Eisenbud, Phys. Rev. 72, 29 (1947).
2. R. Jaffe, Phys. Rev. D15, 267 and 281 (1977).

ON THE EQUATIONS OF STATE IN MANY BODY THEORY*

R. E. Norton

University of California

Los Angeles, California 90024

By generalizing formulae previously employed to study the equations of state in the limit of low temperature, I derive expansions for the equations of state which involve products of full, renormalized (in the sense of relativistic field theory) amplitudes. Higher order terms in the expansions are formally associated with higher powers of the temperature, although in general the effects of composite excitations coming from all orders could dominate at low temperature. The sum of a class of terms in the expansions is shown in the narrow resonance approximation to yield the equations of state of an ideal gas composed of the resonant excitations.

I. INTRODUCTION

Employing a functional expression for the thermodynamic potential, Luttinger[1] and others[2] in the early sixties demonstrated that the equations of state of a normal Fermi system at very low temperature are essentially those of an ideal gas of quasi-particles -- the quasi-particles being the excitations associated with the poles of the full single particle fermion propagators. These works constituted a deri-

---

*Work supported in part by the National Science Foundation.

vation from many-body theory of the predictions of Landau's phenomenological theory of normal Fermi systems.

Somewhat later a similar formalism was applied to boson systems,[3] and the low temperature limit of the equations of state for He^4 were shown[4] to coincide with those implied by the phenomenological theory. More recently,[5] these methods were extended to discuss the non-leading temperature dependent terms in the specific heat of an electron gas.

Although the authors of these works did not apparently have application to relativistic systems in mind, it is easy to see that the formalism can readily be generalized to allow such applications and that the resulting expressions for the equations of state would then involve only renormalized amplitudes (in the sense of relativistic field theory). Partially motivated by this observation, I collaborated[6] in an attempt to derive all the contributions to the equations of state which arise from pursuing these methods. Unfortunately, because of an error in keeping track of the appropriate boundary values of various amplitudes, the expressions for the equations of state presented in reference 6 are incomplete and hence incorrect. In this lecture I will describe the results of my effort over the past few years to obtain the correct generalized expression for the equations of state.

The formulae for the equations of state which emerge from this effort involve products of renormalized n-point amplitudes defined at real frequencies. These amplitudes are analytic continuations of the conventional[7] many body amplitudes involving discrete, imaginary frequencies. In the limit of zero temperature they coincide with the Feynman amplitudes defined as the Fourier transform of the ordinary time ordered product of fields (averaged over the ensemble).

In Section II the various amplitudes are defined and the notation is established. In Section III I discuss aspects of the structure of the n-point amplitudes which are used in subsequent sections, and which are relevant to the analytic continuation to re frequencies. The derivation of the expansion for the entropy density is outlined in Sections IV-VI. In Section VI a class of contributions to the entropy density is displayed and formally summed; in the narrow resonance approximation, the result is shown to yield the entropy density of an ideal gas associated with the composite excitations. Finally, in Section VII, I discuss briefly some aspects of the remaining terms in the expansion for the entropy density, and describe the properties of the analogous expansions for the averages of the conserved charge densities.

Throughout these lectures the discussion is necessarily very sketchy. A more complete presentation of this work will be submitted elsewhere.

II. PRELIMINARIES

The statistical average of any operator A is given by

$$<A> = \frac{1}{Z} \text{Trace}[e^{-\beta(H-\mu^\alpha Q^\alpha)} A] \quad , \qquad (2.1)$$

where Z is the grand partition function

$$Z = \text{Trace } e^{-\beta(H-\mu^\alpha Q^\alpha)} \quad , \qquad (2.2)$$

H is the Hamiltonian (of a relativistic or non-relativistic field theory), and the Q^α are mutually commuting, conserved charges

$$[H, Q^\alpha] = [Q^\alpha, Q^\beta] = 0 \quad . \qquad (2.3)$$

Associated with each conserved charge Q^α is a "chemical potential" μ^α. The grand partition function Z thus depends upon[8] β [equal to $(kT)^{-1}$; k = Boltzmann's constant, T = temperature], the μ^α and whatever other parameters determine the spectrum of states. In the absence of external fields, the only other parameter of this kind is the volume V. The logarithm of Z is proportional to V.

The thermodynamic potential density Ω/V is defined by

$$\frac{\Omega}{V} \equiv -\beta^{-1} V^{-1} \ln Z \quad, \tag{2.4}$$

so that from (2.1) and (2.2) the average charge densities are the chemical potential derivatives of the thermodynamic potential density,

$$\frac{\langle Q^\alpha \rangle}{V} = -\frac{\partial}{\partial \mu^\alpha} \frac{\Omega}{V} \quad. \tag{2.5}$$

The β derivative of Ω/V gives the entropy density S/V,

$$\frac{S}{V} = \beta^2 \frac{\partial}{\partial \beta} \frac{\Omega}{V} \quad. \tag{2.6}$$

A primary objective of this work is to show how the quantities in (2.5) and (2.6) can be expressed as sums of (integrals over) products of renormalized many body amplitudes.

If the "time" dependences of the field operators ϕ are generated by

$$\phi(\underline{x}, x_o) \equiv e^{x_o(H-\mu^\alpha Q^\alpha)} \phi(\underline{x}, 0) e^{-x_o(H-\mu^\alpha Q^\alpha)} \quad, \tag{2.7}$$

the n-point amplitudes $\tilde{T}_n$ are defined by

$$\tilde{T}_n(x_1 \ldots x_n) \equiv \langle T(\phi(x_1) \ldots \phi(x_n)) \rangle \quad, \tag{2.8}$$

where the symbol T indicates an ordering with respect to the times x_{io} (larger values of x_{io} to the left). For an isotropic, homogeneous system the $\tilde{T}_n$ in (2.8) depend only upon the differences of the coordinates x_i. For n=2 the average in (2.8) is the single particle propagator Δ,

$$\Delta^{ij}(x_1 - x_2) \equiv \langle T(\phi^i(x_1) \phi^j(x_2)) \rangle \quad. \tag{2.9}$$

For an ideal gas, where H in (2.1) is simply the Hamiltonian H_o for non-interacting fields, the propagator Δ is given by [suppressing internal indices]

$$\Delta_o(x) = \frac{1}{\beta} \sum_\nu \int \frac{d^3p}{(2\pi)^3} e^{i\underline{p}\cdot\underline{x} - Z_\nu x_o} \frac{1}{\underline{p}^2 - (Z_\nu + \mu^\alpha q^\alpha)^2 + M^2} \qquad (2.10)$$

where the imaginary, discrete frequencies Z_ν are given by[7]

$$Z_\nu = \frac{2\nu\pi i}{\beta} \qquad \text{bosons} \quad (\nu = 0, \pm 1, \pm 2, \ldots)$$
$$= \frac{(2\nu+1)\pi i}{\beta} \qquad \text{fermions}, \qquad (2.11)$$

and where the matrices q^α indicate the charge of Q^α destroyed by the fields

$$[\phi_i, Q^\alpha] = q_{ij}^\alpha \phi_j . \qquad (2.12)$$

When the fields ϕ carry spin there are additional tensor or spinor projection operators in the numerator on the right side of (2.10).

For the interacting system the propagators $\Delta(x)$ have the Fourier transform

$$\Delta(x) = \frac{1}{\beta} \sum_\nu \int \frac{d^3p}{(2\pi)^3} e^{i\underline{p}\cdot\underline{x} - Z_\nu x_o} \Delta(\underline{p}, Z_\nu) , \qquad (2.13)$$

with $\Delta(p, Z_\nu)$ given by

$$\Delta^{-1}(p, Z_\nu) = \underline{p}^2 - (Z_\nu + \mu^\alpha q^\alpha)^2 + M^2 + \Sigma(p, Z_\nu) \qquad (2.14)$$

in terms of the self energy $\Sigma(p, Z_\nu)$. The self energy Σ can be equated to an infinite sum of graphs, where the lines and vertices of the graphs represent the propagators Δ_o and the coupling constants. Associated with each closed loop is an independent momentum integration $(2\pi)^{-3} d^3q$ and frequency summation $\frac{1}{\beta} \sum_\nu$ arranged to conserve momentum and frequency at each vertex.

Similarly, the amplitudes $\tilde{T}_n$ in (2.8) can be equated to a sum of graphs. Associated with each external coordinate x_i in (2.8) is a factor of Δ in the graphical expansion of $\tilde{T}_n$. Let us define the amplitude T_n, without the tilde, as the amplitude obtained by canceling off the factors of Δ associated with the external legs of the graphs for $\tilde{T}_n [\int d^4y \equiv \int d^3y \int_0^\beta dy_0]$:

$$T_n(x_1 \ldots x_n) = \int d^4y_1 \ldots \int d^4y_n \Delta^{-1}(x_1-y_1) \ldots \Delta^{-1}(x_n-y_n) \tilde{T}_n(y_1 \ldots y_n) .$$

(2.15)

Translational invariance implies that the Fourier transform of T_n is proportional to a product of a momentum conserving delta function and a frequency conserving Kronecker delta. Thus, $T_n(p_{\nu_1} \ldots p_{\nu_n})$ is defined by

$$T_n(x_1 \ldots x_n) = \frac{1}{\beta} \sum_{\nu_1} \int \frac{d^3p_1}{(2\pi)^3} \ldots \frac{1}{\beta} \sum_{\nu_n} \int \frac{d^3p_n}{(2\pi)^3} e^{\sum_j (ip_j \cdot x_j - z_{\nu_j} x_{j0})} (2\pi)^3$$

$$\times \delta^3(\Sigma p_i) \beta \delta_{\Sigma \nu_j, 0} T_n(p_{\nu_1} \ldots p_{\nu_n}) , \qquad (2.16)$$

where p_{ν_i} indicates the combination $\underline{p}_i$, Z_{ν_i}. $T_n(p_{\nu_1} \ldots p_{\nu_n})$ is only defined when the total incoming momentum $\Sigma_i \underline{p}_i$ and the total incoming frequency $\Sigma_i Z_{\nu_i}$ are zero.

III. ANALYTIC PROPERTIES OF THE T_n

Suppressing all coupling constants and all reference to three-momenta, the typical graph for $T_n(Z_{\nu_1} \ldots Z_{\nu_n})$ with m internal lines has the form

$$T_n(Z_{\nu_1}\ldots Z_{\nu_n}) = \frac{1}{\beta}\sum_{\alpha_1}\ldots\frac{1}{\beta}\sum_{\alpha_n}\Delta(Z_{\alpha_1})\ldots\Delta(Z_{\alpha_n})$$

(3.1)

$$\times \mathcal{D}(Z_{\alpha_1}\ldots Z_{\alpha_m}, Z_{\nu_1}\ldots Z_{\nu_{n-1}}),$$

where $\mathcal{D}$ is a product of β times a frequency-conserving Kronecker delta for each vertex, except for one, of the graph. In (3.1) the vertex connecting to the line of frequency Z_{ν_n} is omitted from the Kronecker delta product. The Kronecker delta product can be analytically continued away from the discrete frequencies Z_{α_i}. Thus, the frequency sums in (3.1) can be replaced by contour integrals according to the rule [upper sign refers to fermions]

$$\frac{1}{\beta}\sum_\alpha f(Z_\alpha,\ldots) \to \mp\frac{1}{2\pi i}\int_\Gamma \frac{du}{e^{\beta u}\pm 1} f(u,\ldots),$$

(3.2)

where, as shown in Fig. 1, the contour Γ encloses the poles of the Sommerfeld-Watson factors $(e^{\beta u}\pm 1)^{-1}$ at the discrete, imaginary frequencies. The analytically continued $\mathcal{D}(u_1\ldots u_m, Z_{\nu_1}\ldots Z_{\nu_{n-1}})$ is free of singularities[9] for finite u_i, and the product of the factors $(e^{\beta u_i}\pm 1)^{-1}$ and $\mathcal{D}(u_1\ldots u_m, Z_{\nu_1}\ldots Z_{\nu_{n-1}})$ vanishes $[\sim u_i^{-1}]$ as a given u_i goes to infinity in any direction of the complex plane. Thus, the contours Γ can be folded down as indicated in Fig. 1. Since an integral over a folded-down Γ contour is equivalent to an integral over the real axis of the discontinuity of the integrand, the expression in (3.1) is equivalent to

(3.3)

$$T_n(Z_{\nu_1}\ldots Z_{\nu_n}) = \int\frac{du_1}{\pi}\ldots\frac{du_m}{\pi}\frac{A(u_1)}{e^{\beta u_1}\mp 1}\ldots\frac{A(u_m)}{e^{\beta u_m}\mp 1}(U_1\ldots u_m, Z_{\nu_1}\ldots Z_{\nu_{n-1}}),$$

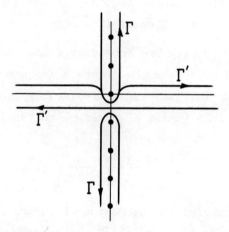

Fig. 1. The contour Γ and the folded down contour Γ'.

where $A(u) \equiv \text{Im } \Delta(u)$ is $(2i)^{-1}$ times the discontinuity of $\Delta(u)$ across the real u axis.

If the Z_{ν_i} in (3.3) were continued to arbitrary complex values Z_i in the same way as the internal frequencies Z_{α_i}, the resulting $T_n(Z_1...Z_n)$ would have no singularities for finite Z_i and would blow up exponentially ($\sim Z_i^{-1} e^{\beta Z_i}$) as any Z_i went to infinity in the right half complex plane. [The $\mathcal{D}$ in (3.3) involves exponential factors $e^{\beta u_i}$ and denominator factors; see below]. The more natural procedure is to select the unique analytic continuation of $\mathcal{D}$ in (3.3) which vanishes as the Z_i approach infinity. This latter continuation is equivalently obtained by taking the former analytic continuation $\mathcal{D}(u_1...u_m, Z_1...Z_{n-1})$ and replacing all the exponential factors $e^{\beta Z_i}$ which appear in the explicit form of this function by the values ∓ 1 which they assume at the discrete frequencies. This analytically continued $\mathcal{D}$ will be denoted by placing a dot over each of the arguments Z_i whose exponential factors are replaced by ∓ 1:

$$\mathcal{D}(u_1...u_m, \dot{Z}_1...\dot{Z}_{n-1}) \equiv \mathcal{D}(u_1...u_m, Z_1...Z_{n-1})\Big|_{e^{\beta Z_i} = \mp 1} . \quad (3.4)$$

The function $\mathcal{D}(u_1...u_m, \dot{Z}_1...\dot{Z}_{n-1})$ has poles involving the Z_i corresponding to every way of cutting internal lines of the graph for T_n which severs off a subgraph having some of the Z_i as external frequencies. The associated pole denominator is the sum of the frequencies entering the separated subgraph. Since these are the only poles of $\mathcal{D}(u_1...u_m, \dot{Z}_1...\dot{Z}_{n-1})$, and since the u_i in (3.3) are real, the contribution to $T_n(Z_1...Z_n)$ obtained by analytically continuing $\mathcal{D}$ in (3.3) in this manner has singularities only when some partial sum of the Z_i is real.

If the Z_i are continued from imaginary to real values by rotating clockwise all the Z_i together, by $90°$, it is easy to see that no partial sum of the Z_i's becomes real during the rotation. Further, as the Z_i approach real values the imaginary part of each denominator factor of $\mathcal{D}(u_1\ldots u_m, \dot{Z}_1\ldots \dot{Z}_{n-1})$ approaches zero with the same sign as the sign of the real part of the partial sum of Z_i appearing in the denominator. [Denominator factors of $\mathcal{D}(u_1\ldots u_m, \dot{Z}_1\ldots \dot{Z}_{n-1})$ involving only undotted frequencies are not associated with poles.] Thus, the analytic continuation of T_n in (3.3) to real frequencies ω_i is given by

$$T_n(\omega_1\ldots\omega_n) = \int \frac{du_1}{\pi}\ldots\frac{du_m}{\pi} \frac{A(u_1)}{\mp e^{\beta u_1}-1}\ldots\frac{A(u_m)}{\mp e^{\beta u_m}-1}(u_1\ldots u_m,\dot{\omega}_1\ldots\dot{\omega}_{n-1})$$

$$+ \ldots \quad , \tag{3.5}$$

where the $+ \ldots$ indicates the sum over all graphs for T_n computed in this manner. The denominator factors of $\mathcal{D}$ in (3.5) involving dotted frequencies ω_i are understood to have the boundary values given by

$$\frac{1}{\Sigma' u_i - \Sigma' \omega_i - i\eta(\Sigma'\omega_i)} \quad ; \quad \eta(\Sigma'\omega_i) \equiv \varepsilon\frac{\Sigma'\omega_i}{|\Sigma'\omega_i|} \quad , \tag{3.6}$$

where ε is a positive infinitesimal. The primes on the sums in (3.6) indicate partial sums over the u_i and ω_i.

In the zero temperature limit the amplitude implied by (3.5) and (3.6) reduces to the Feynman amplitude defined as the usual Fourier transform of the ordinary time-ordered product.

IV. THE EQUATIONS OF STATE

A convenient expression for the thermodynamic potential density is[1,2,6,10] [upper sign refers to fermions]

$$\frac{\Omega}{V} = \frac{\Omega_o}{V} \pm \frac{1}{2} \text{Tr} \frac{1}{\beta} \sum_\nu \int \frac{d^3p}{(2\pi)^3} [\ell n(\Delta(p_\nu)\Delta_o^{-1}(p_\nu)) + \Sigma(p_\nu)\Delta(p_\nu)] + \frac{\Omega'}{V} \quad , \tag{4.1}$$

where Ω_o/V is the thermodynamic potential density for the ideal gas, and where Ω'/V is the sum of all skeleton [i.e. no self-energy insertions] closed linked graphs with internal lines associated with the full propagators Δ. The trace in (4.1) is over the internal indices of the propagators [implicit in (4.1); see (2.9)]. An important property of (4.1) in that it is stationary with respect to variations of the self-energy Σ [the explicit Δ in (4.1) and the Δ associated with each internal line of Ω'/V depend on Σ]. This property is equivalent to

$$\frac{\delta \Omega'/V}{\delta \Delta(p_\nu)} = \mp \frac{1}{2} \Sigma(p_\nu) \ . \tag{4.2}$$

Because of (4.2), the μ^α dependence and the β dependences of Σ can be ignored when the derivatives in (2.5) and (2.6) are applied to (4.1) to obtain the equations of state.

The explicit frequency sum in (4.1) can be replaced by a contour integral over Γ, as in Section III, and the contour Γ folded down to sandwich the real axis and to yield an integral over the discontinuity of the integrand. All the relevant β dependence of this part of (4.1) is then contained in the Sommerfeld-Watson factor $(e^{\beta\omega}\pm 1)^{-1}$. Thus, by applying the derivative in (2.6) to (4.1), and again suppressing reference to the momentum, the entropy density arising from (4.1) becomes

$$\frac{S}{V} = \beta^2 \frac{\partial}{\partial \beta} \frac{\Omega_o}{V} + \mathrm{Tr} \int \frac{d\omega}{2\pi} \frac{\beta^2 \omega e^{\beta\omega}}{(e^{\beta\omega}\pm 1)^2} [\mathrm{Im}\, \ell n\, \Delta(\omega) + \mathrm{Im}\, \Sigma(\omega)\, \mathrm{Re}\, \Delta(\omega)]$$

$$+ \beta^2 \frac{\partial}{\partial \beta} \frac{\Omega'}{V} + \mathrm{Tr} \int \frac{d\omega}{2\pi} \frac{\beta^2 \omega e^{\beta\omega}}{(e^{\beta\omega}\pm 1)^2} A(\omega) \cdot \mathrm{Re}\, \Sigma(\omega)\ , \tag{4.4}$$

where $A \equiv \mathrm{Im}\, \Delta$, and where the terms have been distributed in anticipation of cancellations to come.

Note that the Δ and Σ in the integrand of the first term in (4.4) can be re-interpreted as the corresponding renormalized amplitudes. That is, the renormalized and unrenormalized Δ's differ by a cut-off dependent factor of z; since Im $\ln z$ is zero, and since Im Σ = Im Δ^{-1}, it is easy to see the z dependence of this term in (4.4) disappear. The combination of amplitudes appearing in the last term on the right side of (14.4) does not have this desirable z independence.

Suppressing all dependences on the momenta, coupling constants and numerical factors, the typical graph for Ω'/V with m internal lines has the form

$$\frac{\Omega'}{V} = \text{Tr} \frac{1}{\beta} \sum_{\nu_1} \cdots \frac{1}{\beta} \sum_{\nu_m} \Delta(Z_{\nu_1}) \cdots \Delta(Z_{\nu_m}) \mathcal{D}(Z_{\nu_1} \cdots Z_{\nu_m}) + \cdots \quad (4.5)$$

where $\mathcal{D}$ is the product of β times Kronecker deltas which conserve the discrete frequencies at all the vertices of the graph, except for one. Employing the analytic continuation of $\mathcal{D}$ discussed in Section III, the frequency sums in (4.5) can be replaced by contour integrals over Γ, and the contours Γ folded down to give

$$\frac{\Omega'}{V} = \text{Tr} \int \frac{d\omega_1}{\pi} \cdots \int \frac{d\omega_m}{\pi} \frac{A(\omega_1)}{\mp e^{\beta\omega_1}-1} \cdots \frac{A(\omega_m)}{\mp e^{\beta\omega_m}-1} \mathcal{D}(\omega_1 \cdots \omega_m) + \cdots$$

(4.6)

The relevant β dependence in (4.6) is contained in the explicit exponentials appearing in the Sommerfeld-Watson functions and in the factors of $e^{\beta\omega_i}$ occurring with $\mathcal{D}$. There is thus a β dependence associated with the frequency of each line of the graph, and the total β derivative is the sum of the corresponding partial derivatives. Further, since $\mathcal{D}$ in (4.6) is linear in a given $e^{\beta\omega_i}$, the partial β derivative associated with the frequency ω_i is obtained by differentiating only the function

$$(\mp e^{\beta\omega_i}-1)^{-1}$$

and by replacing the exponential factor $e^{\beta\omega_i}$ in $\mathcal{D}$ by ∓ 1. As dis-

ON THE EQUATIONS OF STATE IN MANY-BODY THEORY

cussed in connection with (3.4), this replacement effectively puts a dot over the argument of ω_i in D. Thus,

$$\beta^2 \frac{\partial}{\partial \beta} \frac{\Omega'}{V} = \pm \int \frac{d\omega_1}{\pi} \frac{\beta^2 \omega_1^2 e^{\beta \omega_1}}{(e^{\beta \omega_1} \pm 1)^2} A(\omega_1) \left[\int \frac{d\omega_2}{\pi} \cdots \int \frac{d\omega_m}{\pi} \right.$$

$$\left. \times \frac{A(\omega_2)}{\mp e^{\beta \omega_2} - 1} \cdots \frac{A(\omega_m)}{\mp e^{\beta \omega_m} - 1} D(\dot{\omega}_1 \omega_2 \cdots \omega_n) \right] + \cdots \quad (4.7)$$

where the $+ \cdots$ indicates the sum of similar terms over all lines of all graphs for Ω'/V.

The individual terms in (4.7) are not really defined until some prescription is given for integrating around the poles of D involving the dotted frequencies. Thus, let us imagine that the ω_i contours are split infinitesimally from each other in some arbitrary, but definite, manner. [Of course the ultimate expressions for the equations of state must be independent of the details of these contour separations.]

Because of (4.2), and because of the structure of the analytically continued amplitudes indicated in (3.5), the bracketed expression in (4.7) appears to be the contribution to $\mp 1/2 \, \Sigma(\omega_1)$ coming from the derivative in (4.2) with respect to $A(\omega_1)$. But the boundary values in D are determined by the contour separations and do not, in general, satisfy (3.6). Nevertheless, the boundary value of each denominator factor can be replaced by the one required by (3.6), if in addition any delta function term which arises from jumping the associated branch cut is also included. If this kind of replacement is made in all the terms by (4.4), the result can be written as $[\Sigma(\omega) \equiv \Sigma(\omega + i\eta(\omega))]$

$$\beta^2 \frac{\partial}{\partial \beta} \frac{\Omega'}{V} = -\text{Tr} \int \frac{d\omega}{\pi} \frac{\beta^2 \omega^2 e^{\beta \omega}}{(e^{\beta \omega} \pm 1)^2} A(\omega) \frac{1}{2} \Sigma(\omega) + \text{P.T.} , \quad (4.8)$$

where the $1/2\ \Sigma(\omega)$ comes from summing up all the contributions with the replaced boundary values, and where the P.T. simply stands for "pulling terms": the sum of all delta function terms coming from pulling the frequencies to the boundary values required by (3.6).

The individual contributions to P.T. in (4.5) coming from the bracketed expression in (4.4) [and from corresponding expressions in all the other terms contained in the + ...] are associated with cuts in the ω_1 channel of the corresponding graph for $1/2\ \Sigma(\omega_1)$. In each of these terms, the delta function associated with a given cut is accompanied by two factors which are, to the extent that they satisfy (3.5), the analytically continued amplitudes for the two parts of the self energy graph separated by the cut. The boundary values of these T amplitudes are determined by the contour separation and do not generally satisfy (3.6). But they can be replaced by boundary values satisfying (3.6) by again adding whatever delta function terms arise from jumping the corresponding branch cuts. Each additional delta function factor is accompanied by one more T-amplitude factor which is linked together with the other T-amplitudes by the lines whose frequencies appear in the arguments of delta functions. When these newly appearing T-amplitudes are "pulled" to the correct boundary values [which conform to (3.6)], new delta function and new T-amplitude factors appear, etc.

In this way the P.T. in (4.5) becomes an infinite sum of terms involving products of successively larger numbers of T-amplitudes [satisfying both (3.5) and (3.6).]. Each contribution to P.T. in (4.5) involving a product of n T-amplitudes is distinguished by a set of lines of a graph for Ω'/V whose frequencies appear within the arguments of the n-1 delta function factors. There is a class of contributions to P.T. associated in this way with each set of lines which link together the T-amplitudes constructed, in accord with (3.5) and (3.6), out of the remaining lines in the graph for Ω'/V. Considering all graphs for Ω'/V, there is an infinite number of other classes of contributions to P.T. which differ from the given class only in the

ON THE EQUATIONS OF STATE IN MANY-BODY THEORY

internal structure of the associated T-graphs. The sum of all the contributions to P.T. belonging to any of these classes can be associated with a T-graph: a graph similar to a graph for Ω'/V except that the vertices represent full T-amplitudes[11] and involve any number of lines. The sum of all contributions to P.T. is the sum of the expressions for all T-graphs.

A few examples of T-graphs are shown in Fig. 2. Since the T-amplitude vertices of a T-graph are connected by lines representing full propagators, the factors of z required to renormalize the T-amplitudes are precisely those which result when each propagator is replaced by z times a renormalized propagator. Thus, T-graphs can be considered to involve products of renormalized amplitudes.

V. EVALUATION OF T-GRAPHS

The T-graph contributions to the entropy density can be evaluated by re-interpreting (4.7) to apply to T-graphs [the T-amplitude factors associated with the vertices being suppressed, and the D function re-interpreted as the D function for the product of Kronecker deltas associated with the T-amplitude vertices] and by extracting the total delta function part of each term, as described below. The $+\ldots$ in (4.7) is re-interpreted to indicate the sum over all lines of all T-graphs.

The total delta function part of (4.7) is computed as follows:

1. $D(\dot{\omega}_1 \omega_2 \ldots \omega_m)$ is a sum of terms, each of which is a product of exponential factors $\exp(\beta \Sigma' \omega_i)$ [$i \neq 1$] divided by a product of denominator factors, some of which involve ω_1. For each subproduct of denominator factors involving ω_1 [$\equiv \Pi_i (a_i - \omega_1)^{-1}$] make the replacement [$\eta(\omega_1)$ defined in (3.6)]

$$\Pi_i \frac{1}{a_i - \omega_1} \, , \, \Pi_i \frac{1}{a_i - \omega_1} = \Pi_i \frac{1}{a_i - \omega_1 - i\eta(\omega_1)} \, . \tag{5.1}$$

The effective, infinitesimal imaginary parts in the denominators in the first term on the right depend upon the contour separations. The

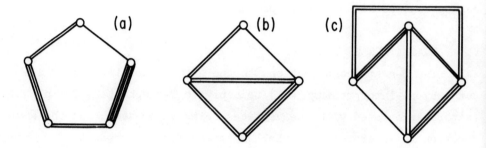

Fig. 2. A few examples of T-graphs. The vertices refer to full, renormalized T-amplitudes[11] and the lines to the imaginary parts of full, renormalized propagators.

contributions coming from the [negative of] the second term on the right of (5.1) give the first term on the right side of (4.8).

2. Expand the results of performing the replacements in 1 in $\mathcal{D}(\dot\omega_1\omega_2\ldots\omega_m)$ as a sum of terms, where each term is a product of exponential factors, delta functions, and denominator factors satisfying (3.6). That is, (3.6) requires the effective iε of each denominator to have a sign equivalent to adding $i\eta(\Sigma'\omega_i)$ to the sum, $\Sigma'\omega_i$, of external frequencies appearing in the denominator. Here the external frequencies in each denominator are the frequencies which appear within the arguments of delta functions multiplying the denominator factor.

3. The "total delta function part" is the sum of all contributions resulting from 2 in which every denominator factor in the product [including those not involving ω_1] has been converted to a delta function [all frequencies $\omega_1\ldots\omega_m$ are thus within the arguments of delta functions]. Terms resulting from 2 which involve fewer delta function factors are effectively included in the pulling terms associated with lower order T-graphs.

As is evident from Fig. 2, a line in a T-graph may occur in parallel with other lines; that is, there may be other lines connecting between the same pair of T-amplitude vertices. Let $\utilde{\omega}_1$ indicate the sum of the frequencies of all lines in parallel with the line of frequency ω_1. The frequency sums a_i in (5.1) always include the negative of the sum of frequencies in parallel with ω_1. Thus, let us re-express the $a_i-\omega_1$ as $b_i-\utilde{\omega}_1$. The right side of (5.1) is equivalent to

$$\left[\prod_i \frac{1}{b_i-\utilde{\omega}_1} - \prod_i \frac{1}{b_i-\utilde{\omega}_1-i\eta(\utilde{\omega}_1)}\right] + \left[\prod_i \frac{1}{b_i-\utilde{\omega}_1-i\eta(\utilde{\omega}_1)} - \prod_i \frac{1}{b_i-\utilde{\omega}_1-i\eta(\omega_1)}\right],$$

(5.2)

and it is easy to check that the second bracket here is equivalent to

$$\frac{1}{2} [\prod_i \frac{1}{b_i - \omega_1 - i\eta(\omega_1)} - \prod_i \frac{1}{b_i - \omega_1 + i\eta(\omega_1)}] - \frac{1}{2} [\prod_i \frac{1}{b_i - \omega_1 - i\eta(\omega_1)} -$$

$$\prod_i \frac{1}{b_i - \omega_1 + i\eta(\omega_1)}] \quad . \quad (5.3)$$

The pulling terms arising from the second bracket on the right side of (5.3) [calculated by applying rates 2 and 3 above] give a contribution to (4.5) which cancels the part of the first term in (4.5) involving Im Σ. The remaining part of the first term in (4.5) cancel against the last term on the right side of (4.4). Thus, substitution of (4.5) into (4.4) gives

$$\frac{S}{V} = \text{Tr} \int \frac{d\omega}{2\pi} \frac{\beta^2 \omega \, e^{\beta\omega}}{(e^{\beta\omega} \pm 1)^2} [\text{Im} \ln \Delta(\omega) + \text{Im} \Sigma(\omega) \cdot \text{Re} \Delta(\omega)] + \frac{S^{(1)}}{V} + \frac{S^{(2)}}{V} ,$$

$$(5.4)$$

where $\frac{S^{(1)}}{V}$ and $\frac{S^{(2)}}{V}$ are the two kinds of pulling contributions to the entropy density obtained by following rules 1-3 above, but with the right side of (5.1) replaced, respectively, by the first term in (5.3) and by the first bracket in (5.2) .

VI.   THE CHAIN PULLING TERMS

A chain T-graph of order n is one in which the connecting set of parallel lines link together n (≥ 3) T-amplitudes in a single channel. An example of a chain T-graph is shown in Fig. 2a.

As described in Section V, the pulling terms involve a sufficient number of delta function factors to conserve the (real) frequencies at each T-amplitude vertex. Thus, when (4.7) is applied to T-graphs, the exponential factors appearing in the D function in (4. can be evaluated on this frequency multi-shell. When evaluated in this way, the D function in (4.7) applied to a chain T-graph of order n satisfies

$$\frac{\mp \omega_1 e^{\beta\omega_1}}{\mp e^{\beta\omega_1} - 1} (\dot{\omega}_1 \omega_2 \ldots \omega_m) = \frac{\mp \omega_1 e^{\beta\tilde{\omega}_1}}{\mp e^{\beta\tilde{\omega}_1} - 1} \prod_{i=2}^{n} \frac{\mp e^{\beta\tilde{\omega}_i} - 1}{\tilde{\omega}_i - \tilde{\omega}_1}, \qquad (6.1)$$

where $\tilde{\omega}_i$ is the total frequency carried by the set of parallel lines comprising the ith link in the chain. The upper or lower sign on the left side of (6.1) applies if the line of frequency ω_1 refers to a fermion or boson, whereas the sign choice on the right side of the equation refers to the fermion or boson character of the channel propagating through the chain.

The chain pulling terms in the entropy density are obtained by applying the pulling rules described in Section V to the denominator product in (6.1). For the first kind of pulling terms, the result is to replace the denominator product in (6.1) according to ($\varepsilon(x) \equiv x/|x|$)

$$\prod_{k=2}^{n} \frac{1}{\tilde{\omega}_k - \tilde{\omega}_1} \rightarrow -\frac{1}{2} \prod_{k=2}^{n} (-2\pi i \, \varepsilon(\tilde{\omega}_k)) \, \delta(\tilde{\omega}_k - \tilde{\omega}_1), \qquad (6.2)$$

where the factor of 1/2 reflects the presence of this factor in the first term on the right of (5.3). The sum of the first kind of chain pulling terms associated with all the parallel lines which contribute to $\tilde{\omega}_1$ is obtained by making the replacement (6.2) in (6.1), substituting the result into (4.7), and replacing the factor ω_1 by $\tilde{\omega}_1$. An equal contribution to $S^{(1)}/V$ in (5.4) arises from the sum of the pulling terms associated with all the parallel lines which contribute to any other $\tilde{\omega}_j$.

There will be a second kind of pulling term associated with the set of lines of total frequency $\tilde{\omega}_i$ only if the frequency contour separations [see Sections IV and V] are arranged so that for $j \neq i$ all denominator factors $(\tilde{\omega}_j - \tilde{\omega}_i)^{-1}$ have effective, infinitesimal imaginary parts equivalent to $(\tilde{\omega}_j - \tilde{\omega}_i + i\eta(\tilde{\omega}_i))^{-1}$ [i.e. the wrong boundary values from the viewpoint of (3.6)]. Regardless of how the contours are separated, one, and only one, of the n ω_i will satisfy this criterion. Thus, the sum of all the second class chain pulling terms is equal to $-2/n$ [n is the order of the T-graph] times the sum of the corresponding first class pulling terms [the factor of 2 arises from the explicit 1/2 in the first term of (5.3)].

The sum of all the chain contributions to $S^{(1)}/V$ and $S^{(2)}/V$ can be conveniently expressed in terms of a matrix $T(\underline{p},\omega)$ defined by [$p \equiv \underline{p},\omega$]

$$<p_1'p_2'\ldots p_n'|T(\underline{p},\omega)|p_1 p_2\ldots p_m> \equiv (2\pi)^4 \delta^3(\underline{p} - \sum_1^m \underline{p}_i)\delta(\omega - \sum_1^m \omega_i)$$
$$\times T_{n+m}(-p_1'\ldots p_n) \quad , \qquad (6.3)$$

where the T_{n+m} amplitude on the right side of (6.3) is the sum of the graphs for T_{n+m} satisfying (3.5) and (3.6). The orthogonality properties of the "states" on the left side of (6.3) are in accord with the completeness relation

$$1 = \sum_n \frac{1}{n!} \int \frac{d^4 p_1}{(2\pi)^4}\ldots\frac{d^4 p_n}{(2\pi)^4} \frac{(\mp e^{\beta \Sigma \omega_i}-1)\epsilon(\Sigma \omega_i)}{\prod_i(\mp e^{\beta\omega_i}-1)} \prod_j (2A(p_j))|p_1\ldots p_n><p_1\ldots$$

(6.4)

where the $(n!)^{-1}$ refers to identical particles. Note that as β approaches infinity ($T \to 0$), the ratio of exponential factors in (6.4) becomes a product of step functions requiring all frequencies ω_i to be of the same sign.

Employing the notation defined in (6.3) and (6.4), the two kinds of chain pulling contributions to the entropy density can be written as

$$\frac{S_{ch}^{(1)}}{V} = -\frac{i}{4} \int \frac{d^3 p}{(2\pi)^3} \int \frac{d\omega}{(2\pi)} \frac{\beta^2 \omega e^{\beta\omega} \epsilon(\omega)}{(e^{\beta\omega}\pm 1)^2} \text{Tr} \sum_{n=3}^{\infty} (-i\, T(p,\omega))^n \qquad (6.5)$$

ON THE EQUATIONS OF STATE IN MANY-BODY THEORY 241

$$\frac{S_{ch}^{(2)}}{V} = \frac{i}{2} \int \frac{d^3p}{(2\pi)^3} \int \frac{d\omega}{2\pi} \frac{\beta^2 \omega e^{\beta\omega} \varepsilon(\omega)}{(e^{\beta\omega} \pm 1)^2} \text{Tr} \sum_{n=3}^{\infty} \frac{1}{n} (-i\, T(p,\omega))^n \quad . \tag{6.5b}$$

The expressions in (6.5) can evidently be formally summed. Performing this sum, taking the real part of the result [the imaginary parts cancel against contributions from non-chain pulling terms], and making use of the symmetry of the integrands under sign reflections of $\underline{p}$ and $\underline{\omega}$, the total chain pulling terms in the entropy density becomes

$$\frac{S_{ch}^{(1)}}{V} + \frac{S_{ch}^{(2)}}{V} = \int \frac{d^3p}{(2\pi)^3} \int_0^\infty \frac{d\omega}{2\pi} \frac{\beta^2 \omega e^{\beta\omega}}{(e^{\beta\omega} \pm 1)^2} \text{Tr}\, [\, \frac{1}{2i} \ln \frac{1+i\, T(p,\omega)}{1-i\, T^*(p,\omega)}$$

$$+ \frac{1}{2} \text{Im}\, (\, \frac{1}{1+i\, T(p,\omega)} - i\, T(p,\omega)\,)\,] \quad . \tag{6.6}$$

The argument of the logarithm in (6.6) is naturally defined as $\exp(4i\delta)$. This term in (6.6) thus involves a sum over the eigenphases $\delta_\lambda(\underline{p},\omega)$. In the narrow resonance approximation (NRA), the δ_λ are constants (zero modulo π) except for jumps of magnitude π at the resonant energies $E_\lambda(p)$:

$$\delta_\lambda^{NRA}(\underline{p},\omega) = \pi\, \Theta(\omega + \mu_\alpha q_\lambda^\alpha - E_\lambda(p)) \quad . \tag{6.7}$$

The q_λ^α in (6.7) are the eigenvalues of Q^α carried by the resonances. When the approximation (6.7) is inserted into (6.6), an integration by parts convert the step function in (6.7) to a delta function, with the result

$$\frac{S_{ch}^{(1)}}{V} + \frac{S_{ch}^{(2)}}{V} \xrightarrow{NRA} \mp \int \frac{d^3p}{(2\pi)^3} \sum_\lambda [1 \mp n_\lambda(p))\, \ln(1 \mp n_\lambda(p)) \pm n_\lambda(p) \ln n_\lambda(p)].$$

$$\tag{6.8}$$

Here

$$n_\lambda(p) = \frac{1}{e^{(E_\lambda(p)-\mu_\alpha q_\lambda^\alpha)} \pm 1} \quad (6.9)$$

is the average occupation number per state for the component λ of an ideal gas. Thus, Eq. (6.6) contains the effects of narrow, composite excitations in much the same way[1-5] as the explicit logarithmic term in (5.4) contains the effects of narrow elementary excitations [i.e. excitations with the quantum numbers carried by the fields].

VII. CONCLUDING COMMENTS

The general structure of the other pulling terms in the entropy density has been established, and certain classes of these terms have been worked out in detail. As yet, however, I have not been able to discover the general rule which would allow one to write down the general pulling term complete in every detail. At least for the sake of completeness, such a rule should be found.

The general pulling contribution to the entropy density involves an ordered product of on-shell T-amplitudes accompanied by factors similar to those appearing in the chain pulling terms discussed in Section VI. In contrast to an opened, ordered chain graph [i.e. simply a stretched, horizontal chain], the general ordered, opened T-graph may have lines which pass by without interacting at a T-amplitude vertex. The general ordered, opened T-graph thus has a complicated topology and is distinguished by a number of channels and subchannels. Each channel has its own frequency $\underset{\sim}{\omega}_i$, and the general pulling term involves a linear combination of these channel frequencies with integral coefficients which depend upon the frequencies' signs. It is the general rule for determining these coefficients which is not yet discovered.

The statistical average of the conserved charge density $<Q^\alpha>$, corresponding to the derivative in (2.5), has an expansion similar to

ON THE EQUATIONS OF STATE IN MANY-BODY THEORY

the expansion for the entropy density symbolized in (5.4). In fact, if the symmetry associated with the charge Q^α is not spontaneously violated, the expansion for $<Q^\alpha>/V$ is obtained by replacing the appearance of each channel frequency $\underset{\sim}{\omega}_i$ in the expansion for the entropy density by β^{-1} times the matrix $\underset{\sim}{q}_i{}^\alpha$ for the total charge carried by the channel [a $\underset{\sim}{q}_i{}^\alpha$ is the direct sum of the q^α in (2.12) associated with all parallel lines which comprise any link in the channel].

There are additional terms in the expansion for $<Q^\alpha>/V$ when the symmetry associated with Q^α is spontaneously violated. The determination of the precise structure of these additional terms is another aspect of this program which has not yet been completed.

The expressions for the equations of state discussed here are, in a sense, expansions in the temperature β^{-1}. Roughly speaking, each line in a T-graph is associated with a factor of β^{-1}, and each frequency conserving delta function is associated with a factor of β. Thus, T-graphs having the fewest number of lines, and therefore having the fewest number of vertices, should be most important at sufficiently low temperatures. In fact, the leading behavior of the equations of state in the low temperature limit appears to arise from the explicit term involving $\ln \Delta$ in (5.4), and from the corresponding terms in the expressions for the conserved charge densities. Such a viewpoint is indeed basic to the works in references 1-5; references 1-4 considered the implications of the $\ln \Delta$ terms typified in (5.4), whereas reference 5 studied the effects associated with the terms with n = 3 in Eq. (6.5). Despite the essential correctness of these works, in general, when there are bound states, the dominant low temperature behavior of the equations of state could arise from terms like the one involving the logarithm in (6.6).

ACKNOWLEDGEMENTS

I would like to thank Professor Gordon Baym for a discussion and for calling my attention to reference 5. I am also appreciative of the hospitality of Professor J. Zinn-Justin and other members of the theory group at SACLAY, where part of this work was performed. Other parts of this work were carried out at the Aspen Center for Physics, and I thank this organization for its hospitality and facilities.

REFERENCES

1. J. M. Luttinger, Phys. Rev. $\underline{119}$, 1153 (1960).
2. J. M. Luttinger and J. C. Ward, Phys. Rev. $\underline{118}$, 1417 (1960); see also the discussion and references in Abrikosov et al., reference 7.
3. W. Götze and H. Wagner, Physica $\underline{31}$, 475 (1965).
4. K. Kehr, Physica $\underline{33}$, 620 (1967).
5. C. M. Carneiro and C. J. Pethick, Phys. Rev. $\underline{B11}$, 1106 (1975).
6. R. E. Norton and J. M. Cornwall, Ann. Phys. $\underline{91}$, 106 (1975).
7. See, for example, A. L. Fetter and J. D. Walecka, <u>Quantum Theory of Many Particle Systems</u>, McGraw Hill, New York, 1971; or A. A. Abrikosov, L. P. Gor'kov and I. Y. Dzyaloshinski, <u>Quantum Field Theoretical Methods in Statistical Physics</u>, Pergammon, New York 1965.
8. The Boltzmann constant k is set equal to unity in the subsequent discussion.
9. See Appendix B of reference 6 and R. E. Norton, forthcoming preprint.
10. C. De Dominicis and P. C. Martin, J. Math. Phys. $\underline{5}$, 14, 31 (19
11. To be more precise, the vertices of a T-graph represent the sum of all single particle irreducible graphs for a T amplitude plus all single particle reducible graphs which do not contain a single line in the same channel when two T amplitude vertices are connected together within the T-graph.

12. For a T-graph with two vertices, replace $\undertilde{\omega}_1$ by ω_1 in (5.2) and (5.3). The pulling contribution from the first bracket in (5.2) then cancel those coming from the first bracket in (5.3). The pulling contributions to the equations of state come from T-graphs with three or more T-amplitude vertices.

DYSON EQUATIONS, WARD IDENTITIES, AND THE INFRARED BEHAVIOR OF YANG MILLS THEORIES*

M. Baker

University of Washington

Seattle, Washington 98195

It is shown that the Schwinger-Dyson equations of Yang-Mills theory in the axial gauge allow for the possibility that the gluon propagator behaves like $(1/q^2)^2$ for small q^2. The possible consistency of such singular low momentum behavior follows from the fact that the vertices appearing in the Schwinger-Dyson equations satisfy the Slavnov-Taylor identities. The general solution of these identities. The general solution of these identities is found and particular solutions relevant to the low q^2 behavior of the gluon propagator are introduced into the Schwinger-Dyson equations. This yields a closed set of integral equations for the gluon propagator which are now being investigated.

In this talk I will describe a general approach to calculate the infrared behavior of Yang Mills theory from the Schwinger-Dyson equations and the Ward identities of the theory. More specifically we eventually want to calculate the low momentum behavior of the running coupling constant $g(q^2)$ as $q^2 \to 0$. This will then give us the nature of the long range force due to gluon exchange in Yang-Mills theory. Yang-Mills theory is defined by the Lagrangian

---

*Work supported in part by the U. S. Department of Energy.

$$L = -\frac{1}{4} G^{\mu\nu a} G^a_{\mu\nu} \quad , \tag{1}$$

where

$$G^a_{\mu\nu} = \frac{\partial}{\partial x^\mu} A^a_\nu - \frac{\partial}{\partial x^\nu} A^a_\mu + gf^{abc} A^b_\mu A^c_\nu \quad ,$$

and f^{abc} are the structure constants of the non-Abelian gauge group underlying the theory.

The outline of this talk is the following:

(1) We write the Schwinger-Dyson equations determining the gluon propagator $D^{ab}_{\mu\nu}(q)$ in terms of the vacuum polarization tensor $\tilde{\pi}^{ab}_{\mu\nu}(q)$. The vacuum polarization tensor in turn is expressed in terms of $D^{ab}_{\mu\nu}(p)$, the three gluon vertex $\Gamma^{abc}_{\lambda\mu\nu}(p,q,r) \equiv \Gamma$, and the four gluon vertex $\Gamma^{abcd}_{4,\lambda\mu\nu\sigma}(p,q,r,s) \equiv \Gamma_4$. In the axial gauge the low momentum behavior of $D^{ab}_{\mu\nu}(q)$ determines $g(q^2)$ as $q^2 \to 0$.

(2) Without specifying Γ and Γ_4 we show that in the axial gauge the vacuum polarization tensor $\tilde{\pi}^{ab}_{\mu\nu}(q)$ has no infrared singularity even if the gluon propagator $D^{ab}_{\mu\nu}(p)$ behaves like $(1/p^2)^2$ as $p^2 \to 0$.

(3) To complete the Dyson equations we must specify the vertices Γ and Γ_4 in terms of $D_{\mu\nu}$. The longitudinal parts of Γ and Γ_4 are determined by the Slavnov-Taylor identities which we write in the axial gauge. We note that any vertices satisfying these identities when inserted into the Schwinger-Dyson equations, yield the structure for $D^{ab}_{\mu\nu}(q)$ demanded by gauge invariance. In particular we show that $\tilde{\pi}_{\mu\nu}(q) \to 0$ as $q^2 \to 0$, even if there is a $(1/p^2)^2$ singularity in the gluon propagators appearing in the Schwinger-Dyson expression for the vacuum polarization tensor. This means that in Yang-Mills theory there is the possibility of a singular long range force corresponding to a running coupling constant $g(q^2)$ behaving like $1/q^2$ for small q.

(4) We present the general solution of the Slavnov-Taylor identities for Γ and Γ_4. We find that the requirement that Γ and Γ_4 contain no kinematic singularities yields transverse parts of Γ and Γ_4 which vanish when any one of the external momenta vanishes. The kinematic singularity free longitudinal parts are determined in te

of the gluon propagator $D^{ab}_{\mu\nu}(q)$.

(5) From section (3) we know that one must use in the Schwinger-Dyson equations vertices Γ and Γ_4 which are exact solutions of the Slavnov-Taylor identities in order to have the possibility of finding a singular long range force. Thus as a first attempt we neglect the undetermined transverse parts of Γ and Γ_4. We then replace Γ and Γ_4 by their kinematic singularity free longitudinal parts. The Schwinger-Dyson equations then become a closed set of equations for the gluon propagator which may be appropriate for computing $D^{ab}_{\mu\nu}(q)$ as $q^2 \to 0$. We discuss some properties of these equations which we are now investigating.

I. THE SCHWINGER-DYSON EQUATIONS FOR THE GLUON PROPAGATOR

The free gluon propagator is

$$D^{(0)ab}_{\mu\nu}(q) = \delta_{ab} D^{(0)}_{\mu\nu}(q) ,$$

where the form of $D^{(0)}_{\mu\nu}(q)$ depends upon the gauge.

In covariant gauges

$$D^{(0)}_{\mu\nu}(q) = \frac{-i}{q^2} \{ g_{\mu\nu} - \frac{q_\mu q_\nu}{q^2} + b \frac{q_\mu q_\nu}{q^2} \} ,$$

where b is a constant which fixes the gauge.

In an axial gauge specified by the gauge condition $n^\mu A^a_\mu = 0$,

$$D^{(0)}_{\mu\nu}(q) = \frac{-i}{q^2} \{ g_{\mu\nu} - \frac{(n_\mu q_\nu + n_\nu q_\mu)}{(n \cdot q)} + \frac{q_\mu q_\nu}{(n \cdot q)^2} n^2 \} . \tag{2}$$

Clearly $n^\mu D^{(0)}_{\mu\nu}(q) = 0$.

The interaction then changes $D^{(0)ab}_{\mu\nu}(q)$ to $D^{ab}_{\mu\nu}(q)$, where $D^{ab}_{\mu\nu}(q) = \delta_{ab} D_{\mu\nu}(q)$. Our task is to calculate the interacting gluon propagator $D_{\mu\nu}(q)$ as $q^2 \to 0$. The bare and interacting gluon propagators are

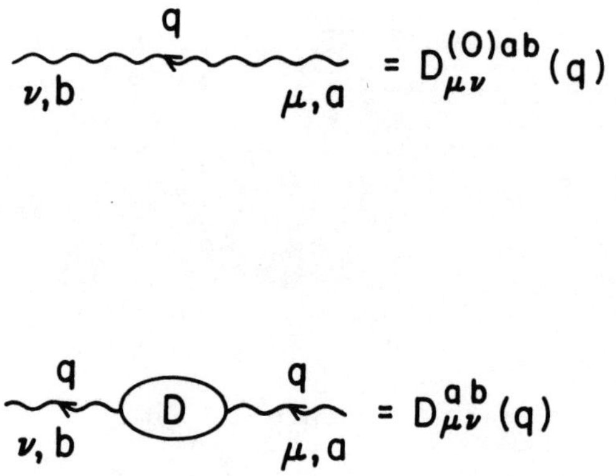

$$\underset{\nu,b \qquad \mu,a}{\overset{q}{\wwbar{}}} = D^{(0)ab}_{\mu\nu}(q)$$

$$\underset{\nu,b \qquad\quad \mu,a}{\overset{q \qquad\quad q}{\wwbar{}\,\boxed{D}\,\wwbar{}}} = D^{ab}_{\mu\nu}(q)$$

Figure 1. Bare and interacting gluon propagators.

DYSON EQUATIONS, WARD IDENTITIES, ETC.

represented diagramatically in Fig. 1.

The Lagrangian (1) contains both cubic and quartic terms in A_μ^a. These give rise to triple and quadruple gluon interactions. The bare triple gluon vertex has the form

$$g\Gamma^{(0)abc}_{\lambda\mu\nu}(p,q,r) = -gf^{abc}[(p-q)_\nu g_{\lambda\mu} + (q-r)_\lambda g_{\mu\nu} + (r-p)_\mu g_{\lambda\nu}], \quad (3)$$

and is represented in Fig. 2(a). The quartic term in L yields the bare quadruple gluon vertex

$$g^2\Gamma^{(0)abcd}_{4,\lambda\mu\nu\sigma}(p,q,r,s) = (-ig^2)\{f^{ade}f^{ecb}(g_{\lambda\mu}g_{\nu\sigma} - g_{\lambda\nu}g_{\mu\sigma})$$
$$+ f^{bae}f^{edc}(g_{\mu\nu}g_{\sigma\lambda} - g_{\mu\sigma}g_{\nu\lambda})$$
$$+ f^{ace}f^{edb}(g_{\nu\sigma}g_{\lambda\mu} - g_{\nu\lambda}g_{\sigma\mu})\}, \quad (4)$$

which is represented in Fig. 2(b).

Due to the interaction these bare vertices $\Gamma^{(0)}$ and $\Gamma_4^{(0)}$ are changed to the interacting vertices Γ and Γ_4 represented by the graphs of Figs. 3(a) and 3(b). The vertices Γ and Γ_4 are one particle irreducible. For example, Γ_4 does not include the graph drawn in Fig. 4.

The Schwinger-Dyson equations then determine the gluon propagator $D_{\mu\nu}$ in terms of the vertices Γ and Γ_4. We first define the vacuum polarization tensor $\tilde{\pi}^{ab}_{\mu\nu}(q)$ by the equations represented by the graph of Fig. 5. Like $D^{ab}_{\mu\nu}$, it is diagonal in color indices so we can write

$$\tilde{\pi}^{ab}_{\mu\nu}(q) = \delta_{ab}\tilde{\pi}_{\mu\nu}(q) .$$

Then $D_{\mu\nu}(q)$ is determined from $\tilde{\pi}_{\mu\nu}(q)$ and $D^{(0)}_{\mu\nu}(q)$ by the equation

$$D_{\mu\nu}(q) = D^{(0)}_{\mu\nu}(q) + D_{\mu\lambda}(q)\tilde{\pi}_{\lambda\alpha}(q)D^{(0)}_{\alpha\nu}(q) . \quad (5)$$

Equation (5) and the equation represented by Fig. 5 then determine

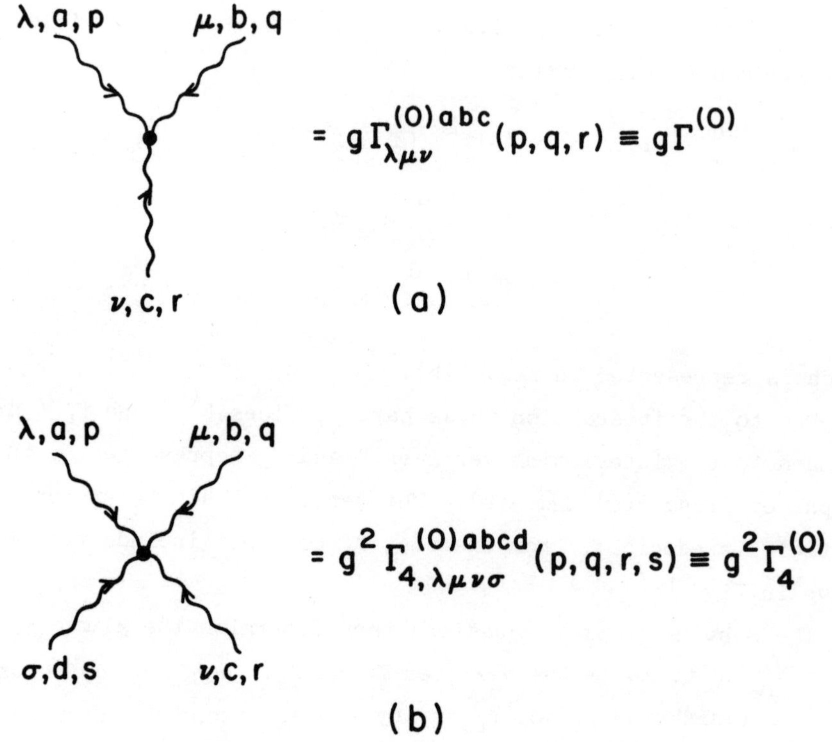
λ, a, p
μ, b, q
ν, c, r
= $g\Gamma^{(0)abc}_{\lambda\mu\nu}(p,q,r) \equiv g\Gamma^{(0)}$
(a)
λ, a, p
μ, b, q
σ, d, s
ν, c, r
= $g^2\Gamma^{(0)abcd}_{4,\lambda\mu\nu\sigma}(p,q,r,s) \equiv g^2\Gamma^{(0)}_4$
(b)

Figure 2. (a) Bare triple gluon vertex.
(b) Bare quadruple gluon vertex.

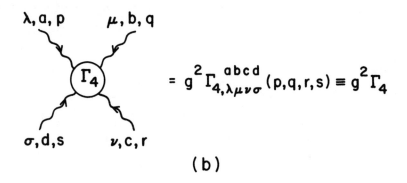

Figure 3. (a) Interacting triple gluon vertex.
(b) Interacting quadruple gluon vertex.

$D_{\mu\nu}$ in terms of the interacting vertices Γ and Γ_4. In covariant gauges $\tilde{\pi}_{\mu\nu}(q)$ also has contributions from internal ghost propagators

These equations were first obtained by Dyson from perturbation theory and by Schwinger from field equations and commutation relations. They are also readily derived from the Feynmann path integral. They are nonperturbative and make no special assumption about the structure of the vacuum.

We have studied these equations mainly in the axial gauge which is ghost free. Furthermore, in this gauge the running coupling constant $g(q^2)$ is determined directly from the gluon propagator $D_{\mu\nu}(q)$, and the Ward identities which we discuss in Section 3 are linear. However, the presence of the external gauge vector n^μ is a complication.

II. INFRARED FINITENESS OF $\tilde{\pi}_{\mu\nu}(q)$

We now ask the question, how singular can $D_{\mu\nu}(p)$ be at small p without generating an infrared singularity in the Schwinger-Dyson equation (Fig. 5) for $\tilde{\pi}_{\mu\nu}(q)$ even at finite q? To answer this question we need only look at the small p region of the p integration. Since q is kept finite, we can neglect p in comparison with q in order to test the convergence at small p for finite q.

First let us look at the expression for $\tilde{\pi}_{\mu\nu}(q)$ given by the second graph in Fig. 5. The small p region of integration then yields a contribution of the form:

$$\tilde{\pi}^{ab}_{\mu\nu}(q) \sim \frac{1}{2!} \int \frac{d^d p}{(2\pi)^d} \Gamma^{(0)bde}_{\nu\lambda\sigma}(-q,q,0) D_{\sigma\rho}(p) D_{\lambda\delta}(-q) \Gamma^{ade}_{\mu\delta\rho}(q,-q,0) \quad ,$$

where for the sake of generality we have written the integral in d dimensional space time. Thus $\tilde{\pi}^{ab}_{\mu\nu}(q)$ will be infrared finite provided

$$\int_0^\varepsilon d^d p \, D_{\mu\nu}(p) < \infty \quad , \tag{6}$$

where the cutoff ε in Eq. (6) restricts the integration to the in-

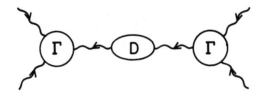

Figure 4. Example of a graph not contained in Γ_4.

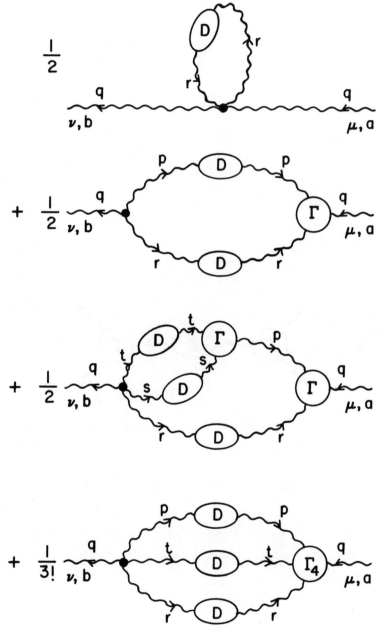

Figure 5. Graphical representation of the equation for the vacuum polarization tensor, $\tilde{\pi}^{ab}_{\mu\nu}(q)$.

tegration to the infrared region.

Let us consider the implication of the convergence criterion, Eq. (6), under the simplifying assumption that $D_{\mu\nu}(p)$ has the tensor structure of the bare propagator $D_{\mu\nu}^{(0)}(p)$. We then set

$$D_{\mu\nu}(p) = \frac{-iZ(p^2)}{p^2} [g_{\mu\nu} - \frac{(n_\mu p_\nu + p_\mu n_\nu)}{(n \cdot p)} + \frac{(n^2 p_\mu p_\nu)}{(n \cdot p)^2}] \quad . \tag{7}$$

Note that if $Z(p^2) \to (1/p^2)^{d/2-1}$ as $p^2 \to 0$, the integral (6) has a potential logarithmic infrared divergence. However, carrying out the integration $d\Omega_p$ over the angles of p, and using the following results valid in any dimension d

$$\int d\Omega_p \frac{(n_\mu p_\nu + p_\mu n_\nu)}{(n \cdot p)} = \frac{2 n_\mu n_\nu}{n^2} \quad ,$$

$$n^2 \int d\Omega_p \frac{p_\mu p_\nu}{(n \cdot p)^2} = - g_{\mu\nu} + \frac{2 n_\mu n_\nu}{n^2} \quad ,$$

we find

$$\int d\Omega_p D_{\mu\nu}(p) = 0 \quad .$$

The vanishing of this angular integral then renders $\tilde{\pi}_{\mu\nu}(q)$ finite even if $Z(p^2) \to (1/p^2)^{d/2-1}$ as $p^2 \to 0$. Thus, in particular, for the physical value d=4 we see that the tensor structure [Eq. (7)] of the axial gauge propagator allows for the possibility that the gluon propagator behaves like $1/p^4$ for small p^2 without producing any unwanted infrared singularity in $\tilde{\pi}_{\mu\nu}(q)$. Since the axial gauge gluon propagator determines the running coupling constant, we conclude that the infrared structure of the Schwinger-Dyson equations does not rule out a singular long range force.

A similar argument applies to each of the graphs of Fig. 5, and we find that a running coupling constant $g(p^2)$ behaving as $1/p^2$ for small p^2 does not produce any infrared singularity in $\tilde{\pi}_{\mu\nu}$.

III. THE SLAVNOV-TAYLOR IDENTITIES FOR Γ AND Γ_4

To use the Schwinger-Dyson equations in a nonperturbative way we must find nonperturbative expressions for Γ and Γ_4. We can obtain certain restrictions on Γ and Γ_4 by making use of the invariance of the Yang-Mills Lagrangian Eq. (1) under the gauge transformation,

$$A_\mu^a(x) \to A_\mu^a(x) + \frac{\partial}{\partial x^\mu} \delta\omega^a(x) + gf^{abc}A_\mu^b(x)\delta\omega^c(x) .$$

This invariance leads to the Slavnov-Taylor[1] identities which in the axial gauge are the following:[2]

$$iq_\mu \Gamma_{\mu\lambda\sigma}^{bac}(q,p,r) = f^{bac}[\pi_{\lambda\sigma}(r) - \pi_{\lambda\sigma}(p)] , \qquad (9)$$

$$iq_\mu \Gamma_{4,\mu\lambda\sigma\rho}^{bade}(q,p,t,r) = -f^{bac}\Gamma_{\lambda\sigma\rho}^{cde}(-(t+r),t,r) ,$$

$$- f^{bdc}\Gamma_{\sigma\lambda\rho}^{cae}(-(p+r),p,r) - f^{bec}\Gamma_{\rho\lambda\sigma}^{cad}(-(p+t),p,t) , \qquad (10)$$

where

$$\pi_{\mu\nu}(q) \equiv i(q^2 q_{\mu\nu} - q_\mu q_\nu) - \tilde{\pi}_{\mu\nu}(q) . \qquad (11)$$

Equations (9) and (10) determine the longitudinal parts of Γ and Γ_4 in terms of $\pi_{\mu\nu}(q)$. Given any solution Γ and Γ_4 of Eqs. (9) and (10), we can add transverse vertices $\Gamma^{(T)}$ and $\Gamma_4^{(T)}$ and obtain a second solution $\Gamma + \Gamma^{(T)}$, $\Gamma_4 + \Gamma_4^{(T)}$. The transverse vertices are any solution of the equations

$$q_\mu \Gamma_{\mu\lambda\sigma}^{(T)bac}(q,p,r) = 0 , \qquad q_\mu \Gamma_{\mu\lambda\sigma\rho}^{(T)bacd}(q,p,t,r) = 0 .$$

The equation for $\tilde{\pi}_{\mu\nu}(q)$ represented by Fig. 5 contains integrations over intermediate momenta p and t. We assume that these integrations are regulated so that the variables p and t can be translated. It can then be shown that if the vertices Γ and Γ_4 in Fig. 5 satisfy Eqs. (9) and (10) then

$$q_\mu \tilde{\pi}_{\mu\nu}(q) = 0 \ . \tag{12}$$

That is, the Schwinger-Dyson equations for $\tilde{\pi}_{\mu\nu}(q)$ and the Slavnov-Taylor identities for Γ and Γ_4 guarantee that $\tilde{\pi}_{\mu\nu}(q)$ is transverse as required by gauge invariance. We can then use any solution of Eqs. (9) and (10) in our expression for $\tilde{\pi}_{\mu\nu}(q)$ and obtain an approximation automatically satisfying Eq. (12).

Differentiating Eq. (12) with respect to q yields

$$\tilde{\pi}_{\mu\nu}(q) = -q_\delta \frac{\partial}{\partial q_\mu} \tilde{\pi}_{\delta\nu}(q) \ . \tag{13}$$

Using the integrals for $\pi_{\delta\nu}(q)$ represented in Fig. 5 and bringing the derivative with respect to q_μ in Eq. (13) under the integral yields an expression for $\pi_{\mu\nu}(q)$ which automatically eliminates the spurious quadratic ultraviolet divergence of perturbation theory. Furthermore, using the identities (9) and (10), we find that differentiation with respect to q_μ produces no new infrared singularity in $\pi_{\mu\nu}(q)$ even if $Z(p^2)$ behaves like $1/p^2$ for small p^2.

We then introduce an ultraviolet cutoff into the expression for $\tilde{\pi}_{\mu\nu}(q)$ obtained from Eq. (13) and Fig. 5. This makes $\tilde{\pi}_{\mu\nu}(q)$ ultraviolet finite and provides a scale. Let us, for simplicity, neclect the dependence of $\tilde{\pi}_{\mu\nu}$ on n^μ. Then from Eq. (12), $\tilde{\pi}_{\mu\nu}(q)$ must have the form

$$\tilde{\pi}_{\mu\nu}(q) = i\tilde{\pi}(q^2)(q^2 g_{\mu\nu} - q_\mu q_\nu) \ , \tag{14}$$

where $\tilde{\pi}(q^2)$ is a dimensionless function of q^2/Λ^2. Inserting Eq. (14) into Eq. (5) then yields an expression for $D_{\mu\nu}(q)$ of the structure of Eq. (7) with

$$Z(q^2) = \frac{1}{1 - \tilde{\pi}(q^2)} \ . \tag{15}$$

Note if $Z(q^2) \to (1/q^2)$ as $q^2 \to 0$, we must have $\tilde{\pi}(q^2) \to 1$ as $q^2 \to 0$.

At first sight, using naive power counting arguments, one might worry that a $1/p^2$ behavior of $Z(p^2)$ at small p^2 would generate via Eq. (13) and Fig. 5 and expression for $\tilde{\pi}(q^2)$ behaving like $1/q^2$ for small q^2.[3] If this were true, there would obviously be no possibility of satisfying the self-consistency condition $\tilde{\pi}(q^2) \to 1$, as $q^2 \to 0$. However, since $\tilde{\pi}(q^2)$ is a dimensionless function of Λ^2/q^2, a $1/q^2$ behavior of $\tilde{\pi}(q^2)$ for small q^2 implies that $\tilde{\pi}$ behaves like Λ^2 for large Λ^2. Since Eq. (13) for $\tilde{\pi}_{\mu\nu}$ has no such quadratic divergence, there can be no $1/q^2$ contribution to $\tilde{\pi}(q^2)$ for small q^2. Now the naive power counting estimate of $1/q^2$ arises from using the limiting form of $Z(p^2) \sim 1/p^2$ over the whole range of the integration in the integral for $\tilde{\pi}(q^2)$. The fact that gauge invariance shows that this $1/q^2$ behavior is absent then means there is a cancellation between the region of integration where $Z(p^2) \sim 1/p^2$ and larger values of p^2 in which this limiting form is not accurate.

It then follows from Eq. (14) that

$$\lim_{q^2 \to 0} \tilde{\pi}_{\mu\nu}(q) = 0 \quad .$$

We thus conclude that a $1/p^2$ behavior of $Z(p^2)$ for small p^2 gives rise not only to an infrared convergent expression for $\tilde{\pi}_{\mu\nu}(q)$ as shown in Section 2, but also to a vanishing $\tilde{\pi}_{\mu\nu}(q)$ as $q^2 \to 0$. Of course, in order to determine the exact behavior of $\tilde{\pi}_{\mu\nu}(q^2)$ for small q^2, we must solve the Schwinger-Dyson equations. However the above considerations show that $\tilde{\pi}(q^2)$ does not have a $1/q^2$ behavior which would make it impossible to satisfy the self consistency condition $\tilde{\pi}(q^2) \to 1$ as $q^2 \to 0$. This potential $1/q^2$ contribution to $\tilde{\pi}(q^2)$ which could arise from the small p region of integration in Fig. 5 is eliminated because the vertices Γ and Γ_4 satisfy the Slavnov-Taylor identities. If we had been dealing with a self coupled field theory which was not a non-Abelian gauge theory, there would be no such mechanism to eliminate a low q^2 singularity in

$\tilde{\pi}(q^2)$ induced by a low p^2 singularity in $Z(p^2)$. There would then be no possibility to generate a self consistent low q^2 singularity.

IV. GENERAL SOLUTION OF THE SLAVNOV-TAYLOR IDENTITIES FOR Γ AND Γ_4

Given $\pi_{\mu\nu}(q)$, we want to determine the most general solutions of Eqs. (9) and (10) for Γ and Γ_4 which satisfy the requirements:

(a) Bose symmetry, e.g.

$$\Gamma^{abc}_{\lambda\mu\nu}(p,q,r) = \Gamma^{bac}_{\mu\lambda\nu}(q,p,r) \quad \ldots \quad,$$

$$\Gamma^{abcd}_{4,\eta\mu\nu\sigma}(p,q,r,s) = \Gamma^{bacd}_{4,\mu\lambda\nu\sigma}(q,p,r,s) \quad \ldots \quad.$$

(b) No kinematic singularities.[4]

These conditions will fix Γ and Γ_4 apart from transverse contributions $\Gamma^{(T)}$ and $\Gamma_4^{(T)}$ which vanish when any one of the external momenta p,q,r, or s goes to zero.

We will explicitly exhibit these solutions under the simplifying assumption that $\pi_{\mu\nu}(q)$ does not depend upon n. That is, we take

$$\pi_{\mu\nu}(q) = iZ^{-1}(q^2)(q^2 q_{\mu\nu} - q_\mu q_\nu) \quad, \tag{16}$$

which from the definition Eq. (11) of $\pi_{\mu\nu}$ is equivalent to Eqs. (14) and (15).

Using Eq. (16) for $\pi_{\mu\nu}(q)$, we can write the solution of Eq. (9) for Γ

$$\Gamma^{abc}_{\lambda\mu\nu}(p,q,r) = \Gamma^{(L)abc}_{\lambda\mu\nu}(p,q,r) + \Gamma^{(T)abc}_{\lambda\mu\nu}(p,q,r) \quad,$$

where

$$\Gamma^{(L)abc}_{\lambda\mu\nu}(p,q,r) = -f^{abc} \{g_{\lambda\mu}[Z^{-1}(p^2)p_\nu - Z^{-1}(q^2)q_\nu]$$
$$+ \frac{[Z^{-1}(p^2)-Z^{-1}(q^2)]}{(p^2-q^2)} [p\cdot q g_{\lambda\mu} - q_\lambda p_\mu](q-r)_\nu\}$$
$$+ \text{cyclic permutations,} \tag{17}$$

and

$$\Gamma^{(T)abc}_{\lambda\mu\nu}(p,q,r) = -f^{abc}\{F(p^2,q^2,r^2)[p\cdot q g_{\lambda\mu}-p_\mu q_\lambda][p_\nu r\cdot q - q_\nu r\cdot p]$$

$$+ G(p^2,q^2,r^2)[g_{\lambda\mu}(p_\nu q\cdot r - q_\nu p\cdot r) + \frac{(r_\lambda p_\mu q_\nu - q_\lambda p_\nu r_\mu)}{3}]\}$$

$$+ \text{ cyclic permutations .} \tag{18}$$

The functions F and G appearing in Eq. (18) are free of kinematic singularities, and G is completely symmetric under the interchange of any pair of variables, while $F(p^2,q^2,r)$ is symmetric under the interchange $p^2 \leftrightarrow q^2$.

We see from Eq. (18) that $\Gamma^{(T)}$, the undetermined part of Γ, contains explicit factors of each of the three external momenta. It thus vanishes linearly when any one of the three momenta approaches zero, since the undetermined functions F and G have no kinematic singularities. Apart from this freedom the kinematic singularity free longitudinal triple gluon vertex is uniquely determined in terms of the gluon propagator (i.e. $Z(p^2)$) via Eq. (18).

In a similar manner we can write

$$\Gamma^{abcd}_{4,\lambda\mu\nu\delta}(p,q,r,s) = \Gamma^{(L)abcd}_{4,\lambda\mu\nu\delta}(p,q,r,s) + \Gamma^{(T)abcd}_{4,\lambda\mu\nu\delta}(p,q,r,s) , \tag{19}$$

where

$$p_\lambda \Gamma^{(T)abcd}_{4,\lambda\mu\nu\delta}(p,q,r,s) = 0 ,$$

and $\Gamma^{(T)}_4(p,q,r,s)$ vanishes when any one of the four momenta p,q,r or s approaches zero. The kinematic singularity free longitudinal part $\Gamma^{(L)}_4$ is given in terms of $Z(p^2)$ according to the equation[5]

$$g^2 \Gamma^{(L)abcd}_{4,\lambda\mu\nu\delta}(p,q,r,s) =$$

$$-ig^2 f^{abe} f^{ecd} \{\frac{g_{\mu\nu}g_{\lambda\delta}}{4} Z^{-1}((p+q)^2) -$$

$$\frac{g_{\lambda\delta}}{2!}(g_{\mu\nu}q\cdot r - r_\nu q_\mu) \frac{[Z^{-1}(q^2)-Z^{-1}(r^2)]}{q^2-r^2}$$

$$+ g_{\mu\nu}[g_\delta(q_\lambda - r_\lambda - s_\lambda)] \frac{[Z^{-1}((p+q)^2) - Z^{-1}(q^2)]}{(p+q)^2 - q^2}$$

$$+ \frac{(g_{\mu\nu}q\cdot r - r_\mu q_\nu)[(r+s)_\lambda - q_\lambda][(p+q)_\delta - r_\delta]}{2![r+s]^2 - q^2} [\frac{Z^{-1}((p+q)^2) - Z^{-1}(r^2)}{(p+q)^2 - r^2} -$$

$$- \frac{Z^{-1}(q^2) - Z^{-1}(r^2)}{q^2 - r^2}]\}$$

+ the remaining 4! permutations of

(a,λ,p), (b,μ,q), (c,ν,r) and (d,δ,s) . (20)

V. THE INTEGRAL EQUATION FOR $Z(p^2)$

We have seen that gauge invariance uniquely determines the vertices Γ and Γ_4 apart from terms which vanish when any of the external momenta approach zero. Assuming that these terms do not play an important role in determining the low momentum behavior of the gluon propagator, we replace Γ by $\Gamma^{(L)}$ and Γ_4 by $\Gamma_4^{(L)}$ in the Schwinger-Dyson equations. They then become a closed set of equations for $Z(p^2)$.

We are now investigating these equations in their original unrenormalized form. As stated in Section 3, we introduce an ultraviolet cutoff which provides a scale for the momentum so that $Z(p^2)$ becomes a function of p^2/Λ^2. Careful treatment of $\pi_{\mu\nu}(q)$ preserves the transverse property, Eq. (12), of $\pi_{\mu\nu}(q)$ in the presence of this cutoff. We then look for the behavior of $Z(p^2/\Lambda^2)$ as $p^2 \to 0$.

We will not write the explicit form of these equations here, but mention some of their general features. The simplifying assumption, Eq. (16), led to solutions for Γ and Γ_4 which did not have any tensors proportional to n_λ, n_μ, n_ν or n_σ. However from Fig. 5 and Eq. (5) we see that Γ and Γ_4 are always contracted with $D_{\mu\nu}^{(0)}$ or $D_{\mu\nu}$. Since $n^\mu D_{\mu\nu} = n^\mu D_{\mu\nu}^{(0)} = 0$, terms proportional to n_μ in Γ or Γ_4 do not contribute to the Schwinger-Dyson equations. Thus only terms in Γ and Γ_4 of the form of Eqs. (17) and (20) are needed to calculate $D_{\mu\nu}(q)$.

The above reasoning does not mean we can neglect the n_μ terms in $\tilde{\pi}_{\mu\nu}$. Such terms induce in $D_{\mu\nu}$ a second invariant structure different from the bare structure and the Schwinger-Dyson equations become a set of coupled equations for Z and the coefficient of the second invariant structure. That is, since $q^\mu \pi_{\mu\nu}(q) = 0$ we can write

$$-i\pi_{\mu\nu}(q) = Z^{-1}(q^2 g_{\mu\nu} - q_\mu q_\nu)$$
$$+ f\left[(n\cdot q)^2 g_{\mu\nu} - (n\cdot q)(n_\mu q_\nu + q_\mu n_\nu) + q^2 n_\mu n_\nu\right], \qquad (21)$$

where in general Z^{-1} and f are functions of the invariants q^2 and $q\cdot n$. Using Eq. (21) the Schwinger-Dyson equations becomes a coupled set of equations for f and Z^{-1}.

We are now studying these equations in order to see whether solutions exist for which $Z(q^2)$ behaves like $1/q^2$ as q^2 approaches zero. We also expect that this investigation will give us insight into the role of the functions F and G of Eq. (18) in determining the infrared behavior of $\tilde{\pi}_{\mu\nu}(q)$.

VI. SUMMARY

We have shown using the Schwinger-Dyson equations and the Slavnov-Taylor identities of Yang Mills theory that no inconsistency arises if the gluon propagator behaves like $(1/p^2)^2$ for small p^2. To see whether the theory actually contains such singular long range behavior, we have formulated a nonperturbative closed set of equations by neglecting the transverse parts of Γ and Γ_4 in the Schwinger-Dyson equations. This simplification preserves all the symmetries of the theory and allows the possibility for a singular low momentum behavior of the gluon propagator. The justification for neglecting $\Gamma^{(T)}$ and $\Gamma_4^{(T)}$ is not evident but we expect that our present study of the resulting equations will elucidate this simplification which leads to a closed set of equations.

This work was carried out in collaboration with R. Anishetty, J. Ball, S. K. Kim, and F. Zachariasen.[6] A related approach is discussed in Ref. 7.

I am indebted to R. Anishetty for his help in preparing this talk.

REFERENCES

1. A. A. Slavnov, Teor. Mat. Fiz. 10, 153 (1972) [Theor. Math Phys. 10, 99 (1972)]; J. C. Taylor, Nucl. Phys. 10, 99 (1971).
2. W. Kummer, Acta Physica Austriaca 41, 315 (1975).
3. In this estimate we have replaced Γ by $Z^{-1}(p^2)p_\mu$, as provided by the Slavnov-Taylor identity (9). (See also Eq. (17) of Section 4.)
4. The requirement of the absence of kinematic singularities is necessary to obtain low energy theorems which follow from taking the $q_\mu \to 0$ limit of Eqs. (9) and (10). For example, letting $q_\mu \to 0$ in Eq. (9) yields the ordinary Ward identity

$$\Gamma_{\mu\lambda\sigma}^{bac}(o,p,p) = if^{bac} \frac{\partial}{\partial p_\mu} \pi_{\lambda\sigma}(p) ,$$

provided

$$q^\mu \frac{\partial}{\partial q_\delta} \Gamma^{bac}_{\mu\lambda\sigma}(q,p,r) \to 0$$

as $q_\mu \to 0$.

5. This result has recently been obtained by S. K. Kim (private communication).
6. Some of our earlier results appear in F. Zachariasen, CERN preprint TH-2601 (1978).
7. R. Delbourgo, University of Tasmania preprint (1978).

INSTANTONS AND CHIRAL SYMMETRY

Robert D. Carlitz*

University of Michigan, Ann Arbor, MI 48109

University of Pittsburgh, Pittsburgh, PA 15260

In this talk I will describe the role that instantons play in the spontaneous breaking of chiral $SU(N) \times SU(N)$ symmetry. My discussion will indicate the physical domain in which instanton effects are likely to be important and will suggest experiments in which some of these effects may be seen.

I will begin with a brief review of what an instanton is and will explain why instantons play an important role in quantum chromodynamics. As pointed out by Belavin, Polyakov, Schwartz and Tyupkin[1] in 1975, there are configurations of the gauge field A_μ for which the field strength tensor $F_{\mu\nu}$ is self-dual or anti-self-dual

$$^*F_{\mu\nu} = \frac{1}{2} \epsilon_{\mu\nu\rho\sigma} F_{\rho\sigma} = \pm F_{\mu\nu} , \qquad (1)$$

and for which the Euclidean action is finite,

$$\frac{1}{4} \int d^4 x_E \, F_{\mu\nu} F_{\mu\nu} = \frac{8\pi^2 n}{g^2} . \qquad (2)$$

---

*Supported in part by the National Science Foundation and the Department of Energy.

These "instantons" and "anti-instantons" provide classical solutions to the QCD field equations in Euclidean space. Such solutions possess a nontrivial topology, since the product $*F_{\mu\nu} F_{\mu\nu}$ is the divergence of a topological current, and has a non-zero integral,

$$\frac{g^2}{32\pi^2} \int d^4x_E \, *F_{\mu\nu} F_{\mu\nu} = n \quad , \tag{3}$$

where n is necessarily an integer.

The meaning of these observations was elucidated[2] when it was pointed out that an instanton describes a quantum mechanical tunnelling process. The amplitude for this process has a non-perbutive structure, $\exp(-8\pi^2|n|/g^2)$. The tunnelling process connects sectors of the vacuum $|m\rangle$ characterized by a topological label m, and forces the true QCD vacuum $|\Omega\rangle$ to be constructed as a superposition of all such sectors,

$$|\Omega(\theta)\rangle = \sum_m e^{im\theta} |m\rangle \quad . \tag{4}$$

The possibility of non-zero values of θ raises interesting questions about the structure of CP violation. In this talk we will deal with massless fermions for which the θ parameter is actually not observable.

The existence of instantons influences the propagation of colored particles in the QCD vacuum and induces interactions among all particles coupled to the color gauge field. I will describe methods[3] for calculating these interactions later in this talk. Here I will note the role they play in influencing the symmetry structure of QCD. The apparent U(N)×U(N) symmetry of a theory with N flavors of massless quarks is too large in comparison with the observed structure of hadronic multiplets. 't Hooft[4] pointed out that instantons reduce this U(N)×U(N) symmetry to U(1)×SU(N)×SU(N). Callan, Dashen, and Gross[5] have suggested further that instanton-induced interactions may drive the spontaneous breakdown of the remaining chiral $SU(N) \times SU(N)$ symmetry, produce the experimentally observed multiplets of $SU(N)$ and generate a set of N^2-1 Nambu-Goldstone bosons. The fact that only N^2-1 (and not N^2) massless bosons result in this scheme is a consequence of 't Hooft's observation[4] that instantons contribute to the anomalous divergence of the axial baryon current.

This talk will emphasize the role of instantons in the spontaneous breaking of $SU(N) \times SU(N)$ symmetry. Our calculations will require only the single instanton ($n = \pm 1$) solutions of Belavin et al.[1] It is, however, worthwhile to note some of the remarkable generalizations of this work that have been obtained in the last four years. The single instanton of Belavin et al. was constructed for the gauge group $SU(2)$. Its existence stems from the decomposition of the rotation group in Euclidean space as $O(4) = SU(2) \times SU(2)$. Instanton solutions with $n = \pm 1$ are described by 8 parameters: four of these refer to the instanton's position, one to its size, and three to its orientation in the group space of $SU(2)$. Solutions with arbitrary n are now known.[8] These are described by 8n parameters,[9] a fact which invites their interpretation as an assembly of n instantons located at various points of space and possessing a variety of sizes and gauge orientations. This interpretation is confirmed by examining the case in which the position parameters are all widely separated (relative to the size parameters). This solution does in fact correspond to an approximate superposition of n individual instantons. Another generalization involves an extension[8] of the gauge group beyond $SU(2)$. These solutions can be interpreted[10] in terms of embeddings of an $SU(2)$ subgroup within the larger group.

Although these generalizations form a fascinating subject in themselves, they will not be relevant for the work I will discuss here. The reason is that for any configuration of n widely separated instantons, the action would be essentially unchanged if one of the instantons were replaced by an anti-instanton. The resulting configuration would no longer be self-dual (or anti-self-dual) and would no longer be an exact solution of the QCD field equations. Nonetheless, since in the path integral quantization procedure each path is weighted by its action, the physical role of these mixed configurations is no smaller than that of configurations with n instantons alone. It follows that we might as well treat these different configurations on the same mathematical footing. Thus for any configuration of widely-separated instantons and/or anti-instantons we write the vector potential A_μ as

$$A_\mu = \sum_i A_\mu^{(i)} , \qquad (5)$$

where $A_\mu^{(i)}$ refers to the potential for a single instanton or anti-instanton. If the $A_\mu^{(i)}$ are specified in a suitable gauge (the "singular" gauge where $A(x) \sim x^{-3}$ for large x), then the field strength tensor will also be an approximate superposition of field strengths arising from the individual instantons.

The field configurations just specified form the basis of a dilute gas approximation.[11] Neglecting interactions between instantons and anti-instantons, we can write the action of a dilute gas configuration as the sum of terms which arise from each individual instanton. This approximation is valid as long as the instantons (and anti-instantons) are on the average widely separated. For questions of chiral symmetry breaking we will argue that this is indeed the case. Other aspects of QCD may involve denser instanton configurations, and Callan, Dashen and Gross[12] have actually suggested that the interactions in a dense instanton fluid may be sufficient to explain color confinement.

Dilute gas calculations require an analysis[3] of quantum fluctuations about the classical field configurations of Belavin et al. (Since the vector potential associated with an instanton configuration is of order g^{-1}, these quantum fluctuations will provide a numerical factor (of order g^0) which multiplies the semi-classical tunnelling amplitude.) The calculations are facilitated by working in the background field gauge.[13] In the gauge, explicit propagation functions have been constructed[14] for the ghost (scalar), quark (spinor) and gauge (vector) particles. These functions are illustrated in Fig. 1. The covariant derivatives D_μ refer to the background instanton field $A_\mu^{(i)}$ in the form

$$D_\mu = \partial_\mu - iA_\mu^{(i)} . \qquad (6)$$

Quantum fluctuations about the instanton background are described by determinants of the operators $-D^2$, $-i\gamma D$ and $(-D^2 \delta_{\mu\nu} - 2F_{\mu\nu})$ for the ghost, quark and gauge fields, respectively. These were calculated by 't Hooft[15] and are illustrated in Fig. 2.

The primes on the operators in Fig. 1 refer to projections into subspaces of non-zero eigenmodes. Zero modes occur[16] for operators describing the quantum fluctuations of any spinning particle in the field of an instanton. Zero modes of the gauge field correspond simply to variations of the 8 parameters which characterize the instanton. These modes are appropriately replaced by collective coordinates.[17] In any calculation these coordinates (namely, the instanton parameters) are integrated over their allowed range of variation. Zero modes of the fermion field have a different physical role. The corresponding eigenvalues are lifted from zero with the introduction of an explicit or dynamically generated mass. If that mass is not too large, the zero modes $|\psi_o\rangle$ will dominate an expansion of the massive fermion propagator,

$$\frac{1}{-i\gamma D + m} = \frac{|\psi_o\rangle\langle\psi_o|}{m} + \frac{1}{-i\gamma D'} + \mathscr{O}(m\rho) . \qquad (7)$$

The instanton size ρ sets the scale for this expansion.

An explicit construction of the propagation functions is facilitated by the fact that the non-zero modes for particles of differing spins are all related.[3,9] Thus, if the scalar propagator $\Delta = -1/D^2$

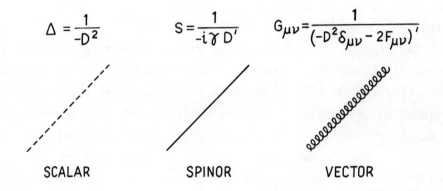

Fig. 1. Propagation functions in the field of an instanton.

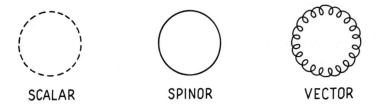

Fig. 2. Determinants resulting from quantum fluctuations in an instanton field.

is known, the spinor and vector propagation functions are given by the formal construction[14]

$$S = \gamma D \, \Delta \left[\frac{1+\gamma_5}{2}\right] + \Delta \, \gamma D \left[\frac{1-\gamma_5}{2}\right] \qquad (8)$$

and

$$G_{\mu\nu} = -q_{\mu\nu\alpha\beta} \, D_\alpha \, \Delta\Delta \, D_\beta \qquad (9)$$

with

$$q_{\mu\nu\alpha\beta} = \tfrac{1}{4} \, \text{tr} \, [\gamma_\mu \gamma_\nu \gamma_\alpha \gamma_\beta (1 - \gamma_5)] \quad . \qquad (10)$$

Equation (9) includes a divergent term proportional to an outer product of zero modes of the gauge field. The resulting ambiguity is of no physical relevance, however, and can be eliminated[18] by a redefinition of the collective coordinates and a corresponding modification in the equation defining $G_{\mu\nu}$. Thus, to construct S and $G_{\mu\nu}$ it suffices to know the form of Δ. That function can be constructed[19] by exploiting the 0(5) invariance[20] of a single instanton Generalizations of Δ for multi-instanton configurations are also known[14,8] but will not be required here.

Let us apply this formalism to the study of chiral symmetry breaking[5-7]. What we seek is the dynamical generation of a quark mass in a theory with N (initially massless) quarks (N>1). If one considers only perturbations about the massless fermion vacuum, then no mass generation will be seen to occur.[21] This, in fact, is just a reflection of the chiral SU(N)×SU(N) symmetry of the theory, realized - if the fermions are massless - in the Wigner mode. In the dilute gas approximation, vacuum tunnelling is suppressed entire by the factor Det $(-i\gamma D)$, which vanishes when zero modes are present The dilute gas is inappropriate for this calculation, since interactions between instantons and anti-instantons do allow tunnelling

INSTANTONS AND CHIRAL SYMMETRY

to occur with the production of correlated pairs of instantons and anti-instantons.[22] There is, however, still no quark mass generated in this process.

The preceding discussion emphasizes the fact that the spontaneous breakdown of chiral symmetry involves a rearrangement of the fermion vacuum. To investigate this effect one can work with a massive fermion vacuum and seek interactions which support the stability of a non-zero but self-consistent mass. The prototype of this approach was described by Nambu and Jona-Lasinio.[23] Its application to instantons[5-7] is straightforward as long as the spontaneously generated mass is small. In that case we can use the expansion of Eq. (7) and write, for the mass shift of a quark moving through a dilute gas of instantons, the expression

$$m(q) \sim \int \frac{d\rho}{\rho^2} [\rho \, m_o(\rho)]^{N-1} \gamma(\lambda\rho) \, f^2\left(\frac{\rho q}{2}\right) [1+o(\rho m)] \quad . \tag{11}$$

This expression is calculated in the massive fermion vacuum. Thus the fermion determinant includes factors of the function

$$m_o(\rho) = \langle \psi_o | m | \psi_o \rangle \quad , \tag{12}$$

which specifies the shift in the eigenvalue of $|\psi_o\rangle$ produced by the momentum dependent fermion mass $m(q)$.

The function $f(\frac{\rho q}{2})$ which appears in Eq. (11) describes the momentum space wavefunction[6] of the mode $|\psi_o\rangle$ which is presumed to dominate the propagation function (see Eq. (7)). The same function enters the definition of $m_o(\rho)$, and Eq. (12) can be rewritten as

$$m_o(\rho) = \rho^2 \int dq \, q \, m(q) \, f^2\left(\frac{\rho q}{2}\right) \quad . \tag{13}$$

Equations (11) and (13) define the self-consistency condition for the mass function $m(q)$. A trivial solution to these equations ($m=0$) corresponds to a realization of chiral SU(N)×SU(N) symmetry in the

Wigner mode, while a nontrivial solution corresponds to its realization in the Nambu-Goldstone mode.

The dependence of Eq. (11) on the running coupling $g(\rho)$ is summarized in the function $\gamma(\lambda\rho)$,

$$\gamma(\rho) = \left[\frac{8\pi^2}{g^2(\rho)} \right]^6 \exp\left[\frac{-8\pi^2}{g^2(\rho)} \right] . \qquad (14)$$

The exponential factor is the familiar semi-classical approximation to the tunnelling amplitude. The factor of g^{-12} arises from the Jacobian associated with the introduction of collective coordinates (for the gauge group SU(3)). The constant λ in Eq. (11) is indeterminate at the one loop level, but is presumably of order one. Note that the condition

$$\rho\ m \ll 1 \qquad (15)$$

needed for zero modes to dominate the expansion of Eq. (7) is precisely the condition needed to insure the validity of a dilute gas approximation[7]. This is because the density of instantons of size ρ is of order $\rho^{-4} [\rho\ m_o(\rho)]^N$.

It is obviously important to understand the behavior of the integrand in Eq. (11) at large ρ. Note that at small ρ, asymptotic freedom guarantees convergence through the vanishing of the running coupling $g(\rho)$ in the limit $\rho \to 0$ (see Eq. (14) for the form of $\gamma(\rho)$) At large ρ, $g(\rho)$ may grow indefinitely; but large values of g produc small values of γ, and indefinite growth of g facilitates convergenc of the integral. An illustration of this phenomenon is given in Fig. 3. Note that even if the qualitative behavior of g changes at large ρ (or if the form of γ is modified when g is large) we have still identified a distinct physical region (at small ρ) which is dynamically selected to yield important instanton effects. This region dominates the dilute instanton gas. Physical effects

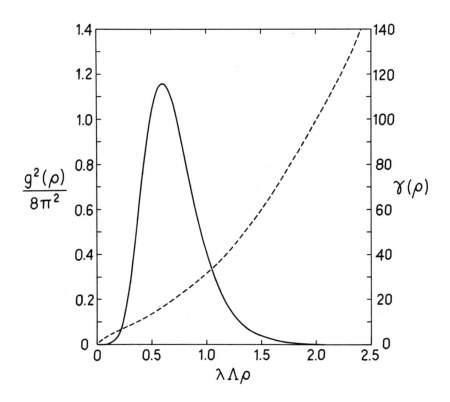

Fig. 3. The running coupling $g^2/8\pi^2$ (dashed line) and the tunnelling factor γ (solid line).

resulting from larger instantons are physically distinct from the effects discussed here and are best left for separate discussion.

The specific value of ρ which maximizes $\gamma(\lambda\rho)$ can be specified by a mass $\mu = \rho^{-1}$, where

$$\frac{g^2(\lambda/\mu)}{8\pi^2} = \frac{1}{6} . \qquad (16)$$

This value is small enough that one can optimistically justify the neglect of the higher order terms in Eq. (14). To actually evaluate the integral in Eq. (11) we need to choose a specific form for $g(\rho)$. Motivated by the idea of a linear potential (and demanding consistency with asymptotic freedom at small ρ), we take[7]

$$\frac{8\pi^2}{g^2(\rho)} = \frac{33-2N}{6} \log\left(1 + \frac{1}{\Lambda^2\rho^2}\right) . \qquad (17)$$

[The potential which results from the Fourier transform of $g^2(1/q^2)/q^2$ is actually quite successful[24] in describing the energy levels and widths of various quarkonium states.] The parameter Λ is fixed by analyses of scaling violations and has the approximate value[25] $\Lambda \approx 500$ MeV. The dominant instanton size μ^{-1} is given by Eqs. (16) and (17) to be

$$\mu = [e^{4/3} - 1]^{1/2} \lambda\Lambda \qquad (18)$$

for $N = 3$. The numerical solution of Eqs. (11) and (13) is given in Fig. 4. Demanding that the dynamically generated mass $m(0)$ have a value of order 300 MeV, we confirm our prejudice that λ should be close to 1 and obtain[26] an estimate

$$\mu \approx 835 \text{ MeV} \qquad (19)$$

The fact that μ^{-1} is no larger than 0.2 fm has a number of important consequences. First of all, it justifies the approxima-

INSTANTONS AND CHIRAL SYMMETRY

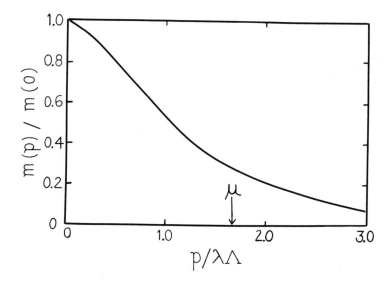

Fig. 4. Self-consistent quark mass function. The dominant instanton size is μ^{-1}.

tion, Eq. (15), which underlies both our dilute gas approximation and our expansion (Eq. (7)) of the massive fermion propagation function. Secondly, it suggests that chiral symmetry breaking does occur at distances smaller than the confinement length. This picture is what underlies a large part of the previous discussion. Were it not valid we could not ignore the interactions between instantons and anti-instantons (or whatever interactions are responsible for confinement) and could not neglect terms of higher order in g^2 throughout our calculations.

The proof of the pudding is, of course, always in the tasting, and it is inviting to look for direct experimental tests of an instanton scale on the order of that suggested by Eq. (19). To this end let us consider what effects instantons might have beyond the generation of (constituent) quark masses. Since mass generation implies the spontaneous breakdown of chiral SU(N)×SU(N) symmetry, the same interaction which produces the quark mass must necessarily bind quark-antiquark pairs to yield a set of N^2-1 Nambu-Goldstone bosons. Explicit calculations[7] verify this general picture and produce Bethe-Salpeter wave functions

$$\Pi_i(p) \sim \frac{m(p) \lambda_i \gamma_5}{p^2 + m^2(p)} . \qquad (20)$$

Note that the large distance ($p \to 0$) behavior of this wavefunction is controlled by the mass $m(0)$. The large momentum region of $\Pi_i(p)$ is however, defined by the parameter μ, as illustrated in Fig. 4.

A similar argument can be applied to the nucleon wavefunction. The quark-quark interaction that instantons induce is strongly attractive[26] for the channel antisymmetric in color, flavor and spi This interaction is strong enough to generate a diquark bound state with a wavefunction similar to that of Eq. (20). The quantum numbe of the diquark allow it to reside in the nucleon but not, for example, in the Δ resonance. This underscores the role that instantons play in breaking the SU(2N) symmetry of the non-relativistic quark model. In the meson sector we have already seen how instantons produce massless pseudoscalar states; the same interactions have no influence at all on vector mesons.

The fact that the instantons which dominate the dilute gas approximation are small relative to the confinement length (~ 1.0 fm) suggests that we picture the nucleon as a quark-diquark bound state.[26] The momenta of quarks within the diquark are of order μ, while the relative momentum of the quark-diquark motion is presumably of order $(1\,\text{fm})^{-1}$, a significantly smaller number. Since the diquark has isospin zero, it follows[26] that there should be differences in the momentum distributions of the up and down quarks within a nucleon. This can be tested experimentally in $\mu^+\mu^-$ production experiments, which can effectively probe the momentum distributions of quarks in the colliding particles. In πp collisions we expect an asymmetry

$$\langle p_\perp^2 \rangle_{\pi^+} - \langle p_\perp^2 \rangle_{\pi^-} \simeq \frac{1}{2} \mu^2 \qquad (21)$$

in the transverse momenta of muon pairs produced with either π^+ or π^- beams. The large value of this difference is a direct consequence of the small size of dominant instantons (see Eq. (19)). The experimental observation of this effect would serve to confirm not only the general picture we have sketched here but even the approximations which we have systematically employed to implement this picture.

We have argued in this talk that instantons may drive the spontaneous breaking of chiral SU(N)×SU(N) symmetry in QCD. If the sizes of the dominant instantons are those specified by Eq. (19), then the region where instanton effects are most important is intermediate between short distances - where perturbation theory is reliable - and long distances - where the confinement mechanism sets in. We have emphasized that this distinction should permit direct experimental observations of instanton-related physical effects.

I would like to thank Dennis Creamer, Choonkyu Lee and Ray Willey for a number of useful discussions. Costumes for my talk were provided by N. Smiley.

NOTES AND REFERENCES

1. A. Belavin, A. Polyakov, A. Schwartz, Y. Tyupkin, Phys. Letters 59B, 85 (1975).
2. C.G. Callan, R.F. Dashen and D.J. Gross, Phys. Letters 63B, 334 (1976); R. Jackiw and C. Rebbi, Phys. Rev. Letters 37, 172 (1976); V.N. Gribov (unpublished).
3. The prototype for all such calculations is found in the work of G. 't Hooft, Phys. Rev. D14, 3432 (1975) and D18, 2199 E (1978).
4. G. 't Hooft, Phys. Rev. Letters 37, 8 (1976).
5. C.G. Callan, R.F. Dashen and D.J. Gross, Phys. Rev. D17, 2717 (1978). This idea has been developed by D.G. Caldi, Phys. Rev. Letters 39, 121 (1977) and by R.D. Carlitz and D.B. Creamer, Refs. 6 and 7.
6. R.D. Carlitz, Phys. Rev. D17, 3225 (1978).
7. R.D. Carlitz and D.B. Creamer, Pittsburgh preprint PITT-199, to be published in Annals of Physics (1979).
8. M.F. Atiyah, N.J. Hitchin, V.G. Drinfeld and Yu. I. Manin, Phys Letters 65A, 185; N.H. Christ, E.J. Weinberg and N.K. Stanton, Phys. Rev. D18, 2013 (1978); E.J. Corrigan, D.B. Fairlie, S. Templeton and P. Goddard, Nuc. Phys. B140, 31 (1978).
9. A.S. Schwartz, Phys. Letters 67B, 172 (1977); R. Jackiw and C. Rebbi ibid, 67B, 189 (1977); L.S. Brown, R.D. Carlitz and C. Lee, Phys. Rev. D16, 417 (1977); M.F. Atiyah, N.J. Hitchin and I.M. Singer, Proc. Natl. Acad. Sci. U.S.A. 74, 2662 (1977).
10. C.W. Bernard, N.H. Christ, A.H. Guth, E.J. Weinberg, Phys. Rev. D16, 2967 (1977).

11. C.G. Callan, R.F. Dashen and D.J. Gross, Ref. 5; N. Andrei and D.J. Gross, Phys. Rev. $\underline{D18}$, 468 (1978); R.D. Carlitz and D.B. Creamer, Ref. 7.
12. C.G. Callan, R.F. Dashen and D.J. Gross, Princeton preprint (1978).
13. J. Honerkamp, Nuc. Phys. $\underline{B48}$, 269 (1972) and G. 't Hooft, Ref. 3.
14. L.S. Brown, R.D. Carlitz, D.B. Creamer, and C. Lee, Phys. Letters $\underline{70B}$, 180 (1977) or $\underline{71B}$, 103 (1977) and Phys. Rev. $\underline{D17}$, 1583 (1977).
15. G. 't Hooft, Ref. 3. A generalization of this work to n-instanton field configurations is given by L.S. Brown and D.B. Creamer, Phys. Rev. $\underline{D18}$, 3695 (1978).
16. S. Coleman (unpublished); J. Kiskis, Phys. Rev. $\underline{D15}$, 2329 (1977); L.S. Brown, R.D. Carlitz and C. Lee, Ref. 9.
17. J.L. Gervais and B. Sakita, Phys. Rev. $\underline{D11}$, 2943 (1975).
18. H. Levine and L.G. Yaffe, preprint SLAC-PUB-2162; C. Lee (unpublished).
19. D.B. Creamer, Phys. Rev. $\underline{D16}$, 3496 (1977).
20. R. Jackiw and C. Rebbi, Phys. Rev. $\underline{D14}$, 517 (1976).
21. J. Hietarinta, W.F. Palmer and S.S. Pinsky, Phys. Rev. Letters $\underline{40}$, 421 (1978).
22. C. Lee and W.A. Bardeen, Michigan preprint UM HE 78-55; W.F. Palmer and S.S. Pinsky, Ohio State preprint COO-1545-243.
23. Y. Nambu and G. Jona-Lasinio, Phys. Rev. $\underline{172}$, 345 (1961).
24. R.Z. Roskies, private communication; R. Levine and Y. Tomozawa, Michigan preprint UM HE 78-41 and work in preparation; J.L. Richardson, preprint SLAC-PUB-2229.
25. See, for example, A. De Rujula, H. Georgi and H.D. Politzer, Ann Phys. $\underline{103}$, 315 (1977); E.C. Poggio, H.R. Quinn and S. Weinberg, Phys. Rev. $\underline{D13}$, 1958 (1976); R. Shankar, Phys. Rev. $\underline{D15}$, 755 (1977); R.G. Moorhouse, M.R. Pennington and G.C. Ross, Nuc. Phys. $\underline{B124}$, 285 (1977).
26. For more details see R.D. Carlitz and D.B. Creamer, Pittsburgh preprint PITT-208 (1979).

QCD AND HADRONIC STRUCTURE

Laurence Yaffe

Princeton University

Princeton, New Jersey 08540

I. INTRODUCTION

Quantum chromodynamics (QCD) is generally believed to be the true theory of strong interactions. It is supposed to predict correctly all observed hadronic behavior. In particular, the inner dynamics of QCD are believed to produce the fundamental properties of confinement and dynamical chiral symmetry breaking. Confinement, or the statement that all observed hadrons are colorless bound states of quarks, is thought to be due to the effective coupling of QCD rising indefinitely as one probes increasing distances. Dynamical chiral symmetry breaking refers to the belief that the true vacuum state of QCD is not chirally invariant (even in the chirally symmetric limit where all light quark masses are neglected.) As a consequence one should find dynamical quark mass generation and formation of composite Goldstone bosons such as the pion. Unfortunately it has proven extremely difficult to derive these properties directly from QCD.

However, on the phenomenological side, one knows that simple quark models, especially the MIT bag model, provide a remarkably good description of the spectrum and static properties of low lying hadrons (except the pion). In such a model one views an ordinary hadron as a bag-like region of space inside of which light quarks are confined. These quarks interact through gluons with a rather weak coupling. The model nicely explains many features of the hadronic spectrum such as, for example, the formation of a string-like bag and consequent linear Regge trajectories for large angular momentum states. It may also be applied to heavy quark states through the use of a Born-Oppenheimer type approximation. Here one considers the quarks momentarily static, then computes the optimum shape of the bag around the quarks, and interprets the resulting energy as a quark potential energy to be inserted into a non-relativistic Schrödinger equation. In general, the bag model provides a very appealing physical view of the structure of hadrons.

I would like to describe an approach to QCD based on semi-classical or instanton methods which results in a physical picture of hadronic structure very similar to that postulated in the bag model. Unfortunately these semiclassical methods are not fully adequate for a quantitative treatment of hadronic properties; however using them to learn what we may from QCD will be seen to provide a very compelling qualitative view of the nature of hadronic structure. Much of what I will describe is due to the work of Curt Callan, Roger Dashen, and David Gross.[1,3]

II. THEORETICAL APPROACHES TO QCD

QCD is specified by the simple Lagrangian

$$L = \frac{1}{4} F^a_{\mu\nu} F^{\mu\nu}_a + \sum_{i=1}^{F} \bar{\Psi}_i [\not{D}+m_i] \Psi_i$$

where $F^a_{\mu\nu}$ is the non-abelian field strength, $F^a_{\mu\nu} = \partial_\mu A_\nu - \partial_\nu A_\mu + g[A_\mu, A_\nu]$, and D_μ is the covariant derivative $\partial_\mu + igA_\mu$. Like all non-abelian gauge theories, QCD possesses the well-known property of asymptotic freedom. This means that the effective coupling of the theory vanishes at short distances or large momenta. Asymptotically $g^2(\rho)/8\pi^2 \sim 1/(11-\frac{2}{3} F) \ln (\rho/\lambda)$ where λ is the renormalizat scale.

Because of this distance dependence of the effective coupling, theoretical approaches to QCD essentially divide into two regimes. When examining the short distance structure of the theory one may apply weak coupling techniques such as ordinary diagrammatic perturbation theory, or semiclassical methods which include the effects of instantons or other non-perturbative fluctuations. In general for weak coupling one considers the contribution of different saddle points (and approximate saddle points) to the functional integral $\int \mathcal{D}A \exp - \frac{1}{g^2} S(A)$. On the other hand, in order to examine the large distance behavior of QCD one must deal with a large coupling and apply some type of strong coupling technique. Essentially one would like to expand in powers of $[\frac{1}{g^2} S(A)]$.

We will strictly concentrate on what may be learned by applying the former weak-coupling semiclassical methods. However, it should be emphasized that neither type of approach is capable of addressing all questions that one might care to ask about QCD. It has unfortunately proven virtually impossible to construct approximation schemes for QCD which are uniformly adequate for all length scales.

With these restrictions in mind we will now approximate the dynamics of the theory by only including the contributions of the dominant small scale fluctuations. Specifically, to calculate the partition function

$$Z = \langle \text{vac}|e^{-HT}|\text{vac}\rangle = \int \mathcal{D}A_\mu(\vec{x}) \exp - \frac{1}{g^2} \int \mathcal{D}^4 x \, \text{Tr} \, F_{\mu\nu} F^{\mu\nu}$$

or any other vacuum expectation value of gauge invariant operators, we will expand about all (constrained) multiple instanton-antiinstanton fields. For more justification and elaboration of this point see ref. 1.

A single instanton is essentially a localized blob with an arbitrary position $(\vec{x}_I)$, size (ρ), and global gauge orientation $(R^{\alpha a})$. The field is given by

$$A^\alpha_\mu = 2\rho^2 \frac{R^{\alpha a}\bar{\eta}^{-a}_{\mu\nu}(x-x_I)^\nu}{(x-x_I)^2[(x-x_I)^2+\rho^2]}$$

and for distances $|x-x_I|$ greater than ρ it approaches the simple form $-\rho^2 R^{\alpha a}\bar{\eta}^{-a}_{\mu\nu}\partial_\nu(1/(x-x_I)^2)$. ($\bar{\eta}^{-a}_{\mu\nu}$ is a certain constant tensor defined by 't Hooft[2].) The single instanton contribution to the partition function is given by the volume of spacetime times $n_0(\rho)$, where

$$n_0(\rho) = \frac{c}{\rho^4}(4\pi^2/g^2)^6 \exp{-8\pi^2/g^2(\rho)}.$$

Here $8\pi^2/g^2$ is simply the classical action of the instanton, the factors of $(1/g^2)$ in front come from the collective coordinate Jacobian, and the constant $c = .1$ (for Pauli-Villars regulators)[1].

The contribution of an arbitrary number of well separated instantons and antiinstantons may be similarly computed, and the resulting approximation to the partition function is given by

$$Z \approx \sum_{n_+,n_-} \frac{1}{n_+!n_-!} \prod_{n=1}^{n_++n_-} [\int d^4x_i \frac{d\rho_i}{\rho_i} dR_i^{\alpha a} n_0(\rho_i) \exp{-S_{INT}(\vec{x}_j,\rho_j,R_j)}].$$

Here $n_{+(-)}$ is the number of (anti)instantons, each instanton and antiinstanton contributes a factor of $n_0(\rho_i)$, and one integrates over the position, size, and orientation of each instanton. S_{INT} is simply the difference between the action of the multiinstanton field and the sum of the individual action for each instanton.

III. THE ANALOG GAS

This result suggests an extremely useful analog gas description since Z is seen to have exactly the form of the grand canonical partition function for a gas of interacting particles. (Here each instanton of a different size and orientation is viewed as a different species of particle.) Such an analogy is useful because it immediately suggests that one may apply many of the techniques used in statistical mechanics to the study of our semiclassical approximation to QCD.

In particular, since the interaction action, S_{INT}, vanishes as the inverse of the fourth power of the interinstanton separation one might begin by considering a perfect gas approximation where S_{INT} is neglected entirely. In this case we see that $n_o(\rho)$ has the physical interpretation of the instanton density. Due to asymptotic freedom it vanishes for small instantons as $\rho^{-4}(\ln(\rho\lambda))^6 (\rho\lambda)^{11}$; however it grows very quickly as one considers increasingly large instantons. Unfortunately, such a perfect gas approximation neglects all collective effects of the instanton gas and is only valid for an extremely dilute system.

In order to learn how to treat the important collective dynamics of the instanton gas, one should consider the form of the instanton field and of the instanton interaction for distances large compared to the core size ρ. What one finds is the remarkable result that in this regime, instantons are equivalent to four dimensional magnetic dipoles. For example, the instanton field has the dipole form

$$A_\mu^\alpha = - D_{\mu\nu}^\alpha \partial_\nu \frac{1}{x^2},$$ where the dipole moment is given by

$$D_{\mu\nu}^\alpha = \rho^2 R^{\alpha a} \bar{\eta}_{\mu\nu}^a.$$ (Remember that in general a magnetic dipole moment is an antisymmetric rank two tensor.) The field strength has the dipole form

$$F_{\mu\nu}^\alpha = - \frac{4}{x^4} T_{\mu\lambda}(\hat{x}) D_{\lambda\epsilon}^\alpha T_{\epsilon\nu}(\hat{x}).$$

Here the conformal matrix $T_{\mu\nu}(\hat{x}) = \delta_{\mu\nu} - 2\hat{x}_\mu \hat{x}_\nu$, and T flips the duality; in other words if the dipole moment is anti-self dual, then the field strength is self dual. Most importantly, for well separated instantons, the interaction is found to equal the dipole-dipole form,

$$S_{INT} = -\frac{8\pi^2}{R^4} \text{Tr}(T(\hat{R})D_1\, T(\hat{R})D_2)$$

and in four dimensions only dipoles of opposite duality interact.

This dipole approximation yields the very useful result that the instanton gas will behave like a four dimensional polar medium. This means that the familiar techniques used to treat dipolar media may be applied here to treat the collective effects of the instanton gas.

IV. HEAVY QUARKS

We would now like to apply these methods to the problem of describing the behavior of heavy quarks in QCD. This is perhaps the simplest physically interesting question one may address in QCD. For convenience we will neglect the modifications due to the presence of light quarks.

For arbitrarily heavy quarks, one need only compute the vacuum expectation value of the ordered loop tr P $\exp \oint_L A \cdot dx$ for a rectangular path of spatial extent R and time T, because in the limit $T \to \infty$

$$\langle \text{tr } P\, e^{\oint A \cdot dx} \rangle \to e^{-V(R)T}$$

where $V(R)$ is the physically relevant static quark potential.

To compute $V(R)$ even in our semiclassical approximation requires evaluating the ordered loop integral for an arbitrary multiinstanton field and then averaging this result over all instanton configurations. This is extremely difficult. However, if instead of attempting the complete calculation we restrict our attention to the large-distance limit of $V(R)$ then a remarkable simplification occurs. In this limit the dominant contribution to the quark potential (or rather its derivative) comes from the numerous instantons far from either leg of the quark loop. These instantons have weak dipole fields at the position of the quark loop and consequently the leading contribution to the quark potential may be obtained by expanding the ordered exponential and retaining the first non-trivial term. After performing the instanton global gauge averaging, one finds that the leading contribution of an instanton to V(R) may be calculated by averaging over all color orientations the interaction of the instanton dipole moment with an effective abelian magnetic field, $F_{\mu\nu}^{ext}$, generated by the quark loop.

Thus, for calculating the large distance behavior of the static quark potential, the non-abelian ordered loop is equivalent to an ordinary current loop generating an external magnetic field. It is important to realize that this effective external field is not an approximation to the full non-abelian gauge field; rather it simply reproduces the leading large distance behavior of the fully gauge invariant quark loop. Corrections to this leading behavior would involve inherently non-abelian interactions induced by the ordered loop.

This result implies that in order to find the true instanton density in the presence of the quark loop (that is, to find which configurations dominate the functional integral), we must examine the behavior of a four dimensional dipolar medium placed in an external field. This is a tremendous simplification since it tells us that we need only generalize ordinary magnetostatics to four dimensions.

Let us focus attention on a region of spacetime far from the quark loop, over which the external field varies slowly (see fig. 1). Within this region we know that the change in energy due to the presence of the external field is given by

$$\varepsilon = \frac{1}{4g^2} F_{\mu\nu}^{\alpha} H_{\mu\nu}^{\alpha} .$$

Here the magnetic induction $H_{\mu\nu}$ is that field which only couples

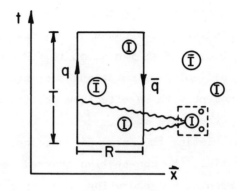

Figure 1a. The quark loop in the presence of a multiinstanton configuration.

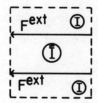

Figure 1b. Expanded view of a small region of spacetime in which the instantons interact with a nearly constant external field.

to the external current $\partial^\mu H_{\mu\nu} = J_\nu^{ext}$ (quark loop), while the magnetic field $F_{\mu\nu} = H_{\mu\nu} + 4\pi M_{\mu\nu}$, where $M_{\mu\nu}$ is the dipole magnetization. Since the external field is weak, we are in a linear response regime and hence $F_{\mu\nu} = \mu H_{\mu\nu}$, where μ is the magnetic permeability. Thus

$$\varepsilon = \frac{1}{4g^2\mu}(F_{\mu\nu})^2$$

and the presence of the medium has the effect of replacing the coupling g^2 with $g^2\mu$. So if the permeability is greater than one then the dipole medium is responding in a paramagnetic or antiscreening manner. This will lead to an increase in the energy of the quarks.

If one applies a perfect gas approximation to the instanton gas and completely ignores the mutual dipole interactions, then one finds that the permeability is given by $\mu_{P.G.} = 1 + \eta[n_o(\rho)]$, where the functional η is essentially the mean square dipole moment,

$$\eta[n(\rho)] \equiv \frac{4\pi^2}{8} \int \frac{d\rho}{\rho} n(\rho)\rho^4 \frac{8\pi^2}{g^2(\rho)} > 0 \quad .$$

And in this approximation the large distance behavior of $V(R)$ is simply a renormalization of the Coulomb potential $- g^2\eta/R$.

To improve upon this naive treatment one must include the effects of the mutual dipole interactions. To do so, consider the free energy F of a gas of fixed density $n(\rho)$ in an external field $F_{\mu\nu}$ (color averaged).

$$F = F_o(n) + F_{INT}(n) - \frac{1}{2} F^2/\mu(n) \quad .$$

Here $F_o = \int \frac{d\rho}{\rho} 2n(\rho) [\ln \frac{n(\rho)}{n_o(\rho)} - 1]$ is the perfect gas free energy, F_{INT} contains the free energy due to the dipole interactions, and the final term is the free energy of the external field. The permeability is of course a function of the density, and a simple calculation shows that

$$\mu(n) = \eta[n] + \sqrt{1+\eta^2[n]} \ .$$

(This differs from the perfect gas result $\mu = 1 + \eta$ due to the fact that we are now taking into account the difference between the local field acting on a dipole and the average background field.)

Minimizing the free energy with respect to $n(\rho)$ will determine the instanton density in the configuration which dominates the functional integral. One finds

$$n(\rho) = n_0(\rho) \exp\left[-\frac{2}{1+\mu^2} \frac{\pi^2 F^2}{8} \rho^4 \frac{8\pi^2}{g^2(\rho)} - \frac{\partial}{\partial n} F_{INT}\right] \ .$$

This is a self consistent equation since μ is itself a function of $n(\rho)$. (See ref. 3 for discussions of the simple mean field treatment of F_{INT}, further consistency conditions and justifications, etc.) Solving this equation for the density $n(\rho;F)$ as a function of the external field, one finds the phase diagram shown in fig. 2. This plots the induction $H = F/\mu$ as a function of the external field F. For large external field one finds a very dilute phase with a very small susceptibility ($\mu-1$). All large instantons are strongly suppressed by the external field and the fraction of space-time occupied by instantons is only a percent or so. As the external field is reduced, the density of instantons and the permeability increase. As the field is further reduced one eventually finds a rather dense phase with a very large permeability; however in between one finds three different branches in the phase diagram.

The situation here is completely analogous to the P-V diagram for a van der Waals fluid (see fig. 3). There one has a first order phase transition from a dilute gas phase to a much denser liquid phase. The condition for thermodynamic stability in the present case is $\partial F/\partial H > 0$ and thus the middle branch of the phase diagram represents unstable states. The condition for phase

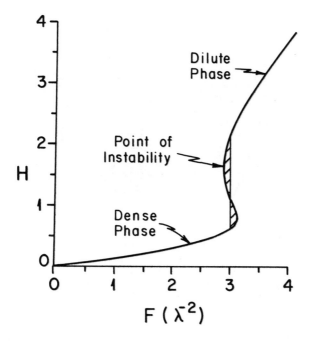

Figure 2. Phase diagram of the instanton gas including the effects of instanton interactions.

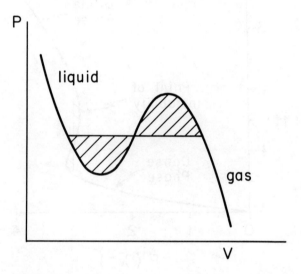

Figure 3. Analogous phase diagram for a liquid-gas transition.

equilibrium requires the free energy in the two phases to be equal. This leads to the familiar Maxwell construction stating that a first order phase transition will occur when the areas of the two shaded regions shown in fig. 2 are equal.

Now, the approximations we have made are not really adequate for a quantitative treatment of the dense phase. However the treatment is adequate all the way through the point of instability. This alone provides very strong evidence for the presence of a first order phase transition in the vacuum structure as a function of the external field.

V. QCD BAG MODEL OF HADRONS

Applying these results for the behavior of the instanton density to the original quark loop, one is immediately led to a bag-like structure because sufficiently close to the quarks the large external field will drive the vacuum into the dilute phase; however, beyond some distance the effective field will be less than the critical field needed for the phase transition and the vacuum will be in the normal dense phase. The four dimensional picture shown in figure 4 is of course equivalent to a three dimensional static bag, inside of which one has a very dilute phase with $\mu \sim 1$, outside of which is the dense, large permeability phase.

If the true permeability of the vacuum were actually infinite this would correspond to perfect paramagnetism or exact confinement. In this case the boundary conditions require $n^\mu F_{\mu\nu} = 0$ (where n^μ is the normal to the boundary) and the effective field of the quarks is completely confined inside the bag. (This of course is just the boundary condition of the MIT bag model). The energy of the quark state contains a volume energy associated with the difference in energy density of the dense and dilute phase, and a field energy of the confined field. Appropriately minimizing this leads to the expected linear quark potential.

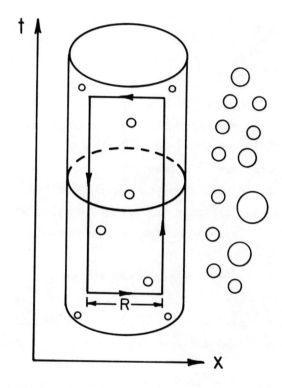

Figure 4. Spacetime view of the resulting heavy quark bag.

Now, including the effects of instantons alone, we found a large but finite permeability for the dense phase. Including the effects of other types of fluctuations such as merons[1] is easily seen to further increase μ. While it is certainly unproven, it seems very likely that if one could adequately treat the effects of all large scale fluctuations, then the permeability would in fact diverge and one would find exact confinement.

It is important to realize that this question of exact confinement is basically irrelevant for the structure of heavy quark states in QCD. If the vacuum permeability were simply very large but finite, then the quark potential would grow linearly until some very large distance at which point the quarks would become unbound. However since the size of heavy quark states decreases with increasing mass, the structure of the heavy quark state is very insensitive to the long distance details of the potential.

We have seen how a consistent treatment of the dynamics of the instanton gas leads to a physical picture of heavy quark states which, in zeroth order, simply reproduces the MIT bag model. There are however, several effects which lead to interesting corrections to the basic bag model. First, there is of course a small density of instantons which remain inside the bag. These will modify $V(R)$ and also contribute to the spin dependent potential. Second, a surface layer of instantons will form due to the attraction of an instanton inside the bag with an image anti-instanton outside, as well as due to the exchange of massless fermions with the outside phase. This surface layer will contribute to the surface tension, cause spin-dependent surface effects, etc.

VI. LIGHT QUARKS

In order to discuss the dynamics of massless fermions, one must treat the breakdown of chiral symmetry at the same time that one considers confinement. This considerably complicates the problem. A brief summary of the resulting picture is sketched below.

In the presence of massless fermions, the contribution of a single instanton to the partition function, or any configuration with unequal numbers of instantons and antiinstantons, vanishes identically. However, this in no way implies that the effects of instantons vanish in this limit. Instead, one finds instanton-antiinstanton pairs bound together through the exchange of massless fermions. The instanton density remains finite and non-zero as the quark masses vanish. (The fundamental point is that the physically relevant limit is the quark masses tending to zero with fixed instanton separations; not the limit of vanishing instanton density with fixed masses.)

As shown by 't Hooft, this gas of instantons induces an effective interaction among the fermions of the schematic form $V \sim \det \bar{\psi}_i (1+\gamma_5) \psi_\partial + (\gamma_5 \rightarrow -\gamma_5)$. This interaction makes the chirally symmetric phase unstable and drives the dynamical chiral symmetry breaking. Consider, for example, the quark propagator in the presence of this instanton gas. One may easily derive a self-consistent integral equation for the modification of the propagator due to the instanton induced interactions.[1] If the instanton density is sufficiently large one does find a chirally asymmetric solution for which the quark mass (self energy) becomes quite large for moderate momenta, but vanishes rapidly at high momenta. One similarly finds in the chirally asymmetric phase a massless pion bound by the instanton interactions. (The need for a self consistent treatment should be clearly understood; one will obviously never find a chirally asymmetric solution in any finite order of perturbation about an initial chirally symmetric vacuum. Equivalently, the relevant configurations in the functional integral are those with a finite density, not a finite number, of instantons and antiinstantons.)

Since a massless fermion propagating through a moderately dense instanton gas will acquire a large dynamical mass, this suggests that in the minimum energy state for a light hadron the instantons will be largely excluded from some region of spacetime surrounding the quarks. Thus, one expects the formation of a mass bag such that inside the bag there is a dilute, chirally symmetric phase, while outside there is a dense phase where the quarks acquire large mass and collective Goldstone excitations such as the pion exist. Including the effects of the dipole interactions one finds that the same bag acts as a color bag much like the case discussed with heavy quarks. Unfortunately it appears very difficult to discuss the dynamics or systematic corrections to this zeroth order bag model.

VII. CONCLUSIONS

We have applied straightforward semiclassical approximations to QCD and, by consistently treating the collective effects of the resulting instanton gas, have derived a qualitative picture of hadronic structure which reproduces, in lowest order, the MIT bag model. In order to significantly improve this description one must learn how to match these weak coupling techniques appropriate inside a hadron with some strong coupling treatment of the large scale fluctuations outside. This is a major theoretical challenge facing us today. If successful, such a combined treatment would hopefully provide one with the understanding of the surface structure and dynamics of the bag needed to calculate interesting hadronic parameters such as the bag constant, the pion coupling (f_π), the mass spectrum, spin-spin and spin-orbit splittings of states, etc. And at that time QCD would finally become a truly quantitative theory of hadronic structure.

REFERENCES

1. C. Callon, R. Dashen, D. Gross; Phys. Rev. D17, 2717 (1978)
2. G. 't Hooft; Phys. Rev. D14, 3432 (1976)
3. C. Callan, R. Dashen, D. Gross; Phys. Rev. D19, (1979), in press.

HIGH ENERGY PREDICTIONS FROM PERTURBATIVE Q.C.D.*

A. Mueller

Columbia University

New York, New York 10027

It has recently become clear that there are many predictions which follow firmly from Q.C.D.[1] This comes about because inclusive processes factor, in some cases, into a term which depends only on the light cone structure of Q.C.D. and a term which depends on the details of on-mass-shell hadrons. Because of asymptotic freedom the term which depends only on the light cone structure is calculable from a Callan-Symanzik equation.

An example of this factorization is the case of $e^+ + e^- \to h(p) + X$ where the outgoing observed hadron has momentum p and the $e^+ e^-$ system has a momentum q. The resulting cross section has invariant amplitudes $\bar{W}_1$ and $\nu\bar{W}_2$ just as in the formally similar case of deeply inelastic electron scattering. Define $\omega = 2p \cdot q/Q^2$; then moments of $\nu\bar{W}_2$ factorize in the following way:

$$\int_0^1 d\omega \, \omega^{\sigma-1} \, \nu\bar{W}_2(\omega, Q^2) \xrightarrow[Q^2 \to \infty]{} \sum_{i=1}^{4} v_\sigma^i \, E_\sigma^i(Q^2) \quad .$$

The E^i obey a Callan-Symanzik equation and can be determined in an

---

*This research was supported in part by the U.S. Department of Energy.

asymptotically free theory. The v^i depend on the details of the hadron $h(p)$. The v^i are analogs of matrix elements of composite operators which appear in deeply inelastic electron scattering. However, the v^i are not simply related to any composite operators.

Nevertheless, the v^i can be defined by a perturbative expansion for elementary external fields. They are what I have elsewhere called cut vertices. The perturbative expansion which defines them is quite similar to the one which defines composite operators in the Wilson expansion. The divergences which appear in the expansion of the v^i are removable by a process similar to the usual BPHZ renormalization procedure. Of course an explicit procedure cannot be given for the computation of v^i in the case of composite external particles and, again, this is the usual situation which occurs in deeply inelastic electron scattering.

All of the arguments leading to factorization in high energy processes are perturbative. It is impossible to confront directly the question as to whether semi-classical effects allow this picture to remain. The difficulty is simply that there is no acceptable formalism for treating instantons in Minkowski space and the process under consideration cannot be directly obtained from a Euclidean space formalism. Nevertheless, it appears that the same mechanism which renders wee gluons ineffective[2] must also make instanton effects small.

I. CLASSIC TESTS OF Q.C.D.

The most straightforward and simple test of Q.C.D. is the process $e^+ + e^- \to$ hadrons. This is a completely inclusive process in that no details of the final-state hadrons are observed. We can express this cross section in terms of the function R by

$$\sigma(Q^2) = \frac{4\pi\alpha^2}{3Q^2} R(Q^2) \quad . \tag{1}$$

R is given by

HIGH-ENERGY PREDICTIONS FROM PERTURBATIVE QCD

$$R = -2\pi \int d^4x e^{-iqx} <j_\mu(x) j_\mu(0)> \quad (2)$$

where $j_\mu(x)$ is the hadronic component of the electromagnetic current. As is well known R satisfies the homogeneous Callan-Symanzik equation

$$(-Q^2 \frac{\partial}{\partial Q^2} + \beta \frac{\partial}{\partial g}) R(Q^2,m^2,g) = 0 \quad (3)$$

when Q^2 is large. This equation can be solved as

$$R = R(Q_0^2, m^2, g(Q^2,Q_0^2,g)) \quad (4)$$

where $g(Q^2,Q_0^2,g)$ is the running coupling constant normalized so that $g(Q_0^2,Q_0^2,g)=g$. Equation (4) means that we can solve the large Q^2 behavior of the cross section in terms of the small coupling (perturbative) expansion in an asymptotically free theory where $g(Q^2,Q_0^2,g) \to 0$ as $Q^2 \to \infty$. We thus obtain the classic result $R \to 3 \sum_a Q_a^2$ where the sum over a goes over the quark degrees of flavor exposed at the particular Q^2 in question. The corrections are only down by $\ln Q^2$ and are calculable.

The other classic test of Q.C.D. involves deeply inelastic scattering. For example, in deeply inelastic electron scattering one has relations like

$$\int_0^1 dx\, x^\sigma \nu W_2 = \int_1^\infty d\omega\, \omega^{-\sigma-2} \nu W_2 = \sum_{i=1}^4 C_{\sigma+2}^{(i)} E_{\sigma+2}^{(i)} (Q^2) \quad (5)$$

where the $E^{(i)}$ obey a Callan-Symanzik equation and the $C^{(i)}$ are directly related to matrix elements of composite operators. In (5) the i=1..4 sum refers to a gluon singlet term, a quark singlet term, and two octet terms. There are thus three independent E's and this makes phenomenological analyses difficult. The most favorable place to confront data is in the F_3 structure function in deeply inelastic neutrino scattering. In this case the singlet operators cannot contribute and so a unique power of $\ln Q^2$ emerges. This process will be discussed elsewhere at this conference so I shall not go into more details here.

II. RECENT RESULTS

A. Jets

Jet results in Q.C.D. were first discussed by Sterman and Weinberg[3] with modified formalism given by Farhi[4] and by Georgi and Machacek.[5] For purposes of illustration let me discuss thrust (T) as given by Farhi. T is defined by

$$T = 2\max \frac{\sum'_i p^i_\parallel}{\sum_i |\vec{p}^i|} . \tag{6}$$

The procedure is as follows. Choose a coordinate system (always in the center-of-mass of the e^+e^- of course) and calculate $\sum'_i p^i_\parallel$. (The $\parallel$ means to take the component of p^i along the positive z axis and the ' on the sum means to take only $p^i_\parallel$ which are positive.) One then varies the coordinate system and defines T to be the maximum value of

$$2 \frac{\sum' p^i_\parallel}{\sum |\vec{p}^i|}$$

obtained. $T(Q^2) \equiv \langle T(Q^2) \rangle$ is the average value of T, the average being taken over experimental events. In the calculation of $T(Q^2)$ it is necessary to include all particles, charged and neutral.

It is widely believed that $T(Q^2)$ obeys the renormalization group equation

$$(- Q^2 \frac{\partial}{\partial Q^2} + \beta \frac{\partial}{\partial g}) T(Q^2) = 0 \tag{7}$$

for large Q^2. I think that it will be very hard to show that $T(Q^2)$ does indeed obey the renormalization group equation though it is likely correct. Then for large Q^2

$$T(Q^2) = 1 - C_1 a(Q^2) \ldots \qquad (8)$$

where the constant C_1 has been calculated. In principle (8) is a precise test of Q.C.D. However, it is a non-trivial task to obtain $T(Q^2)$ reliably from a given experiment and no good test has been made to date.

B. Energy Correlation Functions

The topic of energy correlation functions will be discussed in some detail this afternoon by Lowell Brown, so I shall pass over them in my talk.

C. $e^+ + e^- \to h(p) + X$

The inclusive cross section for $e^+ + e^- \to$ hadron$(p) + X$ has been widely discussed recently.[1] A detailed analysis, including photon spin effects, can be found in Ref. 1. Here I shall discuss the simplified process where the electron angles are integrated out in the center of mass of the $e^+ e^-$. Then one can define moments by

$$\int_0^1 \omega^{\sigma-2} \frac{2E_p \, d\sigma}{d^3 p} \, d\omega = \frac{\alpha^2}{3\pi(Q^2)} F_\sigma(Q^2) \;. \qquad (9)$$

For large Q^2, F_σ takes the factorized form

$$F_\sigma(Q^2) \to \sum_{i=1}^{4} v_\sigma^i \, E_\sigma^i(Q^2) \qquad (10)$$

where the E^i obey a Callan-Symanzik equation and are independent of the particle produced. The v^i depend only on the particle produced and are analogs of the matrix elements of composite operators which appear in the Wilson expansion.

The inclusive production process has a great advantage over the deeply inelastic process in that it is relatively easy to form differences so as to eliminate the singlet terms in (10). Thus, if

$F_\sigma^{\pi^+}$ and $F_\sigma^{\pi^0}$ are the moments for π^+ and π^0 production respectively, the difference

$$F_\sigma^{\pi^+} - F_\sigma^{\pi^0} \to (\ln Q^2)^{-A_\sigma} \times \text{constant}$$

where A_σ is the octet anomalous dimension. This difference, $F_\sigma^{\pi^+} - F_\sigma^{\pi^0}$, has only one dominant term in an asymptotically free theory. It is true, however, that to get the great advantage of eliminating the singlet terms one must first remove those π^+ and π^0, coming from the weak decays of charmed particles and heavy leptons. This separation is a model independent separation; however it is not clear how easy it is to do in practice.

The average multiplicity for a given type of particle is given by

$$\bar{n} = \frac{1}{\sigma_{e^+e^- \to \text{hadrons}}} \int d^3p \frac{d\sigma}{d^3p} . \qquad (11)$$

Thus the average multiplicity is proportional to $F_3(Q^2)$. Unfortunately $F_\sigma(Q^2)$ has a singularity of unknown nature at $\sigma=3$. However, the singularity is purely in the singlet channel so that, for example

$$n^{\pi^+} - n^{\pi^0} \underset{Q^2 \to \infty}{\to} \text{constant} .$$

That is, differences within an isotopic multiplet should have a finite multiplicity for large Q^2.

D. $\underline{h(p_1) + h(p_2) \to \mu^+\mu^-(q) + X}$

In the case of μ-pair production it is useful to talk about two objects

$$F_{\sigma_1\sigma_2}(Q^2,\lambda) = \int \frac{d\sigma}{d^4q} \omega_1^{-\sigma_1} \omega_2^{-\sigma_2} d\omega_1 d\omega_2 \qquad (12)$$

and

$$F_{\sigma_1\sigma_2}(Q^2) = \int d^2\underline{q} F_{\sigma_1\sigma_2}(Q^2,\lambda) \tag{13}$$

separately. In the above $\omega_i = \frac{2p_i \cdot q}{Q^2}$, $\underline{q}$ is the transverse momentum of the µ-pair in the center of mass of the colliding hadrons and $\lambda = \underline{q}^2/Q^2$. When Q^2 is large the F's factorize like

$$F_{\sigma_1\sigma_2}(Q^2,\lambda) = \frac{\alpha^2}{6\pi^2(Q^2)^3} \sum_{i,j} v^i_{\sigma_1} v^j_{\sigma_2} E^{ij}_{\sigma_1\sigma_2}(Q^2,\lambda) \tag{14}$$

and

$$F_{\sigma_1\sigma_2}(Q^2) = \frac{8\pi\alpha^2}{3(Q^2)^2} \sum_{i,j} v^i_{\sigma_1} v^j_{\sigma_2} E^{ij}_{\sigma_1\sigma_2}(Q^2) \ . \tag{15}$$

The E's obey Callan-Symanzik equations which determine the Q^2 dependence in terms of the running coupling and well-defined perturbation expansion. Using asymptotic freedom one recovers something almost like the Drell-Yan formula for $F_{\sigma_1\sigma_2}(Q^2)$ but with the actual non-scaling electroproduction structure functions appearing.

E. Photonic Processes

There are some special simplifications which occur when more than two external currents are involved in the process. As an example let me discuss deeply inelastic scattering off photons.[6] In principle this process can be obtained by considering electron-electron scattering or the two photon contribution to electron-positron scattering. The usual strong interaction operators of the Wilson expansion appear in a discussion of this process. However, in addition the Wilson expansion involves the operator

$$F_{\alpha_1\gamma} \partial_{\alpha_3} \ldots \partial_{\alpha_n} F_{\alpha_2\gamma} = \mathcal{O}_{\alpha_1\alpha_2\ldots\alpha_n}$$

where $F_{\alpha\beta}$ is the electromagnetic field strength. It turns out that the coefficient of this operator dominates the strong interaction operators by a factor of $\ln Q^2$ so that in the leading order, for an asymptotically free theory, the moments can be calculated exactly. That is

$$\int_0^1 x^n \, dx \, F_2^8(x,Q^2) \xrightarrow[Q^2 \to \infty]{} c_n \ln Q^2 \qquad (16)$$

where the c_n are known. There are no anomalous dimensions appearing here because the electromagnetic current is not renormalized.

F. Other Processes

There are many other processes for which exact and precise high energy predictions can be made. The process $e^+ + e^- \to h(p_1) + h(p_2) + X$ is formally very similar to μ-pair production; however the annihilation process has the great advantage that by taking differences of produced hadrons a single factorized term remains for large Q^2. The process $h(p_1) + h(p_2) \to h(p_3) + X$ has been discussed from the Q.C.D. point of view by Feynman, Field and Fox. However, to get clear test here one will need much higher p_3 than has been obtained as of yet. The process $e^- + h(p) \to h(p_1) + X$ is formally similar to $e^+ + e^- \to h(p_1) + h(p_2) + X$ and the predictions look much the same.

REFERENCES

1. A. H. Mueller, Phys. Rev. D18, 3705 (1978) and to be published; R. K. Ellis, H. Georgi, M. Machacek, H. D. Politzer and G. Ross Phys. Letters 78B, 28 (1978) and to be published; S. Libby and G. Sterman, Phys. Letters 78B, 618 (1978) and to be published. D. Amati, R. Petronzio and G. Veneziano, Nucl. Phys. B140, 54 (1978) and to be published.
2. See Ref. 1 and C. DeTar, S. Ellis and P. Landshoff, Nucl. Phys. B87, 176 (1975).
3. G. Sterman and S. Weinberg, Phys. Rev. Letters 39, 1436 (1977).
4. E. Farhi, Phys. Rev. Letters 39, 1587 (1977).

5. H. Georgi and M. Machacek, Phys. Rev. Letters 39, 1237 (1977).
6. E. Witten, Nucl. Phys. B120, 189 (1977); W. Frazer and J. Gunion, to be published; C. H. Llewellyn Smith, to be published.

PERTURBATIVE QCD: AN OVERVIEW *

Stephen D. Ellis

University of Washington

Seattle, Washington 98195

Recent progress in the utilization of perturbative quantum chromodynamics to study lepton-hadron and hadron-hadron reactions is reviewed. Interest is focused on the connections between the various results and the implications for the parton model as a phenomenological tool and as a vehicle to test quantum chromodynamics.

The following discussion is intended to serve as an overview of the recent progress made in using perturbative quantum chromodynamics (QCD) as a phenomenological tool for studying lepton-hadron and hadron-hadron reactions. It is also intended as an introduction to and interpolation between the more specialized companion papers.[1-5]

The basic framework for perturbative QCD arises from a purely pre-QCD notion, the parton model.[6] In this picture the constituents of hadrons, the partons (originally taken to be only quarks), are assumed to suffer confining interactions which are characterized by a (long) time constant of order

$$\tau_L \sim 1/m_h \quad , \tag{1}$$

---

*Work supported in part by the U.S. Department of Energy.

where $m_h \sim$ a few hundred MeV is characteristic of hadron sizes (and masses). Reactions characterized by a momentum Q larger than a few GeV and a correspondingly short time scale

$$\tau_S \sim 1/Q , \qquad (2)$$

($\tau_S \ll \tau_L$) as, for example, in deep inelastic lepton-hadron scattering with Q being the large momentum transfer, may be treated by an impulse approximation scheme. In such hard interactions the partons are considered as essentially free and "on shell" (i.e., with fixed small mass as $|Q^2| \to \infty$) and the full process is assumed to factorize into a convolution of the <u>probability</u> of finding a parton within a hadron with the <u>cross section</u> for the hard process involving the parton. In the "classic" example of lepton-hadron scattering, illustrated in Fig. 1a, the hard process involves the large angle elastic scattering of a lepton and quark. Originally[6] the distributions of partons within hadrons were assumed to be functions of only the fraction x of the hadron's momentum carried by the parton (x = p_q/p_h, the scaling hypothesis) and to be nonzero only for partons collinear with the parent hadron. The parton (quark) distribution F(x) could then be experimentally determined in ep scattering, for example. On the long time scale τ_L, the scattered partons interact and fragment into the observed hadrons. In the naive parton model the resulting hadrons are assumed to be aligned with the parent parton in momentum space and to be described by a fragmentation function D(z), depending only on the ratio z of the hadron's momentum to that of the parent parton (z = p_h/p_q). These fragmentation functions can be determined from the analysis of the annihilation of e^+e^- into hadrons as illustrated in Fig. 1b. Again for this process factorization is assumed such that the quark and antiquark fragment independently. Thus a cross section involving such a hard process assumes a general integral form schematically given by

$$d\sigma_{hadron} \simeq \sum_{\substack{i,j \\ (quarks)}} \iint dx\, dz\, F_i(x) d\sigma_{ij}(x,z,Q^2) D_j(z) . \qquad (3)$$

All of the soft, long-time, confining effects reside in the distributions $F(x)$ and $D(z)$ while $d\sigma_{ij}$ contains all the hard, short-time processes.

While deep inelastic lepton-hadron scattering and the total cross section for e^+e^- annihilation are amenable to more orthodox analyses[7] in terms of the short distance behavior of an operator product expansion, this is not the case for the distribution of hadrons in the final state and the description of this distribution represents a true application of the underlying simple (naive) physical picture of the parton model. This picture was further extended to describe the hadronic production of massive lepton pairs[8] via quark-antiquark annihilation, as in Fig. 1c, and to large p_T hadronic production[9] via quark-quark scattering, as in Fig. 1d. Corrections to the naive picture[6] were presumed to be characterized by the ratio $\langle k_T^2 \rangle / Q^2$, where $\langle k_T^2 \rangle$ is the average squared transverse momentum of quarks in hadrons or hadrons in quarks (jets), assumed originally to be a (few handred MeV)^2. This measure essentially characterizes the "size" of hadrons or the time scale τ_L. Hence the corrections were of approximate magnitude τ_S^2/τ_L^2.

Within this parton model framework it is now possible (and necessary) to study the implications of QCD as the theory of the strong interactions. In the context of QCD, hadrons contain not only quarks but also vector gluons which mediate the interactions between the quarks and, unlike QED, among themselves. Both these new types of partons and the new interactions they imply must be included in the parton model. This is accomplished by <u>assuming</u> that the naive parton model constitutes a zeroth order approximation, calculating perturbative corrections to it, and then demonstrating that the result is consistent with the initial assumption. Typical corrections are illustrated in Fig. 2. Note that there are not only

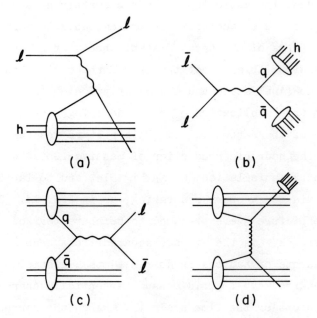

Figure 1: Processes discussed in the parton model:
    (a) Lepton-hadron scattering
    (b) e^+e^- annihilation into hadrons
    (c) Lepton pair production by hadrons
    (d) Large p_T hadron scattering.

truly perturbative corrections due to gluon emission and exchange contributions to the original diagrams of Fig. 1 but also totally new terms arising from gluon-initiated reactions. This program is at first very attractive due to the fact that QCD exhibits asymptotic freedom.[10] Simply stated, the effective coupling of the theory decreases with shorter distances and times or larger momenta as

$$\alpha_s(Q^2) \underset{Q^2 \to \infty}{\approx} \frac{\text{const.}}{\ln(Q^2)} \,. \qquad (4)$$

This renormalization of the coupling already represents the summation of a certain subset of perturbative corrections. Furthermore it suggests that the remaining perturbative corrections will be small at large Q^2. However, the theory also exhibits mass singularities arising from both the collinear fragmentation of, for example, a quark into a quark and a gluon and from the true infrared singularities due to soft gluon emission. This feature of the theory is realized in the perturbation approach as factors of $\ln(Q^2/m_q^2)$ [essentially $\ln(\tau_L/\tau_S)$] which serve to cancel or overwhelm the logarithmic damping inherent in the asymptotically free coupling $\alpha_s(Q^2)$ (which represents the summation of a different set of $\ln(Q^2)$ terms). Thus the factorization Eq. (3) into long time scale effects and short time scale effects which is central to the initial parton picture is threatened. However, recent work of many physicists[1,11] has led to the realization that even in the presence of the logarithms the soft, long-time, quark mass dependent physics still essentially factors from the hard, short-time physics in a <u>universal</u> fashion. Thus the qualitative structure of the parton model remains basically intact but with certain quantitative modifications.

As in the naive model, the asymptotically leading "scaling" contribution to a given process corresponds to the hard scattering of "on shell" (massless) quarks and gluons. However, in order to actually perform the factorization of the long time scale physics from the hard physics, an arbitrary scale, call it M^2, must be introduced. (Care must be exercised to perform this factorization in an

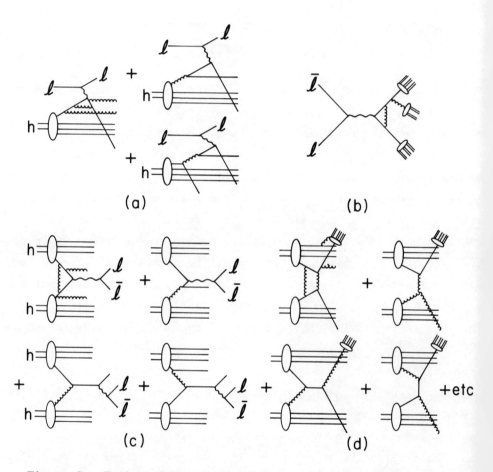

Figure 2: Typical QCD corrections to the naive parton model description of the processes in Fig. 1. Individual processes are as labeled in Fig. 1.

identical fashion for all processes.) The result is then a convolution of distributions for quarks and gluons, evaluated at the scale M^2, with a hard scattering cross section which has a perturbative expansion in terms of $\alpha_s(M^2)$ (for convenience the renormalization point is also taken to be M^2). The hard scattering cross section involves logarithms of both the ratio Q^2/M^2 and the ratios of other large kinematic variables. The schematic form is

$$d\sigma_{hadron} \simeq \sum_{\substack{i,j \\ (quarks,gluons)}} \iint dx\, dz\, F_i(x,M^2)\, d\sigma_{ij}(x,z,\alpha_s(M^2),\theta) D_j(z,M^2) \tag{5a}$$

where θ represents the ratios of other invariants. To insure the rapid convergence of this expression it is useful to choose $M^2 \approx Q^2$ (this is equivalent to using the renormalization group) and thus to express the hard scattering cross section as a series in $\alpha_s(Q^2)$. In this case the distributions will be evaluated at scale Q^2, giving

$$d\sigma_{hadron} \simeq \sum_{\substack{i,j \\ (quarks,gluons)}} \int dx \int dz\, F_i(x,Q^2)\, d\sigma_{ij}(x,z,\alpha_s(Q^2),\theta) D_j(z,Q^2). \tag{5b}$$

For large Q^2 [$\alpha_s(Q^2) \ll 1$] the Q^2 dependence (scaling violation) of the distribution and fragmentation functions is predicted by perturbative QCD and the renormalization group to be as a specific inverse power of $\ln(Q^2)$ [the anomalous dimension]. The present lepton-hadron data are suggestive of this behavior but make no decisive statement, as discussed by L. Abbott[3] in this volume.

In the above factorization analysis the actual singularities associated with "on shell" partons have been hidden in the fully renormalized distribution functions which are assumed to be finite due to nonperturbative confinement physics not calculable by present techniques. It is only the large Q^2 behavior which is amenable to perturbative analysis. It now seems clear that this same noncalculable confinement physics must also induce a nonzero transverse momentum distribution in the functions F and D in addition to that which arises perturbatively[12] from the emission processes illustrated in Fig. 2. Phenomenological studies suggest that the mean of this "primordial" distribution is perhaps as large as 700 MeV/c for quarks in hadrons but more like 350 MeV/c for hadrons resulting from quark fragmentation. Furthermore, applications of this notion, e.g., in μ pair production or large p_T physics, are sensitive to the <u>shape</u> of this noncalculable distribution. The situation is reviewed in the talks of E. L. Berger[4] and J. Owens.[5]

In addition to the power behaved (nonscaling) corrections of the form $<k_T^2>/Q^2$ present in the naive parton model, perturbative QCD predicts further corrections essentially of the form $<m_q^2>/Q^2$ which are <u>process dependent</u> and do not factor into a sum of terms of the above $<k_T^2>$ form.

Finally, the perturbative QCD improved parton model exhibits infrared finite (i.e., genuinely calculable) contributions[12] not present in the naive model which are both process dependent and asymptotically significant in the sense that they do not vanish as a power of $1/Q^2$. There are, for example, the gluon-initiated processes illustrated in Figs. 2c and 2d. For the pair production process (see E. L. Berger[4]) these contributions may be very important for nucleon-nucleon collisions due to their small antiquark content. In the fragmentation process the emission of gluons leads to large k_T within a fragmentation jet, Fig. 3a, which on purely dimensional grounds can be understood to behave as

$$<k_T^2>_{QCD} \sim \alpha_s(Q^2) \cdot Q^2 \sim Q^2/\ln(Q^2) \quad . \tag{6}$$

In the large angle limit this process presumably leads to two distinct jets as illustrated in Fig. 3b (and 2b). Fig. 3 clearly indicates that the physics is the same in the two cases and one must be careful when using this jet language not to treat "fat jets" and

PERTURBATIVE QCD: AN OVERVIEW

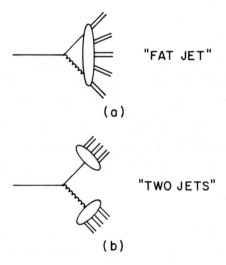

Figure 3: Two descriptions of the result of energetic gluon emission in the fragmentation process.

"double jets" as formally distinct objects.

The successes of this perturbative approach suggest the possibility of performing precise tests to establish whether QCD is indeed the correct theory of the strong interactions. Unfortunately, the nonperturbative, noncalculable contributions discussed briefly above guarantee that, in general, the QCD improved parton model will remain largely a phenomenological tool for the near future. At the same time considerable progress[13] has been made in the direction of avoiding these difficulties by a judicious choice of what experimental measure to study as discussed in the talk by L. S. Brown.[2]

To constitute a satisfactory test of QCD such a measure should satisfy the following constraints:

(a) It should exhibit properties characteristic of QCD, e.g., the presence of both quarks and vector gluons.

(b) It should be reliably calculable, which at present means entirely by the methods of perturbation theory. In order to avoid the necessity of nonperturbative renormalization of infrared singularities, the quantity should be free of explicit infrared logarithms when evaluated perturbatively. It should also be insensitive to corrections due to such nonperturbative confinement effects as the "primordial" k_T distribution.

(c) Finally, it should be accessible to experiment. This condition argues against small effects on a large background or measures which are sensitive to missing neutrals. An example of the latter is a measure which requires <u>event by event</u> determination of an axis.

The clear suggestion is to study various inclusive measures of the final state of e^+e^- annihilation into hadrons where there are no initial hadrons and no reference is made to specific final hadrons. The study of calorimetric, i.e., energy weighted, measures appears to avoid the mass singularities from both the collinear and true infrared divergences (the energy-momentum tensor is not renormalized).

Another area of recent and rapid development is the study[14] of methods actually to sum the perturbative expansion of the hard scattering process discussed above. Since the perturbative expansion typically contains logarithms of ratios of various kinematic variables (the Cheshire cat smile of the mass singularities which have been eliminated by factorization or by a careful choice of the quantity), there are kinematic regions where the logarithms are divergent and thus where a low order expansion is not useful. Successful techniques for summation would serve to enlarge the region of validity of the perturbative approach. A specific example of this approach is discussed in the presentation of L. S. Brown.[2]

Not discussed in any detail in these proceedings, due simply to lack of time, are the perturbative study of spin effects[15] and the interesting comparisons of the structure of gluon initiated jets versus quark initiated jets.[16]

Overall there has been considerable progress in understanding and implementing perturbative QCD. This trend will undoubtedly continue, accompanied by increasing emphasis on the consideration of those quantities which are readily accessible to both theory and experiment. This will in turn result in the clear confirmation (or refutation) of the role of QCD.

ACKNOWLEDGEMENT

Many helpful discussions with H. D. Politzer, G. Sterman, W. I. Weisberger and, especially, L. S. Brown are gratefully acknowledged.

REFERENCES

1. A. H. Mueller, lecture in this volume; Columbia University preprint CU-TP-153 (1979).
2. L. S. Brown, lecture in this volume; University of Washington preprint RLO-1388-777 (1979).

3. L. Abbott, lecture in this volume; SLAC preprint SLAC-PUB-2296 (1979).
4. E. L. Berger, lecture in this volume; SLAC preprint, "Hadroproduction of Massive Lepton Pairs and QCD," (1979).
5. J. F. Owens, lecture in this volume; Florida State University preprint HEP-FSU-790214 (1979).
6. See, e.g., R. P. Feynman, Photon-Hadron Interactions, Benjamin, Reading, MA (1972).
7. See, e.g., K. Wilson, Phys. Rev. $\underline{179}$, 1499 (1969); N. Christ, B. Hasslacher and A. H. Mueller, Phys. Rev. $\underline{D6}$, 3543 (1972).
8. S. D. Drell and T. M. Yan, Phys. Rev. Lett. $\underline{25}$, 316 (1970).
9. For an early discussion of this application see S. D. Ellis and M. B. Kislinger, Phys. Rev. $\underline{D9}$, 2027 (1974).
10. For a review of the properties of QCD see W. Marciano and H. Pagels, Phys. Rep. $\underline{36}$, 137 (1978).
11. A partial summary of recent work in this field includes A. H. Mueller, Phys. Rev. $\underline{D18}$, 3705 (1978); Subhash Gupta and A. H. Mueller, Columbia University preprint CU-TP-139 (1979); C. H. Llewellyn Smith, Oxford University preprint 47/78; W. R. Frazer and J. F. Gunion, University of California preprint UCSD-10P10-194, UCD-78/3 (1978); G. Sterman, Phys. Rev. $\underline{D17}$, 2773 and 2789 (1978); S. Libby and G. Sterman, Phys. Rev. $\underline{D18}$, 3252 (1978); D. Amati, R. Petronzio and G. Veneziano, Nuclear Phys. $\underline{B140}$, 54 (1978) and ibid., $\underline{B146}$, 29 (1978); R. K. Ellis, H. Georgi, M. Machacek, H. D. Politzer and G. G. Ross, MIT preprint CTP-718 (1978) and Caltech preprint CALT 68-684 (1978). See also ref. (1) for further references.
12. See, e.g., G. Altarelli, G. Parisi, and R. Petronzio, Phys. Lett. $\underline{76B}$, 351 (1978) and ibid., $\underline{76B}$, 356 (1978); K. Kajantie and R. Raitio, Nuclear Phys. $\underline{B139}$, 72 (1978); H. Fritzsch and P. Minkowski, Phys. Lett. $\underline{73B}$, 80 (1978).
13. G. Sterman and S. Weinberg, Phys. Rev. Lett. $\underline{39}$, 1436 (1977); C. L. Basham, L. S. Brown, S. D. Ellis and S. T. Love, Phys.

Rev. D17, 2298 (1978), Phys. Rev. Lett. 41, 1585 (1979) and University of Washington preprint RLO-1388-761 (1978); H. Georgi and M. Machacek, Phys. Rev. Lett. 39, 1237 (1977); E. Farhi, Phys. Rev. Lett. 39, 1587 (1977); A. De Rujula, J. Ellis, E. G. Floratos and M. K. Gaillard, Nucl. Phys. B138, 387 (1978); S. Y. Pi, R. L. Jaffe and F. E. Low, Phys. Rev. Lett. 41, 142 (1978).

14. See, e.g., Yu. L. Dokshitzer, D. I. D'Yakonov and S. I. Troyan, Phys. Lett. 78B, 290 (1978), ibid. 79B, 269 (1978) and Proc. XIIIth Winter School of LNPI (Leningrad, 1978) pp. 3–89, part 1 [Engl. trans. SLAC-TRANS-183 (1978)]; K. Konishi, A. Ukawa and G. Veneziano, Phys. Lett. 80B, 259 (1979); R. K. Ellis and R. Petronzio, Phys. Lett. 80B, 249 (1979); C. L. Basham, L. S. Brown, S. D. Ellis and S. T. Love, University of Washington preprint (in preparation).

15. G. L. Kane, J. Pumplin and W. Repka, Phys. Rev. Lett. 41, 1689 (1978); J. Babcock, E. Monsay and D. Sivers, Phys. Rev. Lett. 40, 1161 (1978); C. K. Chen, Phys. Rev. Lett. 41, 1440 (1978).

16. See, e.g., M. B. Einhorn and B. G. Weeks, SLAC preprint SLAC-PUB-2164 (1978); K. Shizuya and S. H. H. Tye, Phys. Rev. Lett. 41, 787 (1978). See also the second and third papers in ref. (14).

TOPICS IN THE QCD PHENOMENOLOGY OF DEEP-INELASTIC SCATTERING *

L. F. Abbott

Stanford Linear Accelerator Center

Stanford University, Stanford, California 94305

ABSTRACT

I review the status of QCD with respect to recent results from deep-inelastic neutrino scattering, emphasizing the theoretical uncertainties coming from effects of non-leading order in $1/Q^2$ and in α_s.

I. INTRODUCTION

During the past year, new results from neutrino experiments[1,2] have been compared with QCD predictions for scaling violations in deep-inelastic structure functions.[1-4] Good agreement between theory and experiment has been found. In what sense do these comparisons test QCD? To answer this question we must examine the various theoretical uncertainties involved in making the QCD predictions. The basic QCD calculations for deep-inelastic scattering,[5] which are discussed in Section II, treat only the lowest-twist operators in the operator product expansion and only leading-order terms in the QCD coupling parameter α_s. As a result, there are corrections of order $1/Q^2$, $1/Q^4$, etc., coming from target mass effects and from higher-

---

*Work supported by the Department of Energy under Contract Number EY-76-C-03-0515.

twist operators, and corrections which are higher order in α_s. Target mass effects can be taken into account,[6-8] but the effects of higher-twist operators remain as a source of uncertainty for predictions at low Q^2. These issues are discussed in Section III. The corrections of second order in α_s to the leading-order QCD results have now been computed.[9] Their phenomenological implications are discussed in Section IV. The work described here was done in collaboration with Michael Barnett.

II. HIGH-Q^2 RESULTS

Two effects which are higher order in $1/Q^2$, elastic scattering contributions and target mass effects, can be directly measured. Using BEBC-Gargamelle data,[1] we find that for analyses of the Q^2-evolution of xF_3, or of moments of xF_3 for $N \leq 6$ both of these effects are smaller than the experimental errors for $Q^2 > 3$ GeV^2. We therefore think it reasonable to assume that other effects of higher-twist operators are also small for $Q^2 > 3$ GeV^2.

In order to compare QCD with neutrino data, I will use the QCD equations for the Q^2-evolution of deep-inelastic structure functions.[5,10] For the non-singlet structure function xF_3 the evolution equation is particularly simple. Defining

$$xF_3(x,Q^2) = F(x,Q^2) \qquad (2.1)$$

we have

$$Q^2 \frac{\partial F(x,Q^2)}{\partial Q^2} = \frac{\alpha_s(Q^2)}{3\pi} \left\{ (3 + 4 \ln(1-x))F(x,Q^2) + \int_x^1 dz \frac{2}{(1-z)} [(1+z^2) F(\frac{x}{z},Q^2) - 2F(x,Q^2)] \right\} \qquad (2.2)$$

where

$$\alpha_s(Q^2) = \frac{12\pi}{(33-2n_f) \ln Q^2/\Lambda^2} \qquad (2.3)$$

QCD PHENOMENOLOGY OF DEEP INELASTIC SCATTERING

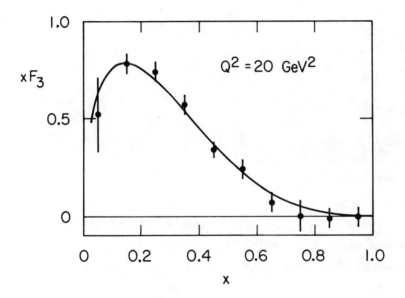

Figure 1. A fit of the form $xF_3 = Ax^a(1-x)^b$ to the combined BEBC-Gargamelle and CDHS data at $Q^2 = 20$ GeV^2.

for n_f quark flavors. In order to proceed, along with Eq. (2.2) one must specify a boundary condition for $F(x,Q^2)$ at some reference point $Q^2 = Q_0^2$. This boundary condition is not completely specified by QCD calculations. We take the standard parameterization

$$F(x,Q_0^2) = Ax^a(1-x)^b \qquad (2.4a)$$

and, following the CDHS group,[2] we use the values

$$\begin{aligned} Q_0^2 &= 20 \text{ GeV}^2 \\ A &= 3.3 \\ a &= .5 \\ b &= 3 \end{aligned} \qquad (2.4b)$$

Then, integrating Eq. (2.2) we obtain the solid curves in Figs. 1-4. The data points shown are a weighted average of BEBC-Gargamelle[1] and CDHS[2] results. Fig. 1 shows the initial fit of Eq. (2.4) to the combined data at $Q^2 = 20$ GeV^2. This initial fit is also shown for reference in Figs. 2-4 by the dashed curve. The curves predicted by QCD on the basis of the initial fit are shown in Figs. 2 and 3 for $Q^2 = 64$ GeV^2 and for $Q^2 = 100$ GeV^2. The agreement between theory and experiment is quite striking. Fig. 4, showing the QCD fit at $Q^2 = 3.9$ GeV^2, is less dramatic but still in reasonable agreement.

In Figs. 1-4, we have combined BEBC-Gargamelle[1] and CDHS[2] data. It should be pointed out that at high Q^2 there is some disagreement between the two data sets, where they overlap, in the range $.5 \leq x \leq .7$. The effects of this discrepancy are minor in the plots I have shown. However, if one takes moments for the two data sets one weights heavily the range of x where the discrepancy occurs and the disagreement for the moments becomes quite large. Another important point is that the Q^2-evolution of $F(x,Q^2)$ at some particular $x = x_0$

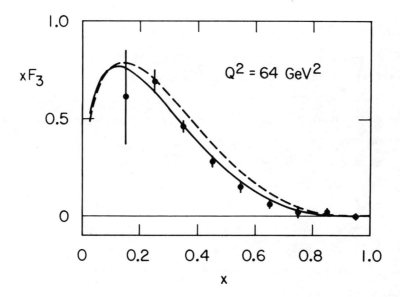

Figure 2. The solid curve is the QCD prediction for xF_3 based on the fit of Fig. 1 which is reproduced here as the dashed curve. The data is combined BEBC-Gargamelle and CDHS results at $Q^2=64$ GeV^2.

depends on $F(x,Q^2)$ only for $x \geq x_0$. Thus, although $F(x,Q^2)$ is not well determined by the data for small x, this uncertainty does not feed through the equations to larger x values.

III. EFFECTS OF HIGHER ORDER IN $1/Q^2$

For $Q^2 < 3$ GeV^2, terms of higher order in $1/Q^2$ <u>are</u> important. This can be seen by the presence of a large elastic-scattering contribution and of sizeable target mass corrections as will be shown below. Terms of higher order in $1/Q^2$ arise from both kinematics and dynamics. Kinematic effects come from the finite target mass, $p^2 = m^2$ where p is the target momentum, and from finite hadronic masses in the final state, $W^2 \geq m^2$ where W is the final-state hadronic mass. Dynamical effects include elastic scattering, resonance formation, diquark scattering, constituent pion scattering, transverse momentum effects and many others. Of these, only the target mass effects can be correctly incorporated into the QCD predictions.

Target mass corrections are made by using the ξ-scaling variable[7,8] or equivalently by taking Nachtmann moments.[6] First, I will discuss the ξ variable within the context of the parton model,[8] and then I will discuss mass corrections in the operator product expansion[7] and Nachtmann moments.[6] The two discussions are equivalent and reflect two ways of looking at the same problem, but it is instructive to state both of them.

Consider the target nucleon momentum parameterized in the convenient form[11]

$$p_{nucleon} = (P + \frac{m^2}{4P}, \vec{0}, P - \frac{m^2}{4P}) . \qquad (3.1)$$

This form guarantees that $p^2_{nucleon} = m^2$.

Then, let the initial momentum of the struck quark be

$$P_{quark} = (xP + \frac{\vec{k}_\perp^2}{4xP}, \vec{k}_\perp, xP - \frac{\vec{k}_\perp^2}{4xP}) . \qquad (3.2)$$

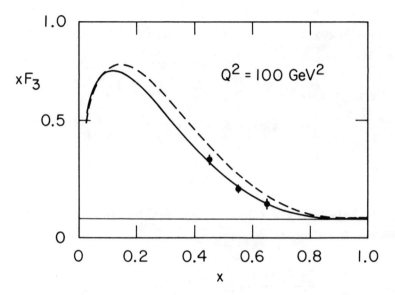

Figure 3. Same as Fig. 2 but at $Q^2 = 100$ GeV^2.

This quark is taken to be massless. Note that

$$x = \frac{(P_0+P_3)_{quark}}{(P_0+P_3)_{nucleon}} \quad . \tag{3.3}$$

The final momentum of the struck quark is $(p_{quark} + q)$. If it too is massless, we have the condition

$$(p_{quark} + q)^2 = 0 \tag{3.4}$$

Solving this equation ignoring $\vec{k}_\perp$ gives the ξ scaling variable

$$x = \xi = \frac{\nu}{m}\left[\sqrt{1 + \frac{Q^2}{\nu^2}} - 1\right] \tag{3.5}$$

where

$$m\nu = q \cdot p_{nucleon} \quad . \tag{3.6}$$

To take target-mass effects into account for xF_3 one would leave Eqs. (2.2 - 2.4) unchanged, but Eq. (2.1) would be replaced by

$$xF_3(x,Q^2) = \frac{x^2}{\xi^2} \frac{1}{(1+\frac{4m^2x^2}{Q^2})} F(\xi,Q^2) + \frac{4m^2}{Q^2} \frac{x^3}{(1+\frac{4m^2x^2}{Q^2})^{3/2}} \int_\xi^1 d\xi' \left[\frac{F(\xi',Q^2)}{(\xi')^2}\right] \tag{3.7}$$

A second derivation of ξ-scaling starts with the operator product expansion of two currents (suppressing vector indices of the currents J)

$$J(x)J(0) = \sum_i \sum_j \left[C_J^i(x^2) x_{\mu_1} \ldots x_{\mu_J} O^{i,J}_{\mu_1 \ldots \mu_J} \right] \tag{3.8}$$

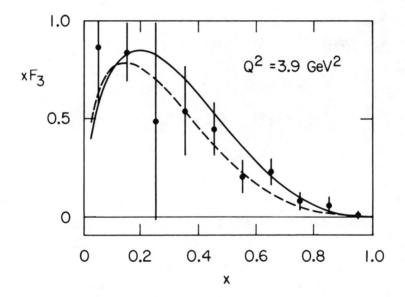

Figure 4. Same as Fig. 2 but at $Q^2 = 3.9$ GeV^2.

where the $O^{i,J}_{\mu_1\ldots\mu_J}$ are local operators of spin J. If we ignore the target mass ($p^2=0$) and write the nucleon matrix elements of the operator $O^{i,J}_{\mu_1\ldots\mu_J}$ as

$$<p| O^{i,J}_{\mu_1\ldots\mu_J} |p> = A^i_J [p_{\mu_1} \ldots p_{\mu_J}] \quad (3.9)$$

then the tensors $[p_{\mu_1} \ldots p_{\mu_J}]$ are symmetric and traceless forming irreducible tensors of O(4). However, if we take the target mass into account ($p^2 = m^2$) these tensors are reducible due to their non-vanishing traces. Since the operator $O^{i,J}_{\mu_1\ldots\mu_J}$ has definite spin we must write

$$<p| O^{i,J}_{\mu_1\ldots\mu_J} |p> = A^i_J [p_{\mu_1}\ldots p_{\mu_J} - \frac{m^2}{4} g_{\mu_1\mu_2}p_{\mu_3}\ldots p_{\mu_J}$$
$$- \text{all other traces}] . \quad (3.10)$$

This modification leads to the ξ scaling variable[7] and to Nachtmann moments[6]

$$\tilde{M}_N = \int_0^1 dx \frac{\xi^{N+1} xF_3(x,Q^2)}{x^3} \left[\frac{1 + (N+1)\sqrt{1+\frac{4m^2x^2}{Q^2}}}{(N+2)} \right] \quad (3.11)$$

to replace the simple moments

$$M_N = \int_0^1 dx \, x^{N-2} \, xF_3(x,Q^2) . \quad (3.12)$$

Although the ξ-scaling scheme correctly accounts for target mass effects, it does not correctly describe the final-state kinematics in deep-inelastic scattering.[12] Kinematics requires that xF_3 vanish at $x = 1$ and quark counting rules[13] suggest a form like $(1-x)^3$ near $x = 1$. When the ξ scaling variable is used, $(1-x)^3$ gets replaced by $(1-\xi)^3$ and at $Q^2 = 1$ GeV^2, for example, x = 1 corresponds to ξ = .64. As a result, xF_3 does not vanish at x = 1 in the ξ-scaling scheme. In the above parton model derivation, we have taken $p^2=m^2$ for the initial nucleon momentum but have ignored the final-state kinematic condition $W^2 \geq m^2$. For example, if we included non-zero final-state hadronic masses we could not have taken the initial quark on mass shell as in Eq. (3.2). Thus, the ξ-scaling scheme assumes that the dominant $1/Q^2$ effects can be removed by correcting the initial-state kinematics <u>only</u> and ignoring final-state kinematics and dynamical effects. This is somewhat supported by the fact that the ξ-scaling variable acts much like the scaling variable of Bloom and Gilman.[14] In overshooting the data near x = 1, one can argue[15] that ξ-scaling accounts for the elastic scattering contribution (and resonance contributions) in the sense that the excess area under the ξ-scaling curve equals the area under the elastic peak at x = 1.

The key issue is then: once target mass corrections have been made, how large are the remaining higher-twist effects? This is an extremely difficult question to answer. We have fit the BEBC-Gargamelle data[1] for moments of xF_3 to the form

$$\tilde{M}_N = M_n^{QCD} [1 + \frac{aN}{Q^2}] \qquad (3.13)$$

where $\tilde{M}_N$ is a Nachtmann moment and M_N^{QCD} is the QCD prediction. The factor of N in the $1/Q^2$ term is suggested by quark counting rules.[13] Fixing the Λ parameter of QCD from the high Q^2 data we find a $\approx .2$ GeV^2. However, if the Λ parameter is not held fixed we find that the data do not distinguish between $1/Q^2$ and $1/\ell nQ^2$ effects, and in fact this question cannot be resolved. The inclusion of other terms like $1/Q^4$, $1/Q^6$, $1/Q^8$, etc., does not change the situation.

Figure 5 shows the effects of elastic scattering and of the target mass at $Q^2 = 1.7$ GeV^2. The dashed curve is the x-scaling prediction of QCD obtained by integrating the QCD evolution equation for xF_3, Eq. (2.2), down from the initial fit of Eq. (2.4) at

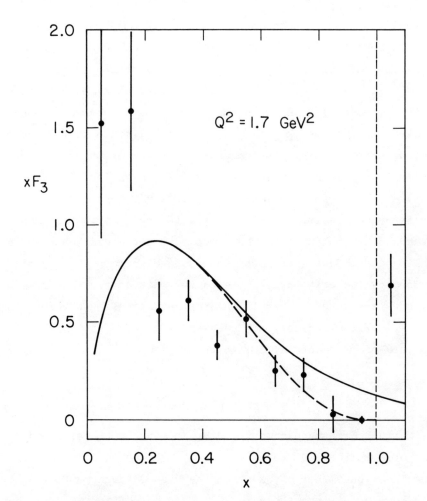

Figure 5. The dashed curve is the x-scaling prediction of QCD based on the initial fit at $Q^2 = 20$ GeV^2 of Fig. 1. The solid curve is the corresponding ξ-scaling prediction. Data is from BEBC-Gargamelle and elastics are shown in a bin from x=1 to x=1.1 where the area under the data point in this bin is equal to the area under the elastic peak at x=1 in the original data.

$Q^2 = 20$ GeV^2, and using Eq. (2.1) to define xF_3. The solid curve is the corresponding ξ-scaling prediction obtained from Eq. (3.7). The elastic data which is actually a sharp spike at $x = 1$ is displayed by adding one extra bin from $x = 1$ to $x = 1.1$. The area under the data point in this bin is equal to the area under the elastic peak in the original data. Note that the elastic contribution is quite large and that the difference between the x-scaling and ξ-scaling curves is substantial. This shows the presence of significant corrections of higher order in $1/Q^2$. The x-scaling (dashed) curve fits quite well in the large x region except for the elastics. This suggests that the simple moments of Eq. (3.12) <u>excluding</u> elastics would agree well with QCD which is in fact true. The ξ-scaling curve overshoots the data at large x but the area between the ξ-scaling curve and the x-scaling curve is approximately equal to the area under the elastic peak. This suggests that the Nachtmann moments of Eq. (3.11) <u>including</u> elastics would also agree well with QCD. Again this is true. In considering which curve in Fig. 5 fits better, it is not at all clear what one means by a good fit at low Q^2. At present the data is not precise enough to distinguish between $1/Q^2$ and $1/\ell n Q^2$ effects. Thus, low Q^2 data cannot be used to test QCD until more is known about higher-twist effects. On the other hand, it does contain information which may be helpful in resolving these issues in the future.

IV. EFFECTS OF HIGHER ORDER IN α_s

Figure 6 shows a comparison of the BEBC-Gargamelle data[1] for the N=3 Nachtmann moment with the leading order QCD prediction

$$\tilde{M}_N = \frac{C_N}{(\ell n \, Q^2/\Lambda^2)^{d_N}} \qquad (4.1)$$

where the d_N are known parameters and the constants C_N and Λ are determined by fitting to the data. We have already seen that the

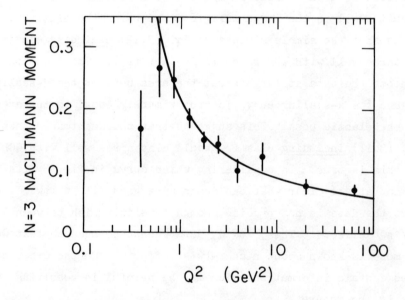

Figure 6. A comparison of the QCD prediction with the N=3 Nachtmann moment from the BEBC-Gargamelle data.

QCD prediction for $Q^2 < 3$ GeV^2 is subject to uncertainties due to effects of higher order in $1/Q^2$. There are also uncertainties coming from terms of higher order in α_s. The corrections of second order in α_s change Eq. (4.1) to[9]

$$\tilde{M}_N = \frac{C_N}{(\ln Q^2/\Lambda^2)^{d_N}} \left[1 + \frac{A_N + B_N \ln\ln Q^2/\Lambda^2}{\ln Q^2/\Lambda^2} \right] . \quad (4.2)$$

The constants A_N and B_N have now been computed.[9] We can then fit Eq. (4.2) to the data of Fig. 6 again varying C_N and Λ to obtain the best fit. We find that although the correction terms are large, the curve obtained using Eq. (4.2) is virtually identical to the curve we found by fitting with Eq. (4.1). This is because the effect of the second-order corrections can almost entirely be absorbed into a renormalization of the parameter Λ.[3] The large second-order corrections change the value of Λ dramatically but have a very small effect on the curve of Fig. 6. This is true for moments over a small range of N (say $2 \leq N \leq 6$) but is no longer valid if we try to fit over a large range of N. Recently, Ross[16] has shown that the large second-order corrections to the QCD curves of xF_3 (Figs. 1-4) occur at large and small x. At large x since xF_3 vanishes this is not too critical and at small x the fit to xF_3 is poorly known anyway. Thus, the second-order corrections and hopefully all corrections of higher-order in α_s do not introduce too much uncertainty into qualitative QCD predictions such as those of Figs. 1-6.

However, the second-order corrections in α_s do present a problem when we go beyond curve fitting and try to give numerical predictions from QCD. For example, the BEBC-Gargamelle group[1] has obtained a parameter r by requiring

$$\frac{d}{dQ^2} \left(\frac{M_N^r}{M_M} \right) = 0 . \quad (4.3)$$

Using the lowest-order form of Eq. (4.1) we see that QCD predicts that Eq. (4.3) can be satisfied at all Q^2 with

$$r = \frac{d_M}{d_N} .$$ (4.4)

To include the second-order corrections in α_s,[4] we must substitute Eq. (4.2) into Eq. (4.3). We find that Eq. (4.3) can no longer be satisfied at all Q^2 but that it can be satisfied at one Q^2 value and then is very nearly satisfied at all other Q^2 values in the experimental range. However, the QCD prediction for r is now quite different from the leading-order prediction of Eq. (4.4). For example, if we choose to satisfy Eq. (4.3) at $Q^2 = 5$ GeV^2 (the values given are not very sensitive to this choice) and take $\Lambda = .5$ Ge we obtain the following results:

TABLE I

QCD Prediction

M/N	r(Leading-Order)	r(Second-Order)
5/3	1.46	1.74
6/4	1.29	1.46
7/3	1.76	2.28

Thus, the parameter r is subject to quite large corrections due to effects of higher order in α_s.[4]

The calculations of the second-order corrections to the moment also serve to point out ambiguities in the definitions of α_s and Λ.[3,17,18] First, α_s can only be defined by specifying a renormalization scheme. Two α_s parameters defined in two different schemes

QCD PHENOMENOLOGY OF DEEP INELASTIC SCATTERING

disagree at second order and are related by

$$\alpha'_s = \alpha_s + \alpha_s^2 + \ldots \quad . \tag{4.5}$$

However, even if the renormalization scheme is fixed, there is still an arbitrariness in the definition of Λ which results in an ambiguity exactly like Eq. (4.5). Consider some function expanded in powers of $1/\ln Q^2$

$$F = \frac{A}{\ln Q^2/\Lambda^2} + \frac{B}{\ln^2 Q^2/\Lambda^2} + \ldots \quad . \tag{4.6}$$

Now define

$$\Lambda_a = \Lambda e^{1/2a} \quad . \tag{4.7}$$

Then,

$$\frac{1}{\ln Q^2/\Lambda^2} = \frac{1}{\ln Q^2/\Lambda_a^2} - \frac{a}{\ln^2 Q^2/\Lambda_a^2} + \ldots \tag{4.8}$$

and we can write

$$F = \frac{A}{\ln Q^2/\Lambda_a^2} + \frac{B-a}{\ln^2 Q^2/\Lambda_a^2} + \ldots \quad . \tag{4.9}$$

Thus, we have the possibility of expanding using different Λ's and getting different values of the constant in the second-order term. If we use the definition of Λ_a in Eq. (4.7) for the moments of Eq. (4.2), we find[3]

$$\tilde{M}_N = \frac{C_N}{(\ln Q^2/\Lambda_a^2)^{d_N}} \left[1 + \frac{A_N - ad_N + B_N \ln\ln Q^2/\Lambda_a^2}{\ln Q^2/\Lambda_a^2} \right] \quad . \tag{4.10}$$

Note therefore, that one cannot define Λ without specifying the constant term in the second-order correction nor can one say how large the second-order term is without specifying what definition of Λ is being used.

In fitting the second-order QCD prediction for the moments to data, one must decide what Λ to use and hence what value of the constant term in the second-order correction to fit with. There are several possible approaches. The calculations of the A_N and B_N were done in the minimal subtraction scheme using dimensional regularization. The A_N contain factors of $\ln(4\pi) - \gamma_E$ coming from expanding around $n = 4$ in the dimensional regularization method. Since these factors are artifacts of the regularization scheme, one can choose[3]

$$a = \ln(4\pi) - \gamma_E \approx 2 \qquad (4.11)$$

in order to remove such factors from the A_N. Another approach is to choose a value of a so that the second-order term is as small as possible over the Q^2 range and for the moments of interest.[3,4] This also leads to an a close to two. Finally, we have fit Eq. (4.10) using C_N, Λ_a, and a all as free parameters. We find a=2.3 ± .6 but this determination can only be made using low Q^2 data where effects of higher order in $1/Q^2$ introduce uncertainties. It is interesting that all three methods lead to roughly the same value of a.

The best definitions of α_s and Λ to use are clearly those which minimize the effects of higher-order corrections. The result of the last paragraph suggest that at moderate Q^2 values, for moments $2 \leq N \leq 6$ the best choice might be $a \approx 2$ in Eq. (4.10). The second-order corrections to the total e^+e^- cross section are now being computed. It will be interesting to see whether the definitions of α_s and Λ which seem to be best in deep-inelastic scattering also minimize the effects of second-order corrections to electron-positron annihilation.

Finally, I would like to point out that in comparing values of Λ or α_s from different processes, one must be sure that the same definitions of these parameters are being used. This can only be d if the second-order corrections are known. In particular, there is no meaning to comparisons of Λ's obtained from different processes using the lowest-order results of QCD.[17]

ACKNOWLEDGMENTS

This work was done in collaboration with Michael Barnett. I am extremely grateful to many colleagues and in particular to R. Blakenbecler, S. Brodsky and F Gilman for helpful conversations.

REFERENCES

1. P. Bosetti et al., Nucl. Phys. B142, 1 (1978).
2. J. G. H. de Groot et al., CERN preprints 79-0168, 79-0132 and 79-0133.
3. W. A. Bardeen, A. J. Buras, D. W. Duke and T. Muta, Phys. Rev. D18, 3998 (1978).
4. R. Barbieri, L. Caneschi, G. Curci and E. d'Emilio, preprint SNS 8/1978.
5. Some reviews are H. D. Politzer, Phys. Reports 14C, 129 (1974); D. J. Gross in Methods in Field Theory, ed. R. Balian and J. Zinn-Justin (North-Holland, 1976); J. Ellis in Weak and Electromagnetic Interactions at High Energy, ed. R. Balian and C. H. Llewellyn Smith (North-Holland, 1977).
6. O. Nachtmann, Nucl. Phys. B63, 237 (1973); B78, 455 (1974); S. Wandzura, Nucl. Phys. B42, 412 (1977).
7. H. Georgi and H. D. Politzer, Phys. Rev. D14, 1829 (1976).
8. R. Barbieri, J. Ellis, M. K. Gaillard and G. G. Ross, Phys. Lett. 64B, 171 (1976); Nucl. Phys. B117, 50 (1976); R. K. Ellis, R. Petronzio and G. Parisi, Phys. Lett. 64B, 97 (1976).
9. W. Caswell, Phys. Rev. Lett. 33, 244 (1974); D. R. T. Jones, Nucl. Phys. B75, 531 (1974); E. G. Floratos, D. A. Ross and C. T. Sachrajda, Nucl. Phys. B129, 66 (1977); Erratum B139, 545 (1978); CERN preprints TH-2566 and TH-2570; W. A. Bardeen, A. J. Buras, D. W. Duke and T. Muta, Phys. Rev. D18, 3998 (1978).

10. G. Altarelli and G. Parisi, Nucl. Phys. B126, 298 (1977).
11. D. Sivers, S. Brodsky and R. Blankenbecler, Physics Reports 23C, 1 (1976).
12. D. J. Gross, F. A. Wilczek and S. B. Treiman, Phys. Rev. D15, 2486 (1976); K. Bitar, P. W. Johnson and W-K Tung, IIT preprint.
13. S. Brodsky and G. Farrar, Phys. Rev. Lett. 31, 1153 (1973); Phy Rev. D11, 1309 (1975); R. Blankenbecler and S. Brodsky, Phys. Rev. D10, 2973 (1974).
14. E. Bloom and F. Gilman, Phys. Rev. Lett. 25, 1140 (1970).
15. A. DeRujula, H. Georgi and H. D. Politzer, Ann. Phys. 103, 315 (1977).
16. D. A. Ross, Caltech preprint 68-699.
17. M. Bacé, Phys. Lett. 78B, 132 (1978).
18. S. Wolfram, Caltech preprint 68-690; W. Celmaster and R. J. Gonsalves, UCSD preprint 10P10-210.

QUANTUM CHROMODYNAMICS AND LARGE MOMENTUM TRANSFER PROCESSES*

J.F. Owens

Florida State University

Tallahassee, Florida 32306

ABSTRACT

A brief review is given of recent progress made in the understanding of high-p_T production of single particles and jets as well as heavy particle production. In each case unresolved questions are discussed and experimental tests are suggested.

I. INTRODUCTION

During the last year progress has been made in justifying the use of perturbative quantum chromodynamics (QCD) to generate predictions for large momentum transfer processes[1,2]. Several questions of principle have now been decided and it appears that factorization of parton distribution and fragmentation functions is established for both leading and non-leading logarithms. Furthermore, at the leading logarithm level, one can perform calculations using a single set of universal, process-independent parton distribution and fragmentation functions. This result has served to stimulate rapid growth in the industry of generating QCD predictions for processes amenable to treatment by perturbative

---

*Work supported in part by the U.S. Department of Energy

techniques, i.e. processes involving large momentum transfers.

In this talk I shall review the status of these predictions in three areas: high-p_T particle production, high-p_T production of hadronic jets, and heavy particle production, by which I mean production of particles containing heavy quarks such as c, b, etc. In each area there remain unresolved questions dealing either with alternative models or with problems encountered while attempting to extract predictions from the theory. I shall endeavor to discuss what I feel are some of the important issues and to point out possible experimental tests.

For some time now there have been two competing models used to describe high-p_T particle production: lowest order QCD[3-5] with scaling violations and parton transverse momentum smearing[6-9] and the constituent interchange model[10] (CIM). In the past year there has been a convergence of these two approaches and it now appears that elements of each are required in order to describe the data[11,12]. Data which support this latter point of view will be discussed.

The same models which are used to obtain predictions for single particle production contain the information necessary for predicting hadronic jet production cross sections. However, making contact with experimental data is made difficult by the fact that gluon jets have not yet been observed. Without knowing the properties of gluon-induced jets it is not possible to estimate the efficiency for detecting them in a given spectrometer. Thus, all measured jet cross sections are subject to a high degree of uncertainty. Calculations have been performed[13-15] which indicate that at very high jet energies gluon and quark jets may possess significantly different properties such as angular size, multiplicity, average charge, etc. However, it is also possible that at the relatively low jet energies which are currently accessible, the gluon and quark jets may look very much alike[16]. The consequences of these two extremes will be discussed with respect to the

QUANTUM CHROMODYNAMICS AND LARGE MOMENTUM TRANSFER

available data.

The question of hadronic charm production cross sections remains a very active area, especially in light of the large cross sections inferred from the various beam dump experiments[17-19]. Some of the different approaches used in estimating these cross sections will be compared.

The above three topics will be discussed in detail in sections II, III, and IV, respectively. A summary and some conclusions will be presented in section V.

II. HIGH-p_T PARTICLE PRODUCTION

As a first step in this section, both the lowest order QCD and the CIM approaches to high-p_T particle production will be discussed. The initial applications of lowest order QCD showed that reasonable agreement with meson production data could be obtained for $\sqrt{s} \gtrsim 5$ GeV/c provided that all of the lowest order subprocesses predicted by QCD were included and the scale violating parton distribution functions were used[3-5]. Subsequent calculations[6,7,9] also showed that the predicted scale violations in the fragmentation functions[20,21] could be accommodated. Lastly, the inclusion of effects due to parton transverse momentum smearing[6-9] resulted in a good description of the data in the region $\sqrt{s} \gtrsim 20$ GeV and $p_T \gtrsim 2$ GeV/c. In order to achieve this agreement, a rather large value for the parton transverse momentum, $<k_T> \simeq 850$ MeV/c, was found to be necessary. The use of such a large value necessitates the use of an arbitrary cut-off which introduces some model dependency in the calculation. Furthermore, it has been argued that using off-shell kinematics for the colliding partons greatly reduces the smearing effect[22]. It must be emphasized, however, that the large smearing calculation was never meant to be exact. Rather, the large value of $<k_T>$ was used in order to approximate the net result of two types of terms[23]. First, there should be an intrinsic (or primordial) component of the parton k_T which results from

the fact that the partons are confined inside the hadron. This
component is then a characteristic of the wave functions of the
colliding hadrons. Furthermore, QCD perturbation theory predicts
that in addition to the lowest order 2→2 subprocesses there will
be higher order terms such as qq→qqg, for example. Portions of
such 2→3 subprocesses are already included in the scale violating
parton distribution and fragmentation functions[24]. In this instance, the contributions from those regions of phase space where
the gluon and one or more of the quarks are parallel have already
been included. However, there are additional terms where the two
quarks and the gluon are each distinct and give rise to separate
jets. Such contributions can mimic the effects which would otherwise be attributed to non-zero parton k_T. Thus, as an initial
attempt at including the effects of such 2→3 subprocesses a large
value of $<k_T>$ was used[6,7,9].

At very large values of p_T, say $p_T \geq 10$ GeV/c, there is general
agreement that the QCD subprocesses are dominant[10,11]. It is only
at lower p_T values that the questions of alternative subprocesses
arise. It has been argued that the effect of smearing on the single
particle distributions is greatly reduced if off-shell kinematics
are used for the colliding partons[22] and, at the same time, that
the other effects, such as the observed degree of acoplanarity, can
still be described adequately. At the lower p_T values the resulting
deficit in the single particle distribution is then filled by new
higher order subprocesses such as those found in the CIM. These
diagrams typically involve the scattering of non-elementary fields
such as mesonic ($q\bar{q}$) and baryonic (qqq) systems contained in the
initial hadrons. There are thus many new subprocesses to be considered and several new distribution functions, as well. As a
result, the question of normalization has always been a weak point
in CIM calculations. Recently, several papers[10,11] have appeared
in which the normalization question was studied in detail. Pre-
scriptions for both the coupling constants and the distribution

functions which are consistent with a wide variety of data were presented. The resulting normalization for high-p_T hadron production is believed to be reliable to within a factor of two or so[11].

The analysis of Ref. (11) goes beyond the CIM and includes the various lowest order QCD subprocesses as well. In this way a successful description of the data is obtained for $\sqrt{s} \geq 20$ GeV and $p_T \geq 2$ GeV/c. In particular, it is possible to study the transition from the region dominated by the CIM to that in which the lowest order QCD subprocesses dominate. Unfortunately, it is difficult to compare these predictions with those of other QCD analyses[3-9] because none of the QCD predicted Q^2 dependences were included. Now, it is not unreasonable to expect that the distributions of color singlet hadrons within color singlet hadrons should scale because such states cannot radiate gluons. However, it can be argued that the other distributions, such as those for the quarks and gluons, should in fact have the logarithmic Q^2 dependences predicted by the theory. Systematically including all of these effects could produce significant changes in many of the contributions of the various subprocesses.

When studying different subprocesses it is often convenient to consider the predictions for the scaling behavior of the invariant cross section at fixed $x_T = 2p_T/\sqrt{s}$ and center-of-mass scattering angle θ. For this the following form is often used:

$$E \frac{d^3\sigma}{dp^3} = p_T^{-n} f(x_T, \theta) \quad . \tag{1}$$

In a scale invariant theory the exponent n depends on the subprocess involved. For meson production in the CIM one has n=8 while for the lowest order QCD subprocess, n=4. Now, inclusion of the QCD predicted scale violation together with the use of the strong running coupling constant, α_s, increases n to about 6.5 in the currently accessible energy region[5-7]. By including parton k_T effects it is then possible to obtain n≈8 over a limited kinematic region[6-9].

Lowest order QCD modified by scaling violations and parton k_T smearing thus predicts that for meson production n≈8 for moderate p_T and that at higher p_T values n will decrease to about 6. This prediction is not modified if CIM terms are present because they too will give n≈8 at moderate p_T values and their influence will die away at larger values of p_T.

Recently, the CERN-Columbia-Oxford-Rockefeller (CCOR) collaboration has obtained data[25] for high-p_T π^0 production between $\sqrt{s}$ = 30 and 62 GeV and $3.5 \leq p_T \leq 14$ GeV/c. Using the $\sqrt{s}$ = 53 and 62 GeV data they find

$$n = 8.0 \pm 0.5 \qquad 3.5 \leq p_T \leq 7.0 \text{ GeV/c}$$

$$n = 5.1 \pm 0.4 \qquad 7.5 \leq p_T \leq 14.0 \text{ GeV/c}.$$

This behavior is in accord with the QCD expectations discussed above. While this result does not directly aid in distinguishing between the two approaches outlined above in the moderate p_T region, it does offer striking evidence for the expected emergence of the lowest order QCD effects.

Recently, it has been stressed that the situation is somewhat different for baryon production[11,12]. Here the CIM subprocesses predict n=12 while the lowest order QCD predictions remain essentially unchanged. Detailed predictions for baryon production are given in both papers. Figs. 1 and 2 show the predictions for p and $\bar{p}$ production from Ref. 12 while Fig. 3 shows the $\bar{p}$ predictions from Ref. 11. As can be seen in Fig. 2, both calculations agree as to the order of magnitude of the lowest order QCD predictions, although there are some differences due to the lack of scaling violations in Ref. 11. In both cases it is concluded however, that as one progresses from meson to antiproton to proton production at fixed s and p_T the lowest order QCD predictions lie increasingly

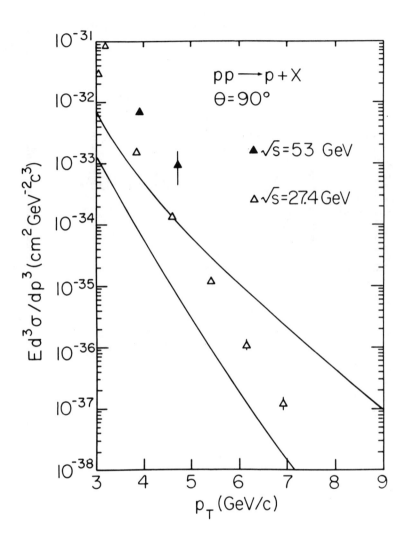

Fig. 1. Lowest order QCD predictions for pp→p+X as obtained in Ref. 12. The data are from Ref. (26).

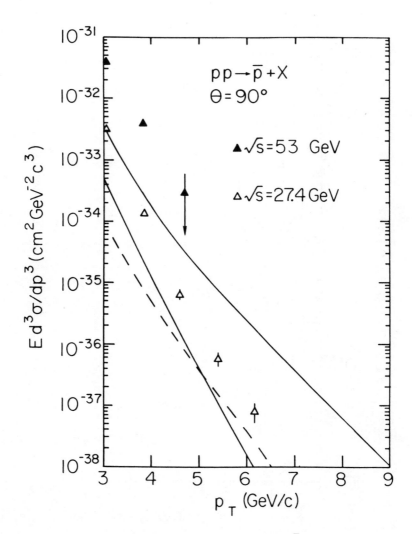

Fig. 2. Lowest order QCD predictions for pp→p̄+X as obtained in Ref. 12. The dashed line is the prediction from Ref. 11 corresponding to lowest order QCD with exact scaling. The data are from Ref. 26.

below the data. This same hierarchy is observed when studying the scaling properties of the invariant cross section. The data[26] at $\sqrt{s} \leq 27.4$ GeV and $p_T \leq 6$ GeV/c show n=12 for p production in agreement with the CIM, n=8 for $\bar{p}$ production which is intermediate between the two models and n=8 for meson production, a result which can be obtained in either model.

The interpretation of this pattern is reasonably simple. The QCD subprocesses all involve partons fragmenting into the observed hadrons while for many of the CIM diagrams, referred to as direct diagrams, the observed hadron takes part in the scattering subprocess. Now, due to the steeply falling p_T distribution the observed hadrons which come from quark fragmentation will tend to take a large share of the parent parton's momentum, i.e. the longitudinal momentum fraction z will be near one. This is the effect known as trigger bias. The parton fragmentation functions behave typically as $(1-z)^m$ with m≃1 for mesons and m≃2-3 for baryons. Therefore, the trigger bias effect tends to suppress the QCD subprocesses more for baryons than for mesons. Thus, any competing subprocesses which are not subject to this effect, e.g. CIM diagrams, should show up more strongly in baryon than in meson production. Such is indeed the case, especially for proton production where significant contributions from the direct diagrams are expected. Finally, for the CIM the direct terms for $\bar{p}$ production are all proportional to small distribution functions (such as sea terms). This explains why the CIM and QCD terms are comparable for $\bar{p}$ production.

The above discussion strongly suggests that in order to understand baryon production at intermediate p_T values one must include the relevant CIM diagrams. In order to be consistent then, it must be that CIM subprocesses contribute at some level to all types of particle production. The basic problem remains one of normalizing all of the various CIM terms.

In order to make further progress in separating the CIM and

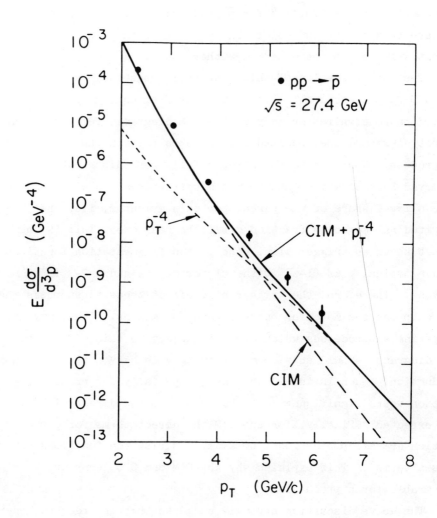

Fig. 3. Predictions for pp→p̄+X from Ref. 11 obtained by combining lowest order QCD with exact scaling and the CIM. The data are from Ref. 26.

QCD subprocess contributions it is necessary to consider two particle correlation predictions. Such calculations have been made by several groups[7,27] using the QCD subprocesses and the results are discussed at length in Ref. 28. Quantitative predictions for the correlation observables have not been made using the CIM, but there have been some qualitative predictions[29].

The results given in Refs. 5, 27, and 28 show that one can understand the away-side momentum distributions (often called x_e or z_p distributions) using lowest order QCD with parton k_T smearing and that the degree of acoplanarity (indicated by the large values of P_{out}) can be approximately described. Also, the relative amounts of positive and negative away-side hadrons for $\pi^{\pm}$ and K^+ triggers are in accord with the data[30]. There is, however, a possible problem with K^- triggers where an excess of positive over negative away-side hadrons has been observed in one experiment[30]. Such an effect is predicted for certain CIM subprocesses[29]. An additional discussion of the impact of these data on various models can be found in Ref. 31.

Another type of experiment which has yielded some interesting results is the measurement of symmetric hadron pairs[32]. This involves a double arm spectrometer which triggers on hadron pairs when the two hadrons have approximately equal and opposite transverse momenta. A simple extension of eq. (1) yields the scaling expression

$$E_1 E_2 \frac{d^6\sigma}{dp_1^3 dp_2^3} = p_T^{-n} f(x_T, \theta) \quad (2)$$

where $x_T = 2p_T/\sqrt{s}$ and $p_T = \frac{1}{2}(|\vec{p}_{T_1}| + |\vec{p}_{T_2}|)$. On simple dimensional grounds n=6 for the lowest order QCD subprocesses and the addition of scaling violations and the use of the running coupling constant increases this to n≈8. Notice that because of the symmetric nature of the trigger the parton k_T smearing has only a small effect and is not expected to change n very much[28]. The experimental results are[32]

$$n = 8.4 \pm 0.4 \quad \pi^{\pm} \text{ pairs}$$

$$n = 9.2 \pm 0.2 \quad h^{\pm} \text{ pairs, } h=\pi, K, p.$$

These results are in nice agreement with the QCD predictions. It would be interesting to have the CIM predictions for comparison.

To summarize this section I would like to make the following comments.

1. The lowest order QCD calculations with scaling violations and parton k_T smearing can describe the data for meson production. Next, the contributions of the higher order 2→3 type subprocesses must be calculated in order to justify the phenomenological ansatz currently being used.

2. Contributions from CIM type diagrams appear to be necessary in order to understand baryon production cross sections. Thus, these types of subprocesses should be present for meson production as well. The dominant question which remains is one of normalization.

3. With regard to point 2, quantitative calculations of CIM predictions for correlation observables and the symmetric pair cross section would be helpful.

4. The extension of the existing QCD calculations to different beam/target/produced hadron combinations will provide new tests with regard to the CIM once the data become available.

III. HIGH-p_T JET PRODUCTION

The study of single particle cross sections is complicated by the presence of the parton fragmentation process and the resulting trigger bias suppression. It has long been realized[33] that a theoretically simpler, though experimentally more complex, quantity would be the high-p_T jet cross section. In this sense a jet is defined as a collection of hadrons, each of which has low p_T with respect to a common (jet) axis. By detecting all of these daughter hadrons one could, at least in principle, reconstruct the original parton four-momentum.

Given the models discussed in the preceding section it is easy to make predictions for such cross sections since all one has to do is to make the replacement

$$D_{C/c}(z) \to \delta(1-z) \qquad (3)$$

where $D_{C/c}(z)$ is the parton fragmentation function and z is the longitudinal momentum fraction. This assumes equal detection efficiencies (of 100%) for all types of jets, an idealization to which I shall return shortly. From this it is clear that the trigger bias discussed above no longer acts as a suppression factor. The QCD predicted jet/particle ratio is thus very large and, for this reason, the QCD subprocesses are expected to dominate over the CIM terms even at moderate p_T values, e.g. $p_T \geq$ 4 GeV/c[10,11]. For this reason I shall concentrate on the QCD predictions for jet production.

Single arm measurements have been reported from two Fermilab experiments: E-260[34] and E-395[35,36]. These data are compared with a set of QCD predictions[37] in Fig. 4. These predictions have been obtained by using the full set of lowest order QCD subprocesses with the QCD predicted Q^2 dependence. Parton k_T smearing has been included using a Gaussian distribution with $<k_T> = 660$ MeV/c. This reduced value of $<k_T>$ is intended to represent only the intrinsic or primordial component discussed above. The order of magnitude of the theoretical predictions appears to be correct, although the curves do lie below the data at $\sqrt{s} = 27.4$ GeV for $p_T \lesssim 5$ GeV/c. Thus, there is room for contributions from higher order terms, e.g. 2→3 subprocesses, as well as some CIM contributions at lower p_T values.

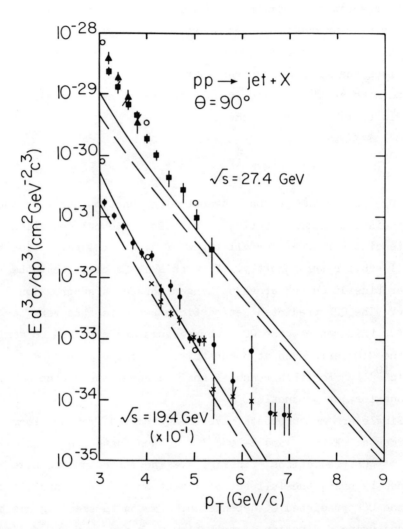

Fig. 4. Lowest order QCD predictions for jet production as obtained in Ref. 37. The dashed curves have no parton k_T smearing while the solid curves have been calculated using a Gaussian parton k_T distribution with $<k_T> = 660$ MeV/c as discussed in the text. The data are from Refs. 34-36.

Evidence for parton k_T smearing and higher order subprocesses may be found by looking at jet-jet correlations. For example, using a double arm spectrometer in a single arm trigger mode allows one to look at the recoiling (away-side) jet. If parton k_T effects and/or 2→3 subprocesses are present then those events in which the parton-parton center-of-mass system is moving toward the detector will be triggered on preferentially. Thus, one expects the average away-side p_T to be less than the trigger p_T. This is shown in Fig. 5 together with the corresponding prediction from Ref. 37. This calculation, which included only the intrinsic k_T component, reproduces the effect seen in the data[35]. The magnitude is, however, slightly less than the data show at the larger p_T values which indicates the possible need for 2→3 type contributions. Extending the data to higher p_T values would be interesting.

Data taken with a double arm trigger can also be enlightening. The Fermilab, Lehigh, Pennsylvania, and Wisconsin collaboration (E-395) have taken data with a "left plus right" (L+R) trigger in which the sum of the magnitudes of the left and right p_T values was required to have a certain value, $|\vec{p}_{T_L}| + |\vec{p}_{T_R}| = 2p_T$. The distribution of p_{T_L} and p_{T_R} about the average, p_T, is then an indication of the momentum imbalance and can be interpreted in terms of parton transverse motion. In this way the flavor averaged rms k_T value has been measured[38] to be about 1 GeV/c at $\sqrt{s} = 27.4$ GeV. Furthermore, there is evidence that this rms value increases with s at fixed p_T and increases with p_T at fixed s. Both of these behaviors would be expected if 2→3 subprocesses were present. For comparison, the intrinsic parton k_T value used in Ref. 37, $<k_T> = 660$ MeV/c, yields $(k_T)_{rms} = 742$ MeV/c using a Gaussian distribution. This is significantly less than the observed value and strengthens the argument for higher order terms.

As in the two particle case, two jet cross sections taken with the "L+R" trigger depend less sensitively on the parton k_T smearing than do the single particle reactions. Such data are therefore

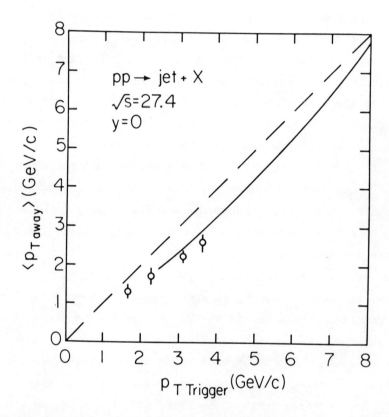

Fig. 5. Predictions from Ref. 37 for the average away-side p_T as a function of the trigger p_T. The data are from Ref. 35.

particularly useful for studying the underlying subprocesses and/or the various parton distributions[37,39].

The theoretical predictions discussed above have all been obtained by assuming perfect efficiency for detecting both gluon and quark induced jets. In actual practice however, this may not be the case. For example, if a jet had a large angular spread, one or more particles could miss the detector and, therefore, the jet would appear to have a lower p_T. It would then form a small background under the much larger cross section existing at the lower p_T values. Correcting for this could boost both the single arm and double arm cross sections significantly. On the other hand, there are particles from the beam and target fragments which could enter the detectors and increase the measured jet momentum. These two effects go in opposite directions and should, therefore, cancel each other to some degree.

These are only two of several effects which must be corrected for when analyzing the data. Unfortunately, the corrections used to determine the jet cross sections depend upon the assumed fragmentation properties of the jets. For example, recent calculations[14,15] have shown that gluon jets may be more spatially extended than quark jets[13], with jet opening angles approximately related by $\delta_{gluon} \simeq \delta_{quark}^{4/9}$. Furthermore, it is expected that gluon jets will be characterized by a higher multiplicity than quark jets[40]. However, it has been shown[15] that the opening angle argument is realistic only for large energies, e.g. jet momenta above 15 GeV/c. Furthermore, a model calculation[16] for both gluon and quark fragmentation shows that for jet momenta on the order of a few GeV, the resulting hadron multiplicities should be comparable for both gluon and quark jets. Thus, for the jet momenta currently under study it may be that the spatial extent and multiplicities associated with quark and gluon jets are essentially the same.

How then does one distinguish between quark and gluon jets at moderate p_T? One way is to look at charge correlations[41]. In a recent CERN experiment the CCHK group used the average charge of the fastest particle in the jet as an indicator of the average charge of the jet. When the jets were in a back-to-back configuration the data indicated an average jet charge of 1/3 as would be expected from quark-quark scattering. However, when both jets were in the forward hemisphere, the trigger side showed 1/3 while the away-side showed zero, consistent with gluon-quark scattering. This subprocess would be expected to dominate since if both jets went forward in the center-of-mass system one parton would be at large x (quark) and the other would be at small x (gluon). The single particle trigger used would favor the quark since the quark fragmentation function is expected to be flatter than that of the gluon. Thus, the opposite side jet should be the gluon. This observation is most interesting since it could be the first evidence for the observation of a gluon jet.

The situation concerning jet cross sections is far from settled and it is clear that much work remains to be done on interpreting these jet experiments. It appears that Monte Carlo calculations using models for quark and gluon fragmentation must be used to unfold the data. Furthermore, it seems likely that the asymptotic predictions for quark and gluon jet behavior will be misleading at the moderate p_T values currently being probed. With regard to these Monte Carlo calculations, it would be useful for experimentalists to quote their results after cutting out particles which have low p_T with respect to the beam. This removes the beam/target fragmentation contamination and the effects of this cut can be corrected for in the theoretical calculation. This restricts the model dependency to just the high-p_T jet fragmentation treatment and removes the uncertainties related to the beam/target fragmentation dynamics.

Already, useful results are coming from the first round of jet experiments. With the advent of larger and more efficient detectors significant gains in our understanding of high-p_T phenomena should be possible.

IV. HEAVY PARTICLE PRODUCTION

Today the bulk of our knowledge concerning charmed particle production has been provided by studies of e^+e^- annihilation and deep inelastic leptoproduction. However, the recent beam dump experiments[17-19], which suggest a charm production cross section between 30 and 100 μb at $\sqrt{s}=27.4$ GeV, have stimulated new calculations for the hadronic production of charmed particles, as well as particles containing other heavy quarks such as the b or t. In the following I will compare and contrast several different mechanisms for heavy particle production.

The first class of calculations to be considered is that based on QCD perturbation theory. Two basic subprocesses are considered: $q\bar{q} \to c\bar{c}$[42-44] and $gg \to c\bar{c}$[42-45]. The charm production cross section is obtained via a convolution with the relevant parton distribution functions:

$$\sigma(A+B \to c\bar{c}+X) \sim \int_{S_L}^{S_H} d\hat{s}\, G_{a/A}(x_a)\, G_{b/B}(x_b)\, \hat{\sigma}(a+b \to c\bar{c}+x) \qquad (4)$$

where $\hat{\sigma}$ represents the relevant subprocess cross section at a center-of-mass energy $\hat{s}$. The integration limits in eq. (4) depend upon the specific reaction being considered and will be specified below. This scheme is closely related to the concept of duality[42, 46-48] used in calculating the production of hidden charm states, e.g. ψ, ψ', etc. In this latter instance it is assumed that integrating from the threshold for $c\bar{c}$ production ($s_L=4M_c^2$) up to the threshold for charm meson production ($s_H=4M_D^2$) yields the sum of the cross sections for all of the $c\bar{c}$ states which exist below the $D\bar{D}$ threshold. The recent analysis presented in Ref. 42 shows that this duality approach works well for ψ, ψ', and T production. Some typical results are shown in Fig. 6.

In order to calculate the charm meson production cross section

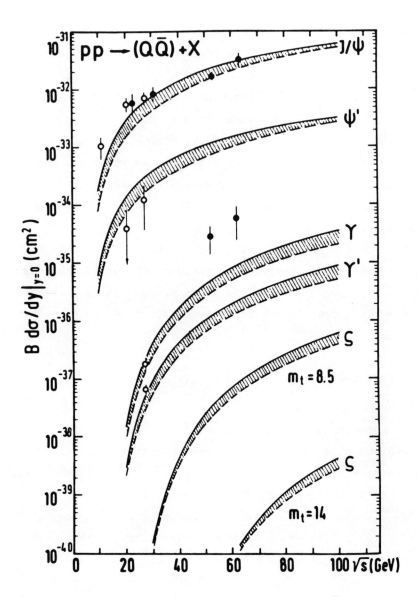

Fig. 6. QCD perturbation theory predictions for $q\bar{q}$ bound state production from Ref. 42. The shaded bands correspond to different choices for the parton distribution functions.

in this approach one simply sets $s_L = 4M_D^2$ and s_H is set at the kinematically allowed upper limit ($\approx s$). The various perturbative calculations agree that the cross section at $\sqrt{s} = 27$ GeV should be on the order of several µb's, i.e. about an order of magnitude below the data.

The authors of Ref. 42 argue that the large cross sections observed in the beam dump experiments can be attributed to coherent nuclear effects which give rise to an increased effective center-of-mass energy. Specifically, one has $s_{eff} \sim A^{1/3} s$ so that for heavy targets, such as the copper ones used in the beam dump, $s_{eff} \gg s$. This in turn yields a large enhancement of the predicted cross section because of the rapid rise near threshold.

Another point of view is stressed in Ref. 49 where the strong dependence on the charm quark mass is discussed. It is argued that potential models for the charmonium spectrum give $m_c < m_\psi/2$ and that by choosing $m_c \approx 1.1$ GeV/c^2 agreement with the beam dump experiments can be obtained. Here the cross section is also enhanced by allowing the integration region to extend below $s=4M_D^2$. It is argued that the $c\bar{c}$ system can interact with the spectator quarks in the beam and target thereby producing a $D\bar{D}$ system even though the original $c\bar{c}$ mass was below the $D\bar{D}$ threshold.

A third charm production cross section calculation[50] avoids the use of perturbation theory altogether. The perturbative calculations do not take into account bound state effects which should be important near threshold. Therefore, the perturbative estimates for charm production may be wrong. Instead, the authors in Ref. 50 relate the charm <u>photo</u>production cross section to the <u>gluo</u>production cross section, $\sigma(gp \to c\bar{c}+x)$. The photoproduction cross section was estimated using vector dominance arguments[51] to be

$$\sigma(\gamma p \to c\bar{c}+x) = \sigma_o(1-s_{Th}/s) \qquad (5)$$

with $\sigma_o = 1.2$µb. This estimate for the asymptotic cross section,

σ_o, agrees closely with a perturbative estimate[51], indicating that the major deviation from the perturbative calculation at intermediate energies is in the modified threshold behavior shown in eq. (5). Given the photoproduction cross section, it is easy to calculate the <u>hadron</u> production cross section by convoluting $\sigma(gp \to c\bar{c}+x)$ with the gluon distribution function $G_{g/p}(x)$. The resulting estimate at $\sqrt{s}$ = 27 GeV is $\sigma(pp \to c\bar{c}+x) \lesssim 170\mu b$, which is compatible with the results of the beam dump experiments.

The key difference between this result and those of the various perturbative calculations lies in the modified threshold behavior given by eq. (5). It is expected that for very large energies, sufficiently far above threshold, the perturbative estimates should again become reliable.

In order to discriminate among the various calculations presented here, detailed measurements of the hadronic charm production cross section will be needed. Specifically, measurements on <u>proton</u> targets will be needed in order to remove the possibility of coherent nuclear effects. Also, measurements spanning a range of energies will be needed in order to establish the correct threshold behavior.

V. SUMMARY

I have attempted to review some of the recent developments concerning the application of perturbative techniques in order to obtain predictions, using QCD, for large momentum transfer processes. In the last year a wide variety of such calculations have been performed and it is fair to say at this stage that no large discrepancies between theory and experiment have been uncovered. However, it should be born in mind that true tests of QCD are difficult to devise using hadron initiated processes because of the problems associated with confinement. As a result, it more often happens that one obtains consistency checks on the theory, and not detailed tests.

There are now, however, hints in the data which suggest possible ways to test certain features of the theory. The behavior of $<k_T>$ as measured in high-p_T jet production is strongly indicative of contributions from higher order, e.g. $2\to3$, subprocesses. Evidence for these terms has also come from studying various high-p_T correlation observables. It is therefore of some importance to pursue the calculation of such higher order terms.

We are now at a stage where much of the basic phenomenology has been done. The goal now should be to devise more precise tests of the theoretical structure contained in QCD.

REFERENCES

1. D. Amati, R. Petronzio, and G. Veneziano, Nucl. Phys. B140, 54 (1978); Nucl. Phys. B146, 29 (1978).
2. R.K. Ellis et al., California Institute of Technology preprint CALT-68-684 (1978).
3. B.L. Combridge, J. Kripfganz, and J. Ranft, Phys. Lett. 70B, 234 (1977).
4. R. Cutler and D. Sivers, Phys. Rev. D17, 196 (1978).
5. J.F. Owens, E. Reya, and M. Glück, Phys. Rev. D18, 1501 (1978).
6. R.D. Field, Phys. Rev. Lett. 40, 997 (1978).
7. R.P. Feynman, R.D. Field, and G.C. Fox, Phys. Rev. D18, 3320 (1978).
8. A.P. Contogouris, R. Gaskell, and S. Papadopoulos Phys. Rev. D17, 2314 (1978).
9. J.F. Owens and J.D. Kimel, Phys. Rev. D18, 3313 (1978).
10. R. Blankenbecler, S.J. Brodsky, and J.F. Gunion, Phys. Rev. D18, 900 (1978).
11. D. Jones and J.F. Gunion, SLAC-PUB-2157 (1978).
12. J.F. Owens, Florida State University preprint FSU-HEP-781220.
13. G. Sterman and S. Weinberg, Phys. Rev. Lett. 39, 1436 (1977).
14. K. Shizuya and S.-H. H. Tye, Phys. Rev. Lett. 41, 787 (1978).

15. M.B. Einhorn and B.G. Weeks, Nucl. Phys. B146, 445 (1978).
16. U.P. Sukhatme, Orsay preprint LPTPE 78/36 (1978).
17. P. Alibran et al., Phys. Lett. 74B, 134 (1978).
18. T. Hansl et al., Phys. Lett. 74B, 139 (1978).
19. P.C. Bosetti et al., Phys. Lett. 74B, 143 (1978).
20. J.F. Owens, Phys. Lett. 76B, 85 (1978).
21. T. Uematsu, Phys. Lett. 79B, 97 (1978).
22. W.E. Caswell, R.R. Horgan, and S.J. Brodsky, SLAC-PUB-2106 (1978); R.R. Horgan and P.N. Scharbach, SLAC-PUB-2188 (1978).
23. R.D. Field, Talk presented at the 1978 Marseille meeting on "Hadron Physics at High Energies", CALT-68-672 (1978).
24. C. Sachrajda, Phys. Lett. 76B, 100 (1978).
25. A.L.S. Angelis et al., Phys. Lett. 79B, 505 (1978).
26. D. Antreasyan et al. Phys. Rev. Lett. 38, 115 (1977); B. Alper et al., Nucl. Phys. B87, 19 (1975).
27. A.P. Contogouris, R. Gaskell, and S. Papadopoulos McGill University preprint (1978).
28. R.D. Field, talk present at the Symposium on Jets in High Energy Collisions, Niels Bohr Institute - Nordita, Copenhagen, July 1978, CALT-68-688 (1978).
29. S.J. Brodsky, talk presented at the VIII Symposium on Multi-particle Dynamics, Kayserberg, France, June 1977, SLAC-PUB-2009 (1977); talk presented at the Symposium on Jets in High Energy Collisions, Niels Bohr Institute - Nordita, Copenhagen, July 1978, SLAC-PUB-2217 (1978).
30. M. Albrow et al., Nucl. Phys. B145, 305 (1978).
31. M.K. Chase, Phys. Lett. 79B, 114 (1978).
32. H. Jöstlein et al., Phys. Rev. Lett. 42, 146 (1979).
33. J.D. Bjorken, Phys. Rev. D8, 4098 (1973).
34. C. Bromberg et al., talk presented at the VIII Symposium on Multiparticle Dynamics, Kayserberg, France, June 1977, FERMILAB-CONF-77/62 (1977).

35. A.R. Erwin, talk presented at the 1978 Vanderbilt Conference, COO-088-36 (1978).
36. M.D. Corcoran et al., talk presented at the 19th International Conference on High Energy Physics, Tokyo (1978), University of Pennsylvania report UPR-0050E (1978).
37. J.F. Owens, Florida State University preprint, FSU-HEP-790122 (1979).
38. M.D. Corcoran et al., talk presented at the Symposium on Jets in High Energy Collisions, Niels Bohr Institute - Nordita, Copenhagen, July 1978, COO-088-63 (1978).
39. M.D. Corcoran et al., Phys. Rev. Lett. $\underline{41}$, 9 (1978).
40. K. Konishi, A. Ukawa, and G. Veneziano, CERN-Ref. TH-2509 (1978).
41. E.E. Kluge, talk presented at the Symposium on Jets in High Energy Physics, Niels Bohr Institute - Nordita, Copenhagen, July 1978, CERN/EP/PHYS 78-35 (1978).
42. M. Glück and E. Reya, Phys. Lett. $\underline{79B}$, 453 (1978).
43. L.M. Jones and H.W. Wyld, Phys. Rev. $\underline{D17}$, 1782 (1978).
44. J. Babcock, D. Sivers, and S. Wolfram, Phys. Rev. $\underline{D18}$, 162 (1978).
45. H.M. Georgi, S.L. Glashow, M.E. Machacek, and D.V. Nanopoulos, Ann. Phys. (N.Y.) $\underline{114}$, 273 (1978).
46. H. Fritzsch, Phys. Lett. $\underline{67B}$, 217 (1977).
47. M. Glück, J.F. Owens, and E. Reya, Phys. Rev. $\underline{D17}$, 2324 (1978).
48. J.F. Owens and E. Reya, Phys. Rev. $\underline{D17}$, 3003 (1978).
49. C.E. Carlson and R. Suaya, SLAC-PUB-2212 (1978).
50. H. Fritzsch and K.H. Streng, Phys. Lett. $\underline{78B}$, 447 (1978).
51. H. Fritzsch and K.H. Streng, Phys. Lett. $\underline{72B}$, 385 (1978).

TESTING QUANTUM CHROMODYNAMICS IN ELECTRON-POSITRON ANNIHILATION AT HIGH ENERGIES*

Lowell S. Brown

Department of Physics, University of Washington

Seattle, Washington 98195

Various measures of the distribution of hadronic energy produced in high-energy electron-positron annihilation provide precise tests of the promising fundamental theory of hadronic physics, quantum chromodynamics. Recent work at the University of Washington on such energy cross sections is reviewed.

Quantum chromodynamics is amenable to precise experimental verification. This happy state of affairs occurs because the theory is asymptotically free. At high energy W it is described by an effective running coupling

$$\bar{g}(W)^2 \sim \frac{1}{\ln W} \tag{1}$$

which vanishes as the energy increases. Therefore, certain suitably selected, high-energy processes can be computed using perturbative methods. The now classic processes for such tests involve deep inelastic lepto-production off nuclear targets. Unfortunately, only the energy variation of these cross sections is predicted, with no prediction being made for their absolute magnitudes and shapes.

---

*Work supported in part by the U.S. Department of Energy

Recently, however, a new method has been developed which provides a more complete description. This is the method of asymptotically free perturbation theory. A large impetus for its development came from the work of Sterman and Weinberg.[1]

The basic ideas of the new method are as follows. Using the total energy W to set the length scale, a partial cross section can be written in terms of a dimensionless function of dimensionless variables,

$$\Delta\sigma = \frac{1}{W^2} F(\frac{W}{\mu}, g_\mu^2, \frac{m_\mu^2}{\mu^2}, x) \quad . \tag{2}$$

Here μ is the mass scale of the renormalization point, g_μ is the value of the renormalized coupling at this renormalization point, and m_μ is the (quark) mass treated as another coupling constant normalized at μ. The symbol x stands for the remaining dimensionless variables such as scattering angles. The physically measurable partial cross section $\Delta\sigma$ must be independent of the value of the arbitrary renormalization point μ. Hence we may take μ = W, turning g_μ^2 into the running coupling $\bar{g}(W)^2$ of the renormalization group. With this choice

$$\frac{m_\mu^2}{\mu^2}$$

is replaced by

$$\frac{m(W)^2}{W^2}$$

which, up to logarithmic corrections, vanishes as $1/W^2$ as the energy W becomes large. Therefore, neglecting terms which are essentially of order $1/W^2$, we secure the high energy limit

$$\Delta\sigma = \frac{1}{W^2} F(1, \bar{g}(W)^2, 0, x) \quad . \tag{3}$$

The existence of the limit (3) requires that the partial cross section be finite in a theory with massless particles. This will generally not be the case. We must restrict our considerations to "proper processes" which are free of infrared mass divergences. This requires that the partial cross section for a "proper process" must be insensitive to soft particle production and to the collinear branching of the massless particles. The partial cross sections will be devoid of mass singularities if they refer only to inclusive energies carried off by the particles. Energy weighting removes divergences which would otherwise arise from soft particle production and the inclusive energy summation removes potential divergences from the collinear branching of the massless particles.

Since the running coupling $\bar{g}(W)^2$ vanishes as $W \to \infty$, the partial cross section (3) can be computed by perturbation theory. Quarks and gluons are, of course, not bound into the observed hadrons in perturbation theory. Hence we can calculate perturbatively only if $\Delta\sigma$ refers to a partial cross section for quark and gluon production and not to the production of a specific hadron. However, there is good empirical evidence to support the assumption that the produced quarks and gluons fragment into hadrons with limited transverse momenta ($p_\perp < 0.3$ GeV). The observed hadrons will therefore follow closely the directions of their parent quark or gluon. Moreover, the energy flow in the observed hadrons will even more closely approximate the energy flow of their parent since a hadron produced at a wide angle relative to its parent direction is soft and weighted little by the energy.

Since the "proper processes" cannot refer to specific, individual hadrons they cannot, in particular, contain hadrons in the initial state; we are restricted to discuss

$$e^+e^- \to \text{hadronic matter.}$$

This reaction is described by a hierarchy[2,3,4] of energy-weighted cross sections. The first member of this hierarchy is simply the ordinary total cross section σ. The next member is the single-energy cross section[2] $d\Sigma/d\Omega$. It describes the "antenna pattern" of the radiated hadronic energy. The single-energy cross section could

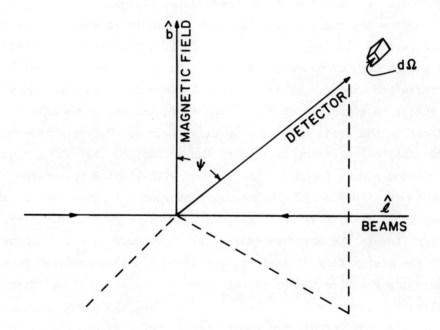

Fig. 1. Geometry for the energy pattern experiment.

TESTINGS QUANTUM CHROMODYNAMICS

be determined, for example, by measurements with a small hadronic calorimeter placed at various angular positions. If the calorimeter collects an energy ΔE in a solid angle $d\Omega$ during a time T, then $d\Sigma/d\Omega = \Delta E/(W\mathcal{L}Td\Omega)$, where $\mathcal{L}$ is the luminosity of the e^+e^- colliding beams and W is the total energy of a e^+e^- pair. The single-energy cross section obeys a sum rule reflecting energy conservation,

$$\int d\Omega \, \frac{d\Sigma}{d\Omega} = \sigma . \tag{4}$$

The third member of the hierarchy is the double-energy or energy-correlation cross section[3,4] $d^2\Sigma/d\Omega d\Omega'$. It can be measured with two calorimeters, one of solid angle $d\Omega$ in the direction $\hat{r}$, the other of solid angle $d\Omega'$ in the direction $\hat{r}'$. The two calorimeters measure the energies dE and dE' which are carried by the hadrons into these solid angles during a single event. The product of the two energies, (dEdE'), is then summed for many similar events with the sum divided by the integrated luminosity $\mathcal{L}T$, times the squared energy of each collision W^2 and the solid angle product $d\Omega d\Omega'$. This procedure defines $d^2\Sigma/d\Omega d\Omega' = \Sigma_{events} (dEdE')/\mathcal{L}TW^2 d\Omega d\Omega'$. Again we have a sum rule reflecting energy conservation,

$$\int d\Omega' \, \frac{d^2\Sigma}{d\Omega d\Omega'} = \frac{d\Sigma}{d\Omega} . \tag{5}$$

Let us first consider the character of the single-energy cross section $d\Sigma/d\Omega$, the hadronic "antenna pattern". The geometry for the measurement of this cross section is illustrated in Fig. 1. The cross section has been calculated[2] to order $\bar{g}(W)^2$. This involves computing the amplitudes corresponding to the Feynman graphs shown in Fig. 2. Although individual amplitudes have mass singularities, the sum of terms which constitute the cross section is finite. The discussion is simplified by considering the case where the e^+e^- beams are perfectly polarized along the magnetic field direction. In this case we have

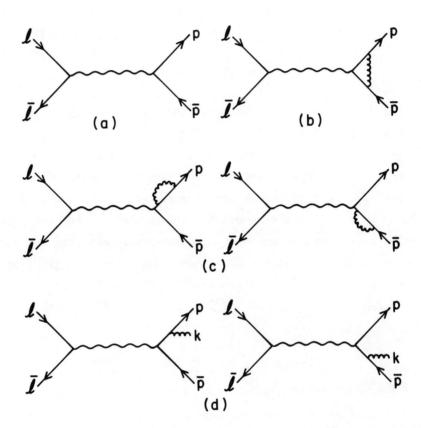

Fig. 2. Zeroth, first, and second order Feynman graphs for electron-positron annihilation.
    (a) Lowest-order graph for $e^+e^- \to \gamma \to q\bar{q}$.
    (b) Vertex modification.
    (c) Self energy insertions.
    (d) Lowest-order gluon emission graphs.

$$\frac{d\Sigma}{d\Omega} = \frac{\alpha^2}{2W^2} \Sigma_f \, 3Q_f^2 \, \{[1 + \frac{\bar{g}(W)^2}{4\pi^2}] \sin^2\psi +$$

$$\frac{\bar{g}(W)^2}{4\pi^2} [3\cos^2\psi - 1] \} \quad , \tag{6}$$

where $\alpha \simeq 1/137$ is the fine structure constant, Q_f is the fractional charge of a quark of flavor f (with the factor 3 accounting for the three quark colors), and ψ is the angle between the detection and magnetic field directions as shown in Fig. 1. At high energy the running coupling is given by

$$\frac{\bar{g}(W)^2}{4\pi^2} = \frac{2}{[11 - \frac{2}{3}N_f]\ln(W/\Lambda)} \quad , \tag{7}$$

where N_f is the number of quark flavors and $\Lambda \simeq 0.5$ GeV is the scale mass determined from deep inelastic lepto-production data. The angular distribution given by Eqs. (6) and (7) is plotted in Fig. 3 for $N_f = 4$. At infinite energy there is a pure $\sin^2\psi$ distribution. As the energy is lowered, the two valleys in this distribution are filled.

The fragmentation of the quarks and gluons into the observed hadrons gives a small effect which can be treated as a correction to the basic quark-anti-quark production [the cross section of Eq. (6) with $\bar{g}(W)^2 = 0$]. We assume that the parent quark or anti-quark yields a number dn of observed hadrons in the momentum interval d^3p) given by the scaling distribution

$$dn = \frac{(d^3p)}{p^0} \, f(\frac{2p_{\parallel}}{W}, p_\perp) \quad , \tag{8}$$

where $p_{\parallel}$ and $p_\perp$ are the momentum components that are respectively parallel and perpendicular to momentum of the parent. The result of this fragmentation correction is simply to replace the coefficient

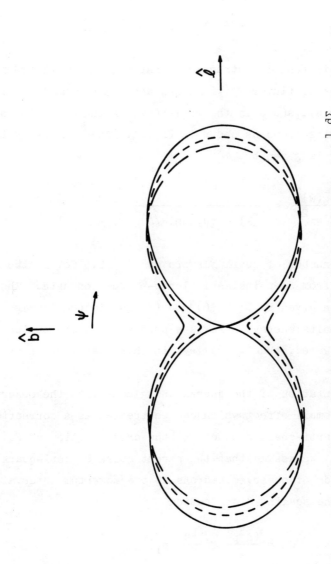

Fig. 3. QCD predictions for the normalized antenna patterns $\frac{1}{\sigma}\frac{d\Sigma}{d\Omega}$ corresponding to perfectly polarized electron and positron beams with various total energies W. The long dashed curve corresponds to W=5 GeV, the short dashes to W=30 GeV, and the solid curve is for infinite W. All unit vectors and angles refer to the geometry displayed Fig. 1.

of $[3\cos^2\psi - 1]$ in Eq. (6) by

$$\frac{\bar{g}(W)^2}{4\pi^2} + \frac{\pi C \langle p_\perp \rangle}{4W},$$

where $C \simeq 2.5$ is the coefficient of the logarithmic rise of the total hadronic multiplicity in e^+e^- collisions and $\langle p_\perp \rangle \simeq 0.5$ GeV is the average transverse momentum of a produced hadron. [It should be noted that, in general, the leading fragmentation corrections to members of the energy cross section hierarchy can be treated in such a simple, analytic manner.] We see that although the fragmentation process also smears the energy pattern, filling in its minima, the fragmentation correction vanishes as $1/W$ as the energy increases. It vanishes much more rapidly than does the perturbative QCD effect which behaves as $1/\ln W$. Figure 4 displays the size of the fragmentation corrections at $W = 10$ GeV and $W = 30$ GeV.

At $W = 30$ GeV the fragmentation corrections are quite small in comparison to the QCD effect. However, as is evident from Fig. 3, the QCD effect is itself small at these energies. We need a test of QCD which is not such a small effect on a large background. This leads us to consider the energy correlation cross section[3,4], $d^2\Sigma/d\Omega d\Omega'$. The geometry of the experimental arrangement is shown in Fig. 5. Since the final hadronic system is produced by an intermediate, virtual photon of spin one, the angular dependence with respect to the beam and magnetic field axes is of a characteristic form. For the case of perfect e^+e^- polarization we have

$$\frac{d^2\Sigma}{d\Omega d\Omega'} = \frac{\alpha^2}{2W^2} \Sigma_f 3Q_f^2 \{A(\chi) [\sin^2\psi + \sin^2\psi']$$

$$+ B(\chi)[\cos\chi - \cos\psi\cos\psi'] + C(\chi)\},$$

(9)

where, as shown in Fig. 5, χ is the angle between the two detectors and ψ, ψ' are their polar angles with respect to the magnetic field direction.

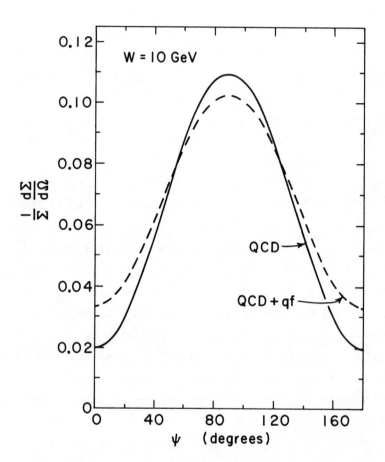

Fig. 4a. QCD and the sum of QCD and quark fragmentation (qf) contributions to the normalized energy pattern $\frac{1}{\Sigma}\frac{d\Sigma}{d\Omega}$ with perfectly polarized electron-positron beams of total energy 10 GeV.

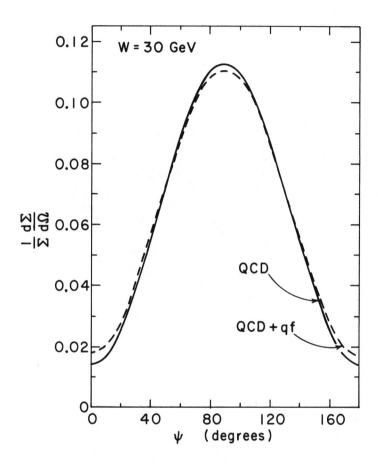

Fig. 4b. QCD and the sum of QCD and quark fragmentation (qf) contributions to the normalized energy pattern $\frac{1}{\Sigma}\frac{d\Sigma}{d\Omega}$ with perfectly polarized electron-positron beams of total energy 30 GeV.

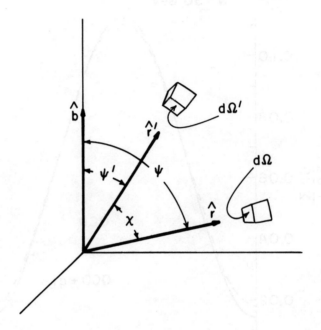

Fig. 5. Geometry for the energy-energy cross section experiment. The two detectors are stationed in directions $\hat{r}$ and $\hat{r}'$ with relative angle χ. These detector directions are respectively at angles ψ and ψ' relative to the magnetic field direction (the polarization direction) denoted by $\hat{b}$.

TESTINGS QUANTUM CHROMODYNAMICS

The calculation of this cross section to order $\bar{g}(W)^2$ again involves the graphs of Fig. 2, and the analytic results are presented in references 3,4. Here we shall simply note that the coefficient $C(\chi)$ vanishes and display the coefficients $A(\chi)$ and $B(\chi)$ in Figs. 6 and 8. Away from $\chi = 0, \pi$, these coefficients are of order $\bar{g}(W)^2$, and the QCD effect is not a small effect on a large background. Figures 6a, b show both the quark fragmentation and the QCD contributions to $A(\chi)$ at $W = 10$ GeV and $W = 30$ GeV. We see that at $W = 30$ GeV the quark fragmentation correction is small over a large angular range. The leading, order $1/W$ fragmentation correction to $A(\chi)$ is symmetrical about $\chi = 90°$. Hence the fragmentation correction to the difference

$$D(\chi) = A(\pi-\chi) - A(\chi) \tag{10}$$

is small, of order $1/W^2$. This difference, which varies over several decades, is plotted in Fig. 7. The coefficient $B(\chi)$ has no leading order $(1/W)$ fragmentation correction. This coefficient is shown in Fig. 8.

We have found that by using the method of asymptotically free perturbation theory we have been able to make predictions for the theory of quantum chromodynamics with essentially no free parameters, predictions which should be accessible to experimental test. Since absolute magnitudes, shapes, and energy variations of several angular distributions are predicted by the theory, their experimental confirmation would be significant. It is, of course, extremely difficult to arrive at a completely objective and precise figure of merit for a test of a theory. An indication of the sensitivity of the energy-correlation cross section can be obtained by calculating it with massless gluons that have spin 0 rather than spin 1. This has been done recently (in fact after this lecture was initially presented) by Basham and Love.[5] Using a coupling $g\bar{q}\lambda_a\phi_a q$ rather than $\bar{g}(W)\bar{q}\gamma^\mu\lambda_a A_{\mu a} q$, they find that with $g = \bar{g}(W)$ the coefficients with

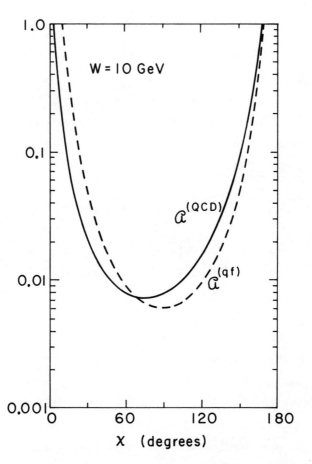

Fig. 6a. The QCD and quark fragmentation (qf) contributions to the A coefficient in Eq. (9) are plotted as a function of χ for total energy W of 10 GeV. Notice that the (qf) contribution is symmetric about $\chi = 90°$ while the QCD behavior is quite asymmetric.

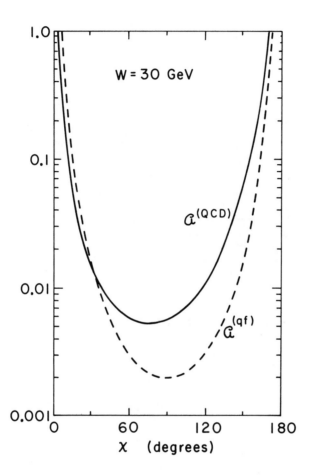

Fig. 6b. The QCD and quark fragmentation (qf) contributions to the A coefficient in Eq. (9) are plotted as a function of χ for total energy W of 30 GeV. Notice that the (qf) contributions is symmetric about $\chi = 90°$ while the QCD behavior is quite asymmetric.

scalar gluons, $A^{(scalar)}(\chi)$ and $B^{(scalar)}(\chi)$, are roughly an order of magnitude smaller than the QCD coefficients $A^{(QCD)}(\chi)$ and $B^{(QCD)}(\chi)$ plotted in Figs. 6 and 8. The shape of $A^{(scalar)}(\chi)$ is roughly similar to that of $A^{(QCD)}(\chi)$ but, as displayed in Fig. (9), the shape of $B^{(scalar)}(\chi)$ is very different from that of $B^{(QCD)}(\chi)$.

Let us conclude with a remark of purely theoretical interest. In perturbation theory, the energy correlation cross section becomes singular for anti-collinear detectors ($\chi \to \pi$). In this region, the leading behavior of the cross section is given by

$$\frac{d^2\Sigma}{d\Omega d\Omega'} \simeq \frac{d\Sigma}{d\Omega} \{1/2\delta(\Omega+\Omega')[1-g^2 c \ln^2 \infty + 1/2 g^4 c^2 \ln^4 \infty - \ldots]$$
$$+ \frac{g^2}{4\pi^2} \frac{1}{3\pi} \frac{1}{1+\cos\chi} \ln\left(\frac{1}{1+\cos\chi}\right) \quad (11)$$
$$- \left(\frac{g^2}{4\pi^2}\right)^2 \frac{2}{9\pi} \frac{1}{1+\cos\chi} \left[\ln\left(\frac{1}{1+\cos\chi}\right)\right]^3 + \ldots \} ,$$

where $\frac{d\Sigma}{d\Omega}$ is the single energy cross section given in Eq. (6) [with $g^2 = 0$], and ... stands for terms of higher order in g^2. The $\delta(\Omega+\Omega')$ term corresponds to the zeroth order production of a quark and an anti-quark which emerge back-to-back. This term is modified by (infrared mass) divergent vertex corrections. With the introduction of proper regulators, these divergences cancel against divergences produced in the remaining terms when the cross section is integrated over a small patch of solid angle covering $\chi = \pi$. Cornwall and Tiktopoulos[6] have proven (to sixth order) that the vertex modification exponentiates as indicated in Eq. (11). The order g^2 angular term previously calculated[3,4] has been recently augmented by a calculation[7] of the leading terms in order g^4 as shown in Eq. (11). Let us suppose that the vertex modifications do indeed exponentiate, multiplying $\delta(\Omega+\Omega')$ by $e^{-\infty}$ and damping it out entirely. And let us further suppose that the remaining terms also exponentia-

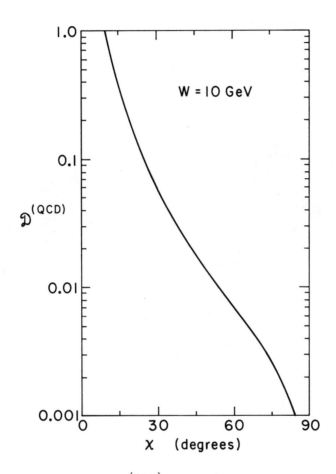

Fig. 7. The difference $D^{(QCD)}$ defined by Eq. (10) as a function of χ for W=10 GeV. The plot for W=30 GeV is obtained by scaling the curve displayed by the factor
$\bar{g}(30)^2/\bar{g}(10)^2 \simeq 0.73$.

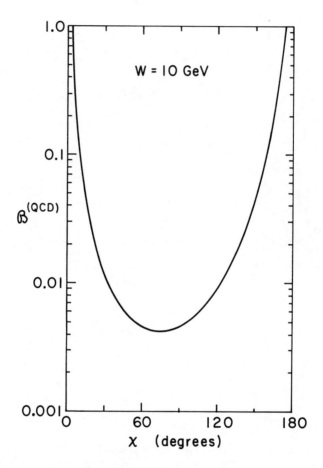

Fig. 8. The coefficient $B^{(QCD)}$ of Eq. (9) as a function of χ for W=10 GeV. The plot for W=30 GeV is obtained by scaling the curve displayed by the factor
$$\bar{g}(30)^2/\bar{g}(10)^2 \simeq 0.73 \quad.$$

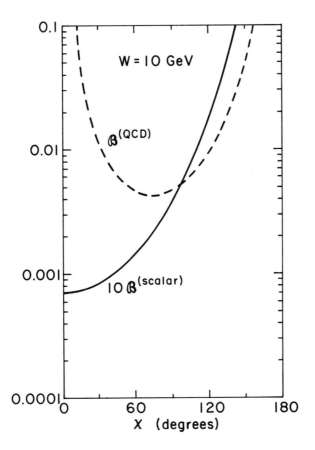

Fig. 9. The QCD and scalar theory results for the B coefficient plotted as a function of χ for W = 10 GeV.

giving near the anti-collinear orientation

$$\frac{d^2\Sigma}{d\Omega d\Omega'} \simeq \frac{d\Sigma}{d\Omega} \frac{g^2}{4\pi^2} \frac{1}{3\pi} \frac{1}{1+\cos\chi} \ln(\frac{1}{1+\cos\chi})$$

$$\exp\{-\frac{g^2}{4\pi^2} \frac{2}{3} \ln^2(\frac{1}{1+\cos\chi})\} \quad (12)$$

$$= \frac{d\Sigma}{d\Omega} \frac{1}{4\pi} \frac{d}{d\cos\chi} \exp\{-\frac{g^2}{4\pi^2} \frac{2}{3} \ln^2(\frac{1}{1+\cos\chi})\} \; .$$

Note that Eq. (12) gives a sharp peak near $\chi = \pi$. With g^2 replaced by the running coupling $\bar{g}(W)^2$, this peak decreases in width, increases in height, and moves toward $\chi = \pi$ as the energy increases.

It is interesting that this conjectured form does, in fact, satisfy a sum rule which must be obeyed by the true $d^2\Sigma/d\Omega d\Omega'$. The sum rule arises as follows: If we integrate $d^2\Sigma/d\Omega d\Omega'$ over a small patch of solid angle covering $\chi = \pi$ we get a well-defined ordinary function which can be developed in a perturbation series in g^2. Accordingly, in view of the sum rule (5), we get

$$\int_{\chi \sim \pi} d\Omega' \frac{d^2\Sigma}{d\Omega d\Omega'} = \frac{1}{2} \frac{d\Sigma}{d\Omega} [1+0(g^2)] \; . \quad (13)$$

[The factor of $\frac{1}{2}$ appears since there is an equal $0(g^0)$ contribution at $\chi = 0$ because the detectors are "transparent" and can detect the same parcel of energy.] Using Eq. (12) we see that Eq. (13) is indeed satisfied. We find that the numerical coefficients of the order g^4 and order g^2 terms in Eq. (11) are related so as to produce precisely the order g^0 result required by Eq. (13).

REFERENCES

1. G. Sterman and S. Weinberg, Phys. Rev. Lett. <u>39</u>, 1436 (1977).
2. C. L. Basham, L. S. Brown, S. D. Ellis, and S. T. Love, Phys. Rev. <u>D17</u>, 2298 (1978).

3. C. L. Basham, L. S. Brown, S. D. Ellis, and S. T. Love, Phys. Rev. Lett. $\underline{41}$, 1585 (1978).
4. C. L. Basham, L. S. Brown, S. D. Ellis, and S. T. Love, Phys. Rev. D (to be published). This paper contains references to related literature.
5. C. L. Basham and S. T. Love, "Energy Correlations in Electron-Positron Annihilation: Sensitivity of Quantum Chromodynamics Tests to Gluon Spin", University of Washington report (unpublished).
6. J. M. Cornwall and G. Tiktopoulos, Phys. Rev. $\underline{D13}$, 3370 (1976).
7. C. L. Basham, L. S. Brown, S. D. Ellis, and S. T. Love, "Energy Correlations in Perturbative Quantum Chromodynamics: A Conjecture for All Orders", manuscript in preparation.

SPIN EFFECTS IN ELECTROMAGNETIC INTERACTIONS

P. A. Souder and V. W. Hughes

(Presented by P. A. Souder)

Yale University, New Haven, Connecticut

I. INTRODUCTION

In this review we discuss a number of experiments measuring spin-dependent phenomena in electromagnetic processes. Emphasis will be on the experimental aspects special to studying spin dependence, both in order to evaluate these experiments and to indicate in what directions progress may be made in this field.

We shall primarily be concerned with electron scattering, but will also mention muons. Targets studied include protons, deuterons, and electrons. Since the electromagnetic interactions are presumably well understood, the experiments focus on what electromagnetic phenomena can tell us about other interactions. The paper is in three sections. First is a discussion of the techniques used to produce polarized electrons and hadrons, second is a description of experiments studying various fundamental symmetries in nature including Lorentz invariance, parity, and time reversal, and third is a report on what has been learned about proton spin structure.

II. EXPERIMENTAL TECHNIQUES FOR ORIENTING SPINS

A large variety of techniques have been developed to produce polarized electrons.[1] At the Stanford Linear Accelerator Center

(SLAC), two different methods have been used for high energy experiments. The first source, called PEGGY I, is based on the photoionization of an atomic beam of polarized lithium.[2] A newer source, PEGGY II, uses polarized light to produce photoelectrons from a GaAs crystal.[3]

Figure 1 is a schematic diagram of the Li source, PEGGY I.[4] An intense beam of ^{6}Li atoms passed through an inhomogeneous six-pole magnet which transmits atoms in only one spin state, atoms of the other state being deflected and forced onto the magnet pole faces. Upon leaving the magnet, the atoms pass into a longitudinal magnetic field provided by the polarizing coil. The spins follow the field lines adiabatically and thus the atoms are in a definite helicity state when they reach the ionization region. Reversing this coil reverses the helicty of the beam. Intense light in the uv range from 1700 Å to 2300 Å is produced by a vortex-stabilized argon flash lamp and is focused onto the beam. The resulting photoelectrons, carrying the spin of the atoms, are accelerated to 70 kV by a static electric field and injected into the accelerator or alternatively into a Mott chamber where the polarization can be measured. Operating characteristics of the source are given in Table I. A major advantage of this source is the large polarization of 85%. The intensity, although considerably less than the maximum that can be accelerated at SLAC, is ample for experiments using polarized targets which suffer from radiation damage.

The GaAs source PEGGY II is shown in Fig. 2.[5] It is conceptually very simple in that one merely shines polarized light on a crystal and accelerates the resulting photoelectrons. The reason for the electron polarization is shown in Fig. 3, where we see that, for GaAs transitions from the $P_{3/2}$ valence band to the $S_{1/2}$ conduction band induced by circularly polarized light favor one spin state over the other. The surface of the crystal is coated with a layer of cesium and oxygen on the order of several atoms thick, and this results in a negative electron affinity for conduction band electrons. A vacuu

SPIN EFFECTS IN ELECTROMAGNETIC INTERACTIONS

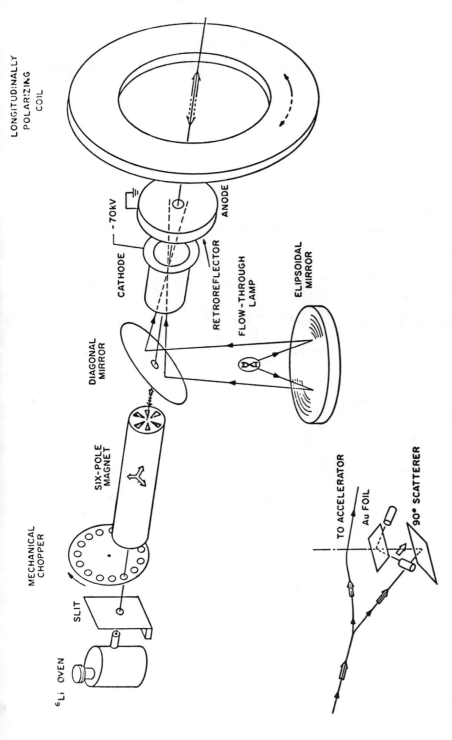

Figure 1. Schematic diagram of PEGGY I showing the layout of the major components.

TABLE I

Characteristics of Polarized Electron Beam

Characteristic	Value
Pulse Length	1.5 sec
Repetition Rate	120 pps
Electron Intensity (at high energy)	~10^9 e^-/pulse
Pulse to Pulse Intensity Variation	<5%
Electron polarization, P_e	0.85 ± 0.08
Polarization reversal time	3 sec
Time between reversals	2 min
Intensity difference upon reversal	<5%
Lifetime of lithium oven load	200 hrs
Time to reload system	36 hrs

SPIN EFFECTS IN ELECTROMAGNETIC INTERACTIONS

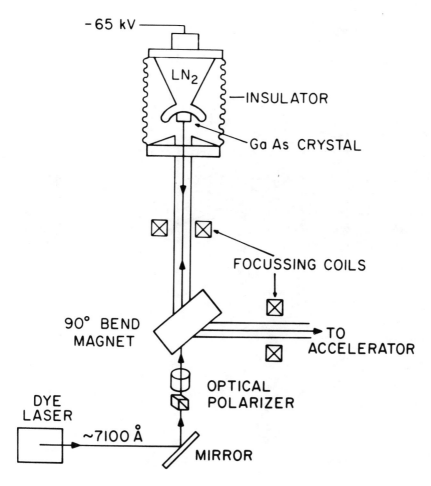

Figure 2. Schematic diagram of PEGGY II, showing the GaAs crystal in the electron gun and the laser optics.

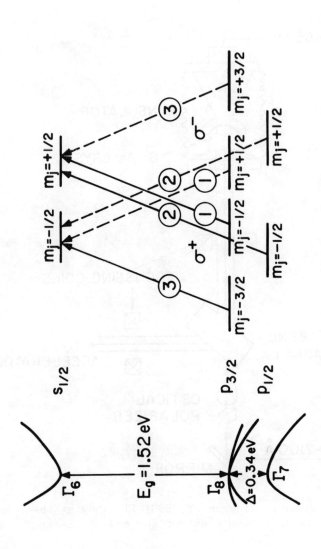

Figure 3. Energy bands of GaAs at the Γ point (left) and transitions between the $S_{1/2}$ levels and the $P_{1/2}$ and $P_{3/2}$ levels. Solid (broken) lines indicate transitions for $\sigma^+(\sigma^-)$ light, and the circled numbers indicate the relative transition strengths.

of 10^{-10} to 10^{-11} torr must be maintained to avoid surface contamination, and the cleaning and coating of the surface is a delicate process.

The operating characteristics of PEGGY II are given in Table II. Two important features are the high intensity, which is nearly the maximum that can be accelerated, and the polarization reversal time, which is less than the time between machine pulses. Polarization reversal is achieved by changing the electric field in a Pockels cell in the laser beam, and to a very high accuracy the only property of the beam changed is its helicity. These features are ideal for the search for very small spin-dependent effects such as those due to parity violation. One limitation of the GaAs source is that the polarization is only about 37%. Present research is directed at finding a crystal that will produce polarization nearer to 100%. This can be achieved by removing the degeneracy of the m_j levels, possibly by using a different material, stressing the crystal, or producing a multilayer structure.[6]

The above sources produce electrons at low energy. The question then arises of whether the RF and static magnetic fields in the accelerator depolarize the beam before it reaches high energy. A theoretically unambigious process, Møller (electron-electron) scattering, is used to measure the polarization of the full energy beam.[7] The cross section and asymmetry for scattering from polarized target electrons are shown in Figure 4. At 90° in the center of mass, the asymmetry is 80% and the cross section is large. The signature in the laboratory frame for a Møller event, an electron of half the beam energy scattered by a small angle, is fairly unique, and a simple single-arm spectrometer is adequate to detect the events. A magnetized iron foil provides the polarized target electrons.

Typical results for Møller scattering using PEGGY I are shown in Fig. 5. The direction of the polarization of the beam depends on the energy because of an energy-dependent spin precession occurring in the 24.5° bend between the accelerator and the experimental

TABLE II

PEGGY II Operating Characteristics

Characteristic	Value
Pulse length	1.5 μsec
Repetition rate	120 pps
Electron Intensity (at high energy)	(1 to 4) x 10^{11} e^-/pulse
Pulse to pulse intensity variation	~3%
Electron polarization	0.37, average
Polarization reversal time	pulse to pulse

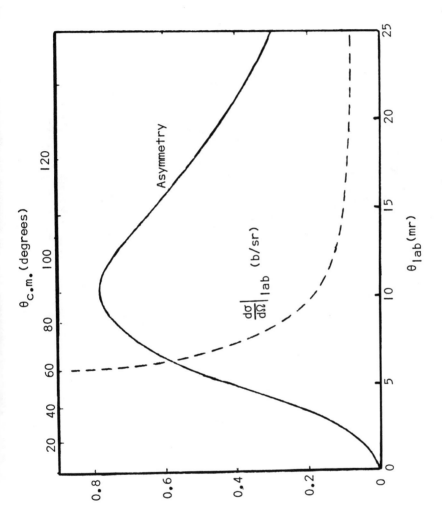

Figure 4. Cross section and polarization asymmetry for Møller (electron-electron) scattering. An incident electron energy of 9.712 GeV is assumed.

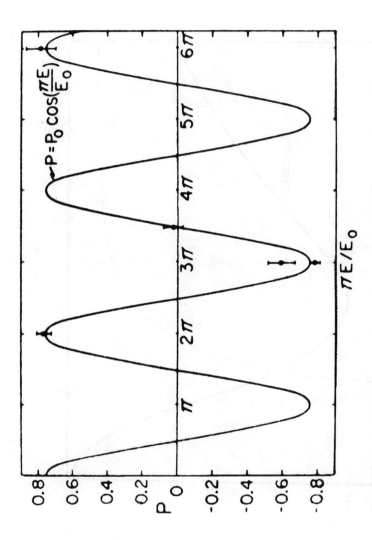

Figure 5. Longitudinal polarization of electrons as a function of energy after the 24.5° bend into the experimental hall, including the effect of the g-2 precession.

hall. This effect will be discussed in more detail in the section on Lorentz invariance. The feature of interest here is the large polarization of 76%, indicating that negligible depolarization occurs during acceleration. Subsequent improvements in PEGGY I, involving filtering out long wavelength light which depolarize the atomic beam in the ionization region, have raised the polarization to 85 ± .8%.[8]

The butanol polarized proton target used for our work at SLAC[9] is based on the usual method of dynamic nuclear orientation.[10] In the target, shown in Fig. 6, butanol is cooled to $1°K$ and immersed in a homogeneous 5T magnetic field produced by superconducting coils. Under these conditions free protons are polarized to about half a percent. A free radical, porphyrexide, is added to the sample, and its quasi-free electrons, having 2000 times the magnetic moment, are nearly 100% polarized. Then 140 GHz microwaves are applied which in effect transfer the spin of the electrons to the protons. Proton polarizations of about 65% are achieved in this fashion. Important operating characteristics of our target are listed in Table III.

Electron scattering experiments usually involved small cross sections, and consequently rather intense beams must traverse the target. This creates a severe problem for polarized targets, which are highly sensitive to radiation damage. Ionizing particles produce stable color centers which presumably serve to depolarize the free protons. Even for the relatively weak 10 na average current of the PEGGY I beam, the target loses half of its original polarization in a few hours. The target can be annealed several times to remove the color centers, but eventually the target material must be completely replaced because the porphyrexide is destroyed. A new target is bright orange; it turns black when irradiated and becomes clear when melted after several annealings.

III. INVARIANCE PRINCIPLES

Invariance principles have become a central theme in modern physics. Polarization phenomena have made important contributions,

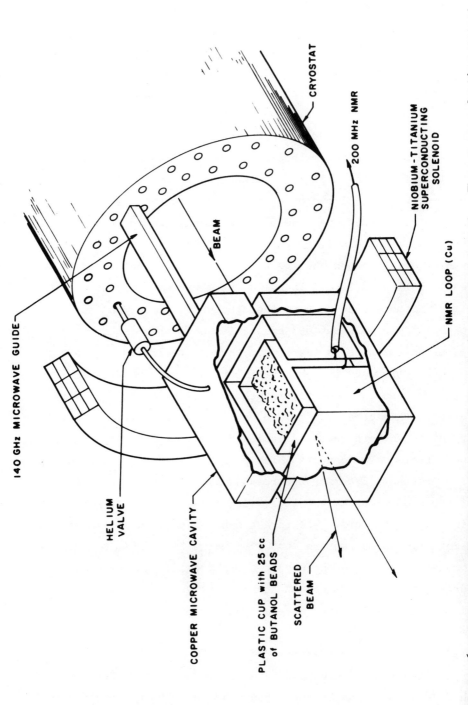

Figure 6. Schematic diagram of the SLAC-Yale polarized proton target. The superconducting magnet

TABLE III

Operating Characteristics of Polarized Proton Target

Characteristic	Value
Magnetic field (longitudinal field of superconducting magnet)	50 kG
Temperature	1.05° K
Target material	25 cm^3 of butanol-porphyrexide beads (~1.7 mm diam)
Initial polarization of free protons[a]	0.50 to 0.65
Depolarizing dose (1/e)	~3 x 10^{14} e$^-$/cm^2
Polarizing time (1/e)	~4 min
Anneal or target change time (including polarizing)	45 min

[a] Improvements in target operation gave the larger polarization values in the later parts of the experiment.

including sensitive tests of Lorentz invariance, time-reversal invariance, and parity invariance.

A test of Lorentz invariance at high energy was made at SLAC by studying the precession of the spins of the polarized electrons relative to their momenta occurring in the $24.5°$ bend between the en of the linear accelerator and the experimental hall.[11] If the g-factor of the electron were exactly 2, the electron momentum and spi vectors would precess at exactly the same frequency, and a beam woul maintain its helicity in a bending magnet. However, g is slightly greater than 2, and the spin precesses relative to the momentum by an angle θ_a given by

$$\theta_a = \theta_s - \theta_c = \gamma a \theta_c ,$$

where θ_s is the spin precession angle, θ_c the angle the beam is bent γ is the usual Lorentz factor and $a \equiv (g-2)/2$ is the g-factor anomal Although a is only about 10^{-3}, γ is on the order of 10^4 at SLAC energies, and an electron precesses three full cycles in the $24.5°$ bend at the highest energy of about 20 GeV. The g-2 precession is evident in Fig. 5, and from that data we have measured the anomaly a with a precision of 2%. Table IV gives this number along with results from high-precision measurements of the anomaly measured at low velocites.[12,13] The fact that the high γ value agrees with the low energy values verifies the predictions of special relativity spanning a range in γ of four orders of magnitude.

Although there are no fashionable theories for the breakdown of Lorentz invariance, we can make the above observation quantitati by using the analysis of Newman et al.[14] They noted that the cyclo tron frequency $\omega_c = eB/\tilde{\gamma} m_o c$ uses a "dynamic" $\tilde{\gamma} \equiv \frac{p}{m_o} \frac{dp}{dE}$ and that th spin precession frequency $\omega_s = \frac{eB}{2mc} g + (1-\omega)\omega_s$ uses a "kinematic" $\gamma \equiv (1-\beta^2)^{-1/2}$. Experiments measure the difference between the two $\omega_a = \omega_s - \omega_c = (g-2\gamma/\tilde{\gamma})eB/2m_o c$. Values of g-2 at different velocities may thus be viewed as measurements of the velocity dependenc of $\gamma/\tilde{\gamma}$. To obtain a figure of merit for evaluating different exper

TABLE IV

Electron g-2 Measurements

γ	$a \times 10^3$	Reference
$1 + 10^{-9}$	1.159 652 41 (20)	12
1.2	1.159 657 70 (350)	13
2.5×10^4	1.162 2 (200)	11

iments, we expand

$$\frac{\gamma}{\tilde{\gamma}} = 1 + c_1(\gamma-1) + \ldots$$

so that $\gamma=\tilde{\gamma}=1$ in the Newtonian limit and all higher terms vanish if $\gamma=\tilde{\gamma}$ as predicted by special relativity. By comparing our data with either of the other low energy experiments and by assuming that the c_1 term dominates, we establish the limit that $c_1 < 10^{-9}$.

Time reversal invariance of the electromagnetic interactions has been tested in two experiments[15] which studied the inelastic scattering of unpolarized electrons by polarized protons. That data covered the resonance region with Q^2 between 0.2 and 1.0 $(GeV/c)^2$. Within the 2% to 15% experimental errors, no violation of T invariance was seen.

The final symmetry we shall discuss and the one that has created the most excitement recently is parity invariance. The highly successful Weinberg-Salam model[16] of the weak and electromagnetic interactions as well as many other gauge theories predict that parity is violated in electron-hadron scattering due to an interference between the ordinary electromagnetic current and the neutral weak current. Deep inelastic scattering of polarized electrons (or muons) is a particularly clean and useful test of these effects.[17] Any helicity dependence in the differential cross section (only the momentum of the electron is determined in the final state) can only be due to the presence of a $\vec{\sigma}_e \cdot \vec{p}_e$ term which is manifestly parity violating. Here $\vec{p}_e$ is the electron momentum vector. In contrast to the situation for the T-invariance test mentioned above, two-photon exchange and other higher order effects cannot simulate parity violation.

The quantity measured is polarized lepton parity experiments is

$$A = \frac{d\sigma^+ - d\sigma^-}{d\sigma^+ + d\sigma^-}$$

where the signs refer to the helicity of the incident electron. Fo most models predicting non-zero values of A, A is proportional to the square of the four-momentum transfer, Q^2, and typically $A \simeq (10^{-5}$ to $10^{-4})Q^2/M_p^2$. The asymmetries may also depend on the kinematic variable y, which is the ratio of the energy lost by the electron during the scattering to the initial energy.

Before we discuss in detail the recent SLAC-Yale experiment which was able to measure the asymmetry with a precision sufficient to test these predictions, we will survey some earlier, less sensitive experiments, the results of which are summarized in Table V.

The first high energy test of parity violation in electron-nucleus scattering came as a byproduct of measurements at SLAC of asymmetries in elastic and deep inelastic scattering of polarized electrons from PEGGY I by the polarized proton target.[18] A sensitivity of about $10^{-3} \, Q^2/M_p^2$ was achieved.

A second experiment at SLAC using PEGGY I was dedicated to the search for parity violation.[19] The precision of this experiment was limited by a possible correlation between beam energy and helicity, which is magnified in the asymmetry by the large sensitivity of the cross section to incident energy. The identification of this systematic error stimulated the development of improved beam monitoring devices which proved important in later, more sensitive experiments at SLAC.

A third pre-1978 experiment listed in Table 5 is a muon scattering experiment performed at Serpukhov.[20] The helicity of the muons was controlled by accepting either forward or backward pion decays. Data over a wide range of Q^2 from 1 $(GeV/c)^2$ to 11 $(GeV/c)^2$ were obtained and limits of 5% to 40% were established at each point. No parity violation was seen at this level.

In the spring of 1978 a new parity experiment at SLAC used the polarized electron source PEGGY II and was able to make a significant test of the weak interaction theories.[21] This achievement was based on a number of interesting techniques which were combined to minimize both statistical and systematic errors.

TABLE V

Results on Search for PNC Prior to 1978

SLAC E80 Results (95% CL) (Ref. 18)

$|A| < 3 \times 10^{-3}$, H, Elastic, $Q^2 = 0.77$ $(GeV/c)^2$

$|A| < 3 \times 10^{-3}$, C, DIES, Q^2 between 1 and 4 $(GeV/c)^2$

SLAC E95 Results (95% CL), (Ref. 19)

$|A| < 2 \times 10^{-3}$, D, DIES, $Q^2 = 1.2$ $(GeV/c)^2$

$|A| < 10^{-2}$, D, DIES, $Q^2 = 4.2$ $(GeV/c)^2$

SERPUKHOV MUON DEEP INELASTIC SCATTERING RESULTS (95% CL) (Ref. 20)

$|A| < 1.6 \times 10^{-2}\, Q^2$, Fe

$[p_\mu \simeq 20\ GeV/C;\ 1 < Q^2 < 11\ (GeV/c)^2]$

A general schematic diagram of the major components of the experiment is shown in Figure 7. At the injection end of the accelerator is the GaAs polarized electron source PEGGY II, which has been described above. PEGGY II is matched to the duty factor of the accelerator at SLAC, producing 1.5 sec pulses 120 times each second. The electrons, after being accelerated to high energy, passed through an elaborate system of beam monitors and then into the experimental hall. The polarization of the electrons was measured during the run with a Møller scattering spectrometer. For the parity measurement, the electrons scattered from a liquid deuterium or hydrogen target and were detected by a new large-acceptance spectrometer.

A beam monitoring system, shown in detail in Figure 8, was developed to keep systematic errors in the experiment well below 10^{-5}. It measured the beam's intensity, position, direction, and energy. The central component was a resonant position monitor (RPM),[22] which is a microwave cavity that is excited in a position-dependent way by the microstructure of the traversing beam. This device can measure the position of the primary beam to 10 μm during an individual 1.5 μsec beam pulse. The intensity of the incident beam was measured with sensitivity of 0.02% each pulse by two toroids, T_1 and T_2, which are shown in Figure 8.

A new spectrometer, comprised of magnets from the SLAC 8-GeV and 20-GeV spectrometers, was assembled for this experiment. It was oriented at a fixed angle of $4°$ relative to the incoming beam. The spectrometer was designed to maximize the acceptance (see Figure 9) in order to obtain sufficient statistics. Two independent devices, a gas Cerenkov counter and a lead-glass total absorption counter, detected the scattered electrons. Since the instantaneous rate was about 10^9/sec, signals from the detectors had to be integrated and digitized instead of counted. Nevertheless, the cross section was measured to a precision of 3% each pulse, approximately as would be predicted by counting statistics.

One strength of this experiment was the ability to reverse the

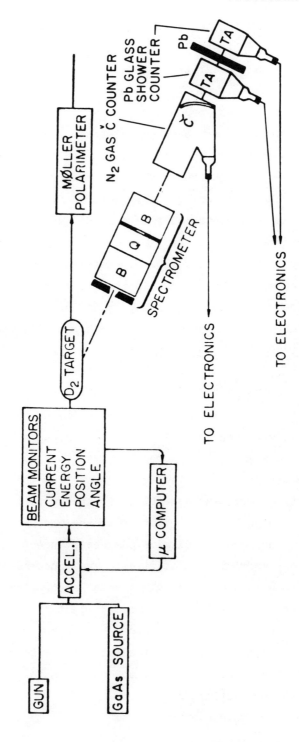

Figure 7. Block diagram of the apparatus used to detect parity nonconservation.

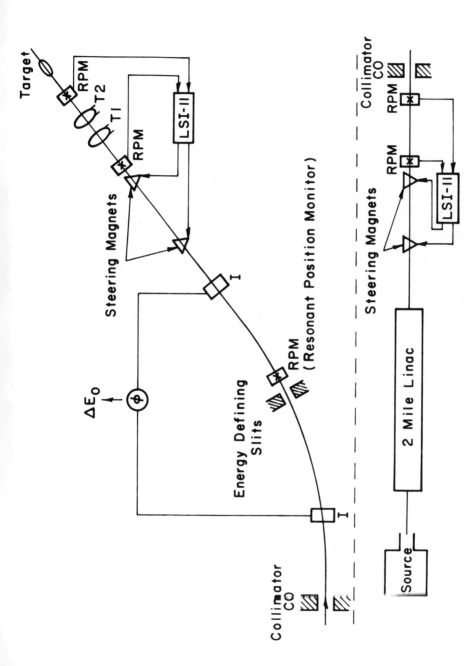

Figure 8. Detail of the beam monitoring system, including resonant position monitors, ΔE monitor, and toroids T_1 and T_2.

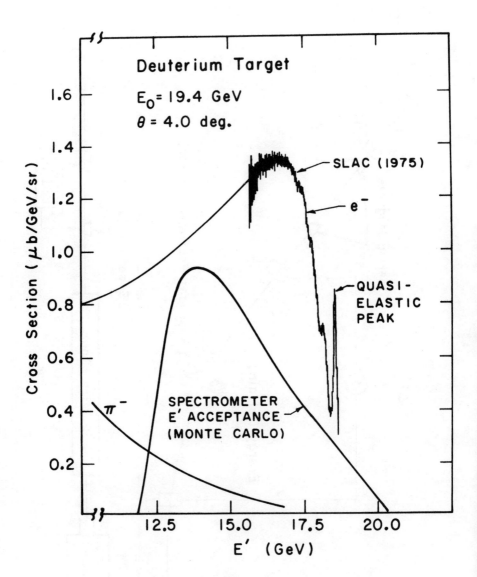

Figure 9. Acceptance of the new high rate spectrometer used for the parity experiment.

beam helicity by several independent methods. The first and most
important method was changing the electric field in the Pockels cell
in the laser beam. In this way the helicity of the laser light and
in turn the helicity of the photoelectrons could be reversed. No
other beam properties, intensity, phase space, etc., are expected to
be affected significantly by this operation. Moreover, this method
of reversal could change the helicity of the beam randomly on a pulse-
to-pulse basis in order to eliminate the effects of slow drifts.
Basically, each beam pulse provided a complete and independent
measurement of the cross section for a given helicity.

A second method of polarization reversal was to rotate a cal-
cite prism in the laser beam. When oriented at zero degrees, the
prism does not change the polarization of the light, whereas at $90°$
the polarization is reversed. In addition, a systematic test of the
apparatus can be made by orienting the prism at $45°$ so that the light
becomes linearly polarized and the photoelectrons are unpolarized.
Data obtained with the Pockels cell in operation but with the prism
in this orientation is shown in Figure 10. As required for unpolarized electrons, there is no asymmetry. Results for polarized electrons are also shown in Figure 10. The data exhibit a nonzero asymmetry which changes sign when the prism is rotated. This is what
is expected if there is a parity-violating asymmetry in the cross
section.

The fact that the asymmetry behaves correctly when the prism
is rotated is strong evidence that spurious asymmetries are absent.
However, still another method of spin reversal is possible, the g-2
precession in the switchyard. By varying the beam energy between
16.2 and 22.2 GeV the spin of the electrons can be varied by $360°$.
Figure 11 shows the asymmetry versus energy, and the behavior is
exactly as expected for a true parity-violating effect.

The final experimental result[21] is given in Table VI for both
deuterium and hydrogen. The largest systematic errors are the 5%
uncertainty in the beam polarization and a 3% uncertainty due to a

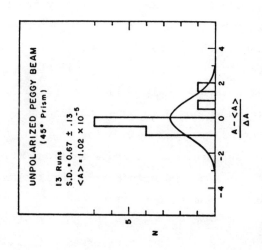

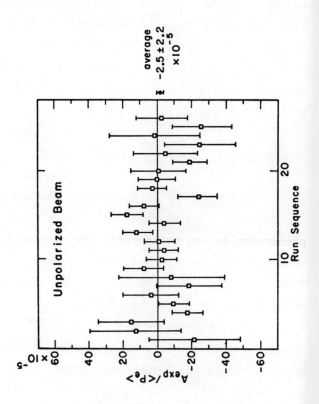

SPIN EFFECTS IN ELECTROMAGNETIC INTERACTIONS

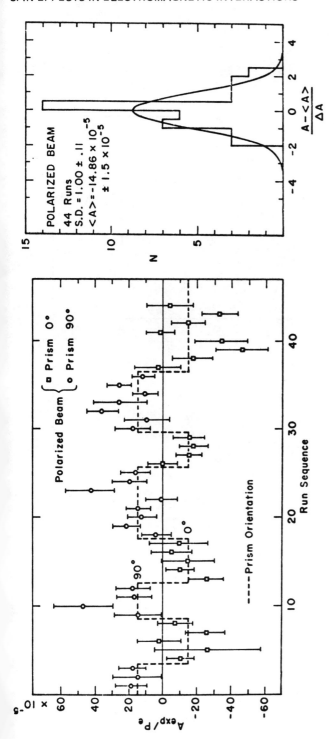

Figure 10. Top Left: Asymmetry for unpolarized electrons (prism at 45°) versus run number.

Bottom Left: Asymmetry for polarized electrons versus run number. Broken line is fit to the data assuming that the asymmetry reverses sign when the electron helicity is reversed by rotating the prism.

Top right: Residuals for unpolarized data.

Bottom right: Residuals for polarized data.

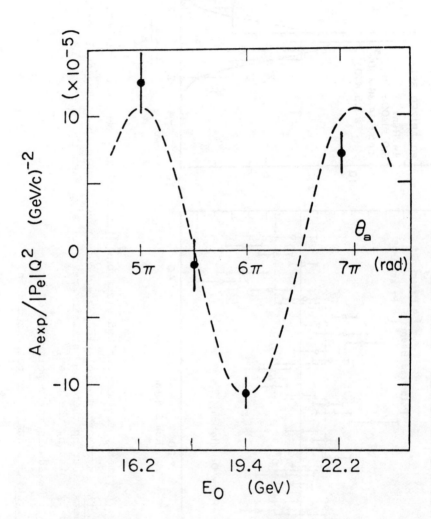

Figure 11. Experimental asymmetry as a function of beam energy. Broken line was calculated assuming that the theoretical asymmetry is independent of energy but that the beam helicity changes in the 24.5° bending magnet due to the g-2 precession.

TABLE VI

Experimental Results

$e^- + D \to e^- + X$ (DIES)

$\dfrac{A}{Q^2} = (-9.5 \pm 1.6) \times 10^{-5}$ $(GeV/c)^{-2}$

[Statistical error = 0.86×10^{-5}]

[Systematic error $\simeq 0.7 \times 10^{-5}$]

$<Q^2> = 1.6$ $(GeV/c)^2$

$<y> = 0.21$

$e^- + P \to e^- + X$ (DIES)

$\dfrac{A}{Q^2} = (-9.7 \pm 2.7) \times 10^{-5}$ $(GeV/c)^{-2}$

[Statistical error = 1.6×10^{-5}]

[Systematic error = 1.1×10^{-5}]

$<Q^2> \simeq 1.6$ $(GeV/c)^2$

$<y> = 0.21$

small correlation between beam helicity and energy detected by the beam monitor system.

The deuterium point is plotted in Figure 12 along with the predictions of some gauge theories. The line A=0 is predicted by theories with no parity violation in the neutral current sector but is clearly excluded by the data. The "hybrid model,"[23] shown as dotted lines, is also excluded except for values of $\sin^2\theta_w$ which are inconsistent with neutrino data. The solid lines are the predictions of the Weinberg model, and the agreement with the data is good for $\sin^2\theta_w = 0.20 \pm 0.03$.

Values of $\sin^2\theta_w$ for a large number of different experiments[24] are shown in Figure 13. The consistency of these values is truly impressive. Although there are other weak interaction models which are still viable, the fact that this wide and varied body of data on neutral currents is described well by a one-parameter model developed even before the discovery of neutral currents must stand as a major achievement.

The discovery of parity nonconservation in deep-inelastic scattering has opened up the new field of the weak interaction physics of electron scattering. From the usual phenomenological point of view, the parity violating Hamiltonian at low energy is given by[25]

$$\mathcal{H}^{PV} = \Sigma\, \varepsilon^{AV}_{ab}\, \bar\psi_a \gamma_\mu \gamma_5 \psi_a\, \bar\psi_b \gamma^\mu \psi_b$$

where the ε's are coupling constants and where a and b refer to electrons or quarks. For electron-hadron scattering, there are four independent coupling constants,

$$\varepsilon^{AV}_{eu},\ \varepsilon^{AV}_{ed},\ \varepsilon^{VA}_{eu},\ \text{and}\ \varepsilon^{VA}_{ed}.$$

One linear combination of these constants, approximately $2\varepsilon^{AV}_{ed} - \varepsilon^{AV}_{eu}$, has been measured by the SLAC-Yale experiment.[26] A new run studying the y-dependence of the asymmetry has been completed, and the result

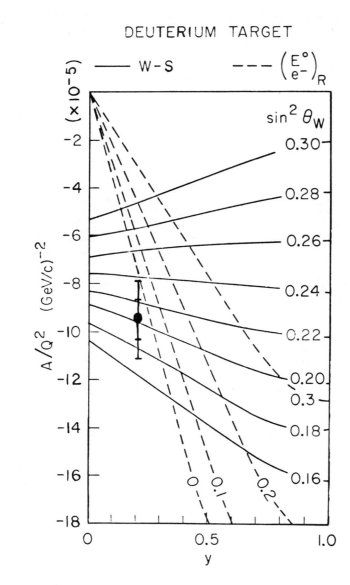

Figure 12. A/Q^2 for the parity experiment plotted together with the prediction of the Weinberg-Salam model (solid lines) and the hybrid model (broken lines).

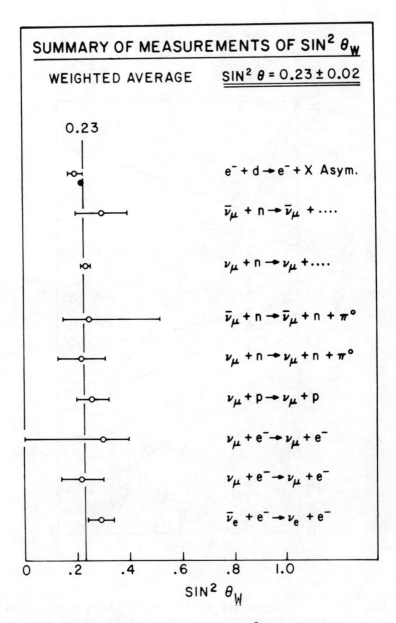

Figure 13. Symmary of measurement of $\sin^2\theta_w$ obtained from electron and neutrino scattering.

will be published soon. These data will provide information about the axial-hadronic coupling constants.

Useful information can also be provided by experiments at lower Q^2 in which polarized electrons are scattered coherently from nuclei. This is in contrast with the incoherent scattering from individual quarks characteristic of deep-inelastic scattering. A particularly clean example is elastic scattering from carbon. Since the spin and isospin of this nucleus are both zero, there is only one form factor, and it cancels out in the calculation of the asymmetry.[27] The asymmetry is proportional to $\varepsilon_{eu}^{AV} + \varepsilon_{ed}^{AV}$, a combination of the coupling constants nearly orthogonal to the one measured in the SLAC-Yale experiment. Another interesting possibility is elastic scattering from hydrogen, particularly in the backward direction.[28]

The cross sections for coherent reactions are large only for small values of Q^2. Since the asymmetries are proportional to Q^2, rather small effects, on the order of 10^{-5} to 10^{-6}, must be measured. However, a level of precision on the order of 10^{-7} has already been achieved in a parity experiment involving the scattering of polarized protons.[29] Thus it is not unreasonable to attempt similar precision in electron experiments, and indeed these difficult experiments have been proposed at several laboratories.[30]

At even lower Q^2, parity violation involving the same coupling constants can be studied in atoms. Reviews of this field appear elsewhere.[31] We should also mention the interest at CERN in studying parity nonconservation in deep-inelastic muon scattering.[32] According to muon-electron universality, the coupling constants involved in this reaction should be equal to the ones for electron scattering. Finally, Møller scattering can provide information about parity violation in the electron-electron interaction.

Looking ahead, the most critical test of the ideas behind the Weinberg model will come with the search for the intermediate vector bosons. However, we feel that it is also very important to measure carefully all of the low energy coupling constants, and to achieve

IV. HADRON SPIN STRUCTURE

The final topic we shall discuss is a series of experiments in which polarized electrons are scattered from polarized protons.[18,33] These experiments measure a term in the cross section proportional to $\vec{\sigma}_e \cdot \vec{\sigma}_p$ which of course violates no symmetry principle. However, it is quite sensitive to the details of proton structure, especially in the quark-parton model. According to this model, the proton is comprised of elementary spin 1/2 fermions called quarks, and deep inelastic scattering involves scattering from a single quark. Since fermion-fermion scattering is extremely spin-dependent (large angle scattering occurs only for antiparallel spins), spin dependent scattering is a good probe of the spin-dependence of the quark wavefunction.

The detailed kinematics required to determine proton structure by electron scattering are fairly complicated. The relevant variables are defined in Figure 14. As indicated in Figure 15, the cross section may be expressed in terms of four structure functions, W_1 and W_2, which are measured in unpolarized scattering, and G_1 and G_2, which appear only for spin-dependent scattering. The physical interpretation of the structure functions is simplified if they are expressed in terms of four virtual-photon absorption cross sections $\sigma_{1/2}$, $\sigma_{3/2}$, σ_L, and σ_{TL}. The quantity $\sigma_{1/2}(\sigma_{3/2})$ is the cross section for the absorption of a virtual photon with helicity antiparallel (parallel) to the proton spin vector. For elementary spin 1/2 particles like quarks, $\sigma_{3/2} = 0$. Longitudinal photons are absorbed with a cross section σ_L, and σ_{TL}, which may be negative, is a spin-dependent interference term involving longitudinal photons.

The kinematic points where we have obtained data are plotted in Figure 16. The data are concentrated in the deep inelastic region, covering a range of ω between 2 and 10 and a range of Q^2 between 1 and 4 $(GeV/c)^2$. Data have also been obtained for elastic scattering

POLARIZED ELECTRON–PROTON SCATTERING

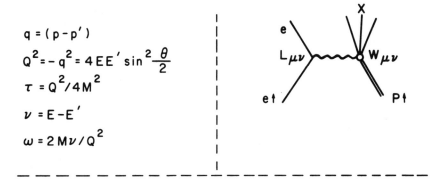

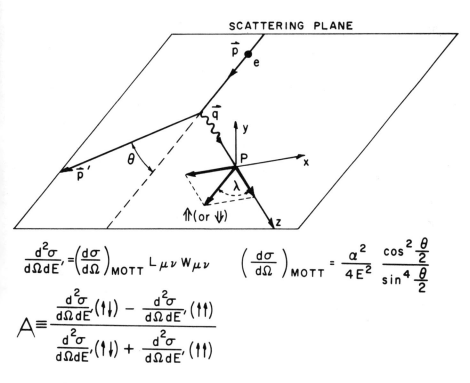

Figure 14. Kinematic variables used to describe the asymmetries for deep inelastic electron-proton scattering. The initial electron energy is E, the final energy is E', and the proton mass is M.

$$\frac{d^2\sigma}{d\Omega dE'} = \left(\frac{a^2\cos^2\frac{\theta}{2}}{4E^2\sin^4\frac{\theta}{2}}\right)\left[W_2 + 2\text{TAN}^2\frac{\theta}{2} W_1 \pm 2\text{TAN}^2\frac{\theta}{2}(\epsilon + E'\cos\theta)MG\right.$$
$$\left. \pm 8 EE'\text{TAN}^2\frac{\theta}{2}\sin^2\frac{\theta}{2} G_2\right]$$
+(A)
−(P)

$$\frac{d^2\sigma}{d\Omega dE'} = \left(\frac{d\sigma}{d\Omega}\right)_M \left(\frac{1}{\epsilon(1+\nu^2/Q^2)}\right) W_1 \left\{1 + \epsilon R \pm (1-\epsilon^2)^{1/2}\cos\psi A_1\right.$$
$$\left. \pm \left[2\epsilon(1-\epsilon)\right]^{1/2}\sin\psi A_2\right\}$$

$$\epsilon = \left[1 + (1+\nu^2/Q^2)\text{TAN}^2\frac{\theta}{2}\right]^{-1}$$

$$R = \sigma_L/\sigma_T; \quad \sigma_T = (\sigma_{1/2} + \sigma_{3/2})/2$$

$$A = \frac{d\sigma(\uparrow\downarrow) - d\sigma(\uparrow\uparrow)}{d\sigma(\uparrow\downarrow) + d\sigma(\uparrow\uparrow)}$$

$$A = D(A_1 + \eta A_2)$$

$$D = \frac{E - E'\epsilon}{E(1+\epsilon R)} = \frac{(1-\epsilon^2)^{1/2}\cos\psi}{(1+\epsilon R)}$$

$$\eta = \frac{\epsilon(Q^2)^{1/2}}{E-E'\epsilon} = \left(\frac{2\epsilon}{1+\epsilon}\right)^{1/2}\text{TAN}\psi \simeq \text{TAN}\psi$$

$$A_1 = \frac{\sigma_{1/2} - \sigma_{3/2}}{\sigma_{1/2} + \sigma_{3/2}}$$

$$A_2 = \frac{2\sigma_{TL}}{\sigma_{1/2} + \sigma_{3/2}}$$

$$|A_1| \leq 1; \quad |A_2| \leq \sqrt{R}$$

Figure 15. Summary of formulae used to describe the cross section and asymmetries for polarized electron-proton scattering

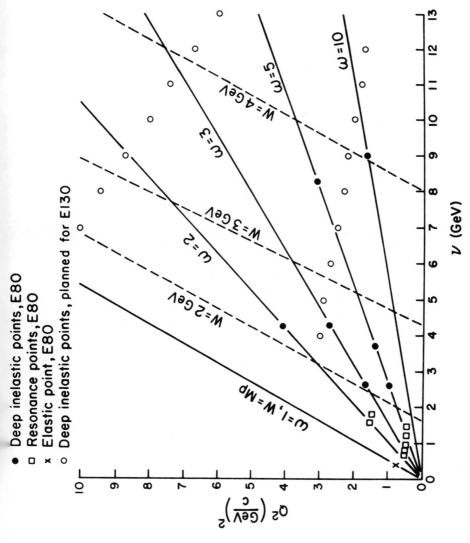

Figure 16. Kinematic range of data points obtained in experiment E80 and also points planned for SLAC experiment E130.

which serves as a check of the experimental method, and in the resonance region.

The results of our measurements are given in Figure 17. The vertical error bars are predominantly statistical, and th horizontal bars reflect the kinematical smearing due to radiative effects. The asymmetry is always large and positive, as expected in the simplest quark-parton models. The quantity A_1 is expected to scale (be a function of $x \equiv 1/\omega$ only and not of Q^2), and our data are consistent with this prediction.

A very general prediction about A_1 is given by the Bjorken sum rule:[34]

$$\int_0^\infty \frac{d\omega}{\omega} [\nu W_2^p A_1^p - \nu W_2^N A_1^N] = \frac{1}{3} g_A/g_V$$

where the superscript P(N) refers to the proton (neutron) and where g_A and g_V are the weak interaction coupling constants. If we neglec A_1^N, which is small in most models, we can use our data to evaluate the sum rule as shown in Figure 18. About half of the sum rule is saturated in the region between $\omega=2$ and $\omega=10$. If we extrapolate ou data to high ω by assuming that $A_1 \propto 1/\sqrt{\omega}$ as predicted by Regge theory[35] the sum rule is satisfied.

Finally, our data can be used to test predictions of various quark-parton and other models, a number of which are graphed in Figure 19. The models[36-41] are generally consistent with our data and indeed do not differ from each other very much except for $x > 0$ The $x=1$ limit is interesting because QCD predicts that the asymmetr should be 100%.[42]

Soon we will take more data with an improvement in precision of a factor of three to five. A new large-acceptance spectrometer will be used. The planned kinematic points, shown as open circles in Figure 16, will cover a larger range in ω and Q^2 than the presen data. We shall also study A_1^N by using polarized deuterons. These new data will lead to imporved tests of scaling, the Bjorken sum

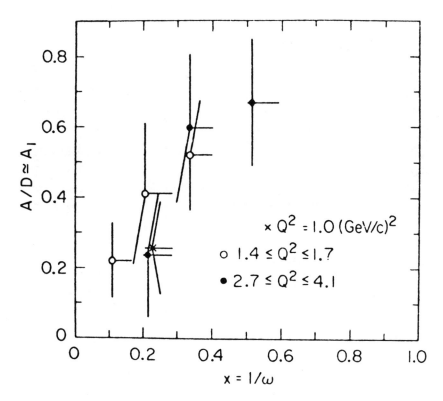

Figure 17. Experimental asymmetries for polarized electron-proton scattering as a function of the scaling variable x and also of Q^2.

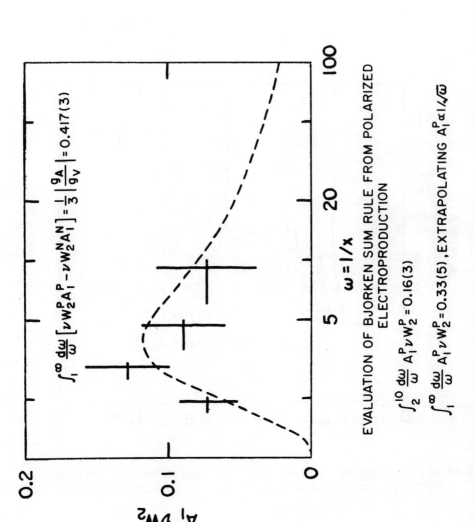

Figure 18. Plot of $A_1\nu W_2$ used to evaluate the Bjorken sum rule. Dashed curve is the fit

SPIN EFFECTS IN ELECTROMAGNETIC INTERACTIONS 433

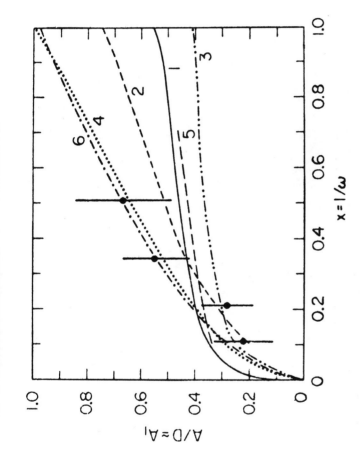

Figure 19. Experimental values of $A/D \simeq A_1$ compared to various theoretical predictions: (1) a relativistic symmetric valence-quark model (Ref. 36); (2) a model using the Melosh transformation (Ref. 37); (3) a model with nonvanishing orbital angular momentum for the quarks (Ref. 38); (4) an unsymmetric model in which the spin of the nucleon is carried by a single quark at x=1 (Ref. 39); (5) The MIT bag model (Ref. 40); (6) source theory (Ref. 41).

TABLE VII

$\Delta\nu_H$ (Theory) and Comparison with Experiment

$$\Delta\nu_{theor} = [\tfrac{16}{3}\alpha^2 cR_\infty(\mu_p/\mu_B^c)][1+(m_e/m_p)]^{-3}$$

$$\times [1 + \tfrac{3}{2}\alpha^2 + a_e + \varepsilon_1 + \varepsilon_2 + \varepsilon_3 + \delta_p]$$

where

$$a_e = \tfrac{\alpha}{2\pi} - 0.328\,48\,\tfrac{\alpha^2}{\pi^2} + (1.181\pm0.015)\tfrac{\alpha^3}{\pi^3};\quad \varepsilon_1 = \alpha^2(\ell n 2 - \tfrac{5}{2})$$

$$\varepsilon_2 = -\tfrac{8\alpha^3}{3\pi}\ell n\alpha(\ell n\alpha - \ell n 4 + \tfrac{281}{480});\quad \varepsilon_3 = \tfrac{\alpha^3}{\pi}(18.4 \pm 5)$$

δ_p = Proton recoil and proton structure term

$\delta_p = \delta_p(\text{rigid}) + \delta_p(\text{polarizability})$

$\delta_p = -34.6(9) \times 10^{-6} + \delta_p(\text{pol})\ [|\delta_p(\text{pol})|<4\text{ ppm}]$

$\Delta\nu_{expt} = 1\,420\,405\,751.766\,7\,(10)$ Hz

$\delta_p(\text{pol}) = -0.25(1.2)$ ppm

CONTRIBUTION OF PROTON POLARIZABILITY
TO HYDROGEN HFS INTERVAL $\Delta\nu$

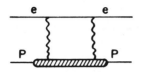

$$\delta_P(\text{inelastic}) = \frac{\alpha}{\pi} \frac{m_e}{M} \frac{1}{2(1+\mu_A)} \int_0^\infty \frac{d(-q^2)}{(-q^2)} \left[\Delta_1(q^2) + \Delta_2(-q^2) \right]$$

$$\Delta_1(q^2) = \frac{9}{4}\left[F_2(q^2)\right]^2 + 5M^3 \int_{\nu_I(q^2)}^\infty \frac{d\nu}{\nu} \beta_1\left(\frac{\nu^2}{-q^2}\right) G_1(\nu, q^2)$$

$$\Delta_2(q^2) = 3M^2 \int \frac{d\nu}{\nu^2} \beta_2\left(\frac{\nu^2}{-q^2}\right) q^2 G_2(\nu, q^2)$$

$$\beta_1(x) \equiv \frac{4}{5}(-3x + 2x^2 + 2(2-x)\sqrt{x(x+1)}\,) \longrightarrow 1 - \frac{1}{2x} \cdots \text{ as } x \to \infty$$

$$\beta_2(x) \equiv 4x + 8x^2 - 8x\sqrt{x(x+1)} \longrightarrow 1 - \frac{1}{2x} \cdots \text{ as } x \to \infty$$

$F_2(q^2)$ = Pauli form factor ; $F_2(0) = \mu_A$

$$\nu_I(q^2) = m_\pi + \frac{(m_\pi^2 - q^2)}{2M}$$

Present theoretical bound without using any asymmetry
data: $|\delta_P| < 4$ ppm

Figure 20. Formulae relating the proton polarizability for the hydrogen hyperfine interaction to asymmetries in inelastic electron-proton scattering.

rule, and quark-parton models.

In the more distant future, we are planning to construct a target which can orient the proton spins perpendicular to the electron beam and in the scattering plane. With this geometry, we can measure a new asymmetry which is approximately proportional to A_2. The quantity A_2 is small for naive quark-parton models but plays an important role in more complete descriptions of the proton.

Since this review began with a discussion of an atomic beam device used to produce polarized electrons, it is appropriate to conclude with atomic physics. The hyperfine structure interval $\Delta\nu$ for hydrogen has been measured to twelve significant digits.[43] This is the highest precision experiment ever performed. However, the theory[44] as outlined in Table VII is valid to only six digits. There are many limits to the theoretical expression, including high order QED corrections and knowledge of the fundamental constants. However the largest uncertainty is in a term $\delta p^{(pol)}$ which arises from the spin dependent polarizability of the proton.

It turns out that the information needed to calculate $\delta p^{(pol)}$ is exactly what we have measured at SLAC.[45] The relation between $\delta p^{(pol)}$ and G_1 and G_2 is given in Fig. 20. This formula together with positivity bounds has set a limit on δp of <4 ppm.[34,35] The inclusion of our data should yield a useful value. It is striking that studying the scattering of 10 GeV electrons sheds light on an atomic energy level shift on the order of 10^{-20} GeV!

Research supported in part by the Department of Energy under Contract No. EY-76-C-02-3075.

REFERENCES

1. M. S. Lubell, in Atomic Physics 5, edited by R. Marrus, M. Prior and H. Shugart (Plenum, New York, 1977), p. 325.
2. V. W. Hughes, et al., Phys. Rev. A 5, 195 (1972).
3. M. Campagna, et al., Advances in Electronics and Electron Physics 41, 113 (1976).

4. M. J. Alguard et al., "A Source of Highly Polarized Electrons at the Stanford Linear Accelerator Center," SLAC-PUB-2244. To be published in Nuclear Instruments and Methods.
5. C. K. Sinclair et al., in *High Energy Physics with Polarized Beams and Targets*, Proceedings of the Argonne Symposiu, ed. by M. L. Marshak (AIP, New York, 1971), p. 424.
6. R. C. Miller, et al., Inst. Phys. Conf. Ser. No. 43, p. 1043 (1979).
7. P. S. Cooper, et al., Phys. Rev. Lett. $\underline{34}$, 1589 (1975).
8. M. J. Alguard et al., Phys. Rev. A $\underline{16}$, 209 (1977).
9. W. W. Ash, in *High Energy Physics with Polarized Beams and Targets*, Proceedings of the Argonne Symposium, ed. by M. L. Marshak (AIP, New York, 1976), p. 485.
10. M. Borghini, et al., Nucl. Instrum. Meth. $\underline{84}$, 168 (1970).
11. P. S. Cooper et al., Bull. Am. Phys. Soc. $\underline{24}$, 72 (1979); P. S. Cooper, et al., to be published in Phys. Rev. Lett.
12. R. S. Van Dyck, et al., Phys. Rev. Lett. $\underline{38}$, 310 (1977).
13. J. C. Wesley and A. Rich, Phys. Rev. A $\underline{4}$, 1341 (1971).
14. D. Newman et al., Phys. Rev. Lett. $\underline{40}$, 1355 (1978).
15. J. R. Chen et al., Phys. Rev. Lett. $\underline{21}$, 1279 (1968); S. Rock et al., Phys. Rev. Lett. $\underline{24}$, 748 (1970).
16. S. Weinberg, Phys. Rev. Lett. $\underline{19}$, 1264 (1967), A. Salam in "Elementary Particle Theory; Relativistic Groups and Analyticity," Nobel Symp. No. 8, ed. N. Svartholm (Almquist and Wiksell, Stockholm, 1968), p. 367.
17. Y. B. Zel'dovich, Sov. Phys. JETP $\underline{36}$, 682 (1959); E. Derman, Phys. Rev. D $\underline{7}$, 2755 (1973); S. M. Berman and J. R. Primack, Phys. Rev. D $\underline{9}$, 2171 (1974); D $\underline{10}$, 3895 (1974); W. W. Wilson, Phys. Rev. D $\underline{10}$, 218 (1974).
18. M. J. Alguard, et al., Phys. Rev. Lett. $\underline{37}$, 1261 (1976); M. J. Alguard, et al., Phys. Rev. Lett. $\underline{41}$, 70 (1978).
19. W. Atwood, et al., Phys. Rev. D $\underline{18}$, 2223 (1978).
20. Y. B. Bushnin, et al., Soviet J. Nucl. Phys. $\underline{24}$, 279 (1976).

21. C. Y. Prescott, et al., Phys. Lett. 77B, 347 (1978).
22. Z. D. Farkas, et al., SLAC-PUB-1823 (1976).
23. A. DeRujula, et al., Rev. Mod. Phys. 46, 391 (1974).
24. C. Baltay, Proceedings of the 19th Intl. Conf. on High Energy Physics, Tokyo, 1978, p.
25. M. A. B. Beg and G. Feinberg, Phys. Rev. Lett. 33, 606 (1974).
26. J. D. Bjorken, Phys. Rev. D 18, 3239 (1978).
27. G. Feinberg, Phys. Rev. D 12, 3575 (1975).
28. E. Reya and K. Schilcher, Phys. Rev. D 10, 952 (1974); D. Cuthiell and J. N. Ng. Phys. Rev. D 16, 3225 (1977); E. Hoffmann and E. Reya, Phys. Rev. D 18, 3230 (1978); Frederick J. Gilman and Thomas Tsao, Phys. Rev. D 19, 790 (1979).
29. J. M. Potter, et al., Phys. Rev. Lett. 33, 1307 (1974); J. M. Potter, et al., in High Energy Physics with Polarized Beams and Targets, Proceedings of the Argonne Symposium, edited by M. L. Marshak (AIP, New York, 1976), p. 266.
30. P. A. Souder, et al., "Search for Parity Violation in the Elastic Scattering of Polarized Electrons from Nuclei," Bates Proposal 77-14 (1977); E. W. Otten, et al., "Proposal for Searching for Parity Violating Neutral Currents in Elastic e-d Scattering," 1976 (Univ. of Mainz).
31. See for example P.G.H. Sandars, in Proceedings of the 1977 International Symposium on Lepton and Photon Interactions at High Energies, edited by F. Gutbrod (Deutsches Elektronen Synchrotron DESY, Germany, 1977), p. 599; or G. Feinberg, Unification of Elementary Forces and Gauge Theories, edited by David B. Cline and Frederick E. Mills (Harwood Academic Publishers Ltd., London 1977), p. 117.
32. E. Gabathuler, et al., "Proposed Experiments and Equipment for Program of Muon Physics at the SPS," by European Muon Collaborators, CERN, July 1974.
33. M. J. Alguard, et al., Phys. Rev. Lett. 37, 1258 (1976); M. Bergstrom, et al., Bull. Am. Phys. Soc. 23, 529 (1978).

34. J. D. Bjorken, Phys. Rev. D $\underline{1}$, 1376 (1970).
35. R. L. Heimann, Nucl. Phys. $\underline{B64}$, 429 (1973).
36. J. Kuti and V. W. Weisskopf, Phys. Rev. D $\underline{4}$, 3418 (1971).
37. F. E. Close, Nucl. Phys. B $\underline{80}$, 269 (1974).
38. G. W. Look and E. Fischbach, Phys. Rev. D $\underline{16}$, 211 (1977); L. M. Sehgal, Phys. Rev. D $\underline{10}$, 1663 (1974).
39. R. Carlitz and J. Kaur, Phys. Rev. Lett. $\underline{38}$, 673, 1102 (E); J. Kaur, Nucl. Phys. B$\underline{128}$, 219 (1977).
40. R. J. Hughes, Phys. Rev. D $\underline{16}$, 622 (1977); R. L. Jaffe, Phys. Rev. D $\underline{11}$, 1953 (1975).
41. J. Schwinger, Nucl. Phys. B$\underline{123}$, 223 (1977).
42. G. R. Farrar, et al., Phys. Rev. Lett. $\underline{35}$, 1416 (1975).
43. E. R. Cohen and B. N. Taylor, J. Phys. Chem. Ref. Data $\underline{2}$, 663 (1973).
44. S. J. Brodsky and S. D. Drell, Ann. Rev. Nucl. Sci. $\underline{20}$, 147(1970) B. E. Lautrup, et al., Phys. Rep. Phys. Lett. $\underline{3C}$, 193 (1972).
45. C. K. Iddings, Phys. Rev. $\underline{138}$, 446 (1965); E. de Rafael, Phys. Lett. $\underline{37B}$, 201 (1971).

VERY RECENT SPIN RESULTS AT LARGE $P_\perp^2$ *

A. D. Krisch

Randall Laboratory, University of Michigan

Ann Arbor, Michigan 48109

I would like first to give you a general impression of the nature of proton-proton elastic scattering at large $P_\perp^2$ by showing Fig. 1. This is a plot of spin-averaged proton-proton elastic scattering cross-sections measured with an unpolarized beam and an unpolarized target. I have plotted the spin-averaged cross-section $\langle d\sigma/dt \rangle$ against my favorite[1,2] scaled variable, $\rho_\perp^2$. This $\rho_\perp^2$ variable removes most of the energy dependence or "shrinkage" that occurs when elastic cross-sections are plotted against $-t$ or $P_\perp^2$. I have plotted here essentially all existing p-p data from 3 GeV/c to 2000 GeV/c.

The most striking feature of this plot is that it contains several quite distinct regions or components. At small $\rho_\perp^2$, or forward angles, is the best known region, which is usually called the "diffraction peak". Its slope of $10(\text{GeV/c})^{-2}$ corresponds to diffraction scattering from the outside of the proton which has a size of about 1 fermi. This is followed by an "intermediate-$\rho_\perp^2$ region" which is important at low energy but disappears by 100 GeV/c.

---

*Supported by a Research Grant for the U.S. Department of Energy.

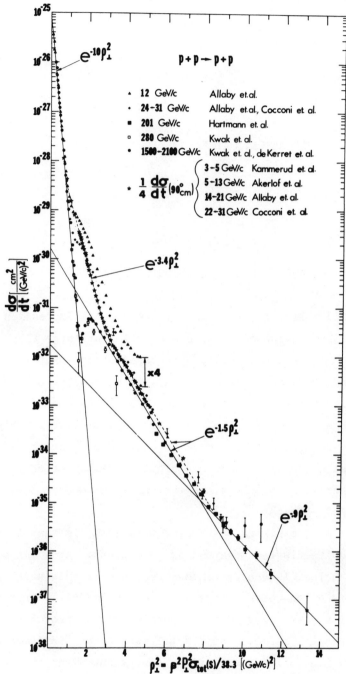

Figure 1. The differential elastic proton-proton scattering cross-section is plotted against the scaled $P_\perp^2$ variable, $\rho_\perp^2$.

Next comes the $e^{-1.5\rho_\perp^2}$ component at large $\rho_\perp^2$. This is the "hard-scattering component" which may be caused by the direct scattering of the proton's constituents. There is also a hint of an even harder-scattering component at still larger $\rho_\perp^2$ but its existance is not yet established.[3]

The other remarkable feature of this graph is the nature of the energy independence of p-p elastic scattering when it is plotted against this $\rho_\perp^2$ variable, which comes from the Lorentz-contracted geometric model:[1,2]

$$\rho_\perp^2 = \beta^2 P_\perp^2 \, \sigma_{tot}(s)/38.3 \quad . \tag{1}$$

In the diffraction peak and in the hard scattering region $d\sigma/dt$ is independent of incident energy. However, at intermediate $\rho_\perp^2$, there is an enormous amount of structure. There is a strong $e^{-3\rho_\perp^2}$ component at 10 GeV/c which starts to disappear at about 20 to 30 GeV/c. By 100 GeV/c it has disappeared and by 1500 GeV/c it has dropped well below the dip which now appears near $\rho_\perp^2 = 1.5(\text{GeV}/c)^2$. Thus there are 3 distinct regions, each with a characteristic behavior. The intermediate $\rho_\perp^2$ region has a very strong energy dependence (~1/s) and is probably direct (non-diffractive) elastic scattering possibly due to some exchange mechanism. The small angle diffraction peak is totally independent of energy, which has been known for some years. The recent surprise is that the large-$P_\perp^2$ hard scattering region is also independent of energy when plotted against $\rho_\perp^2$. This suggest that the "hard-scattering" is also diffraction scattering from some objects. The slope of $e^{-1.5\rho_\perp^2}$ indicates that these objects must have a size of about 1/3 fermi. I like to call these objects inside the proton "constituents". Other people give them more specific names such as "quarks" or "partons". On occasion I have talked about "cores" and "onions". I prefer the general name constituent to emphasize that we really don't know much about these objects. However, I certainly believe that this p-p elastic data and other experiments such as deep inelastic e-p scattering do show that the proton contains constituents. The session this afternoon will concentrate on direct experimental evidence for the nature of these constituents from scattering experiments.

I will now turn to the spin dependence of proton-proton scattering. I will begin by defining a pure-4-spin cross-section, which is a cross-section where both initial spins (beam and target protons) and both final spins (scattered and recoil protons) are measured. We will consider cross-sections when the spins are measured transverse to the horizontal scattering plane; such cross-sections are called "transversity" cross-sections. There are $2^4 = $ such pure-4-spin cross-sections, but the conservation laws reduce the number of independent transversity cross-sections to five:

$$d\sigma/dt \ (\uparrow\uparrow \to \uparrow\uparrow) \qquad \text{up non-flip}$$
$$d\sigma/dt \ (\downarrow\downarrow \to \downarrow\downarrow) \qquad \text{down non-flip}$$
$$d\sigma/dt \ (\uparrow\uparrow \to \downarrow\downarrow) \qquad \text{parallel double-flip} \qquad (2)$$
$$d\sigma/dt \ (\uparrow\downarrow \to \uparrow\downarrow) \qquad \text{antiparallel non-flip}$$
$$d\sigma/dt \ (\uparrow\downarrow \to \downarrow\uparrow) \qquad \text{antiparallel double-flip.}$$

The spin-average cross-section, $<d\sigma/dt>$, is a complex average of these 5 pure-spin cross-sections. This averaging may well obscure some of the most interesting and important aspects of strong interactions. Of course this will only occur if there is some spin dependence. If all cross-sections are totally independent of spin, then you gain nothing by knowing the spin; all cross secitons would then equal the spin-averaged cross-section. But if spin effects are large, then you will completely miss a very important aspect of strong interactions if only unpolarized beams and unpolarized targets are available. From a more formal point of view, the pure-4-spin cross-sections are the simplest and most precisely defined quantities which you might someday hope to calculate directly from a theory of strong interactions.

Before turning to our experimental data I would like to list my colleagues who participated in these experiments:

<u>University of Michigan</u>:

    K. Abe, D. G. Crabb, P. H. Hansen, A. J. Salthouse, B. Sandler, T. Shima, K. M. Terwilliger.

<u>Argonne</u>:

    E. A. Crosbie, L. G. Ratner, P. F. Schultz, G. H. Thomas.

<u>AUA</u>:

    J. R. O'Fallon

<u>Abadan</u>:

    A. Lin

<u>Coral Gables</u>:

    A. Perlmutter

I have already discussed our experimental apparatus several times[4] and I don't think it would be appropriate to give another detailed discussion of this equipment. Let me just remind you that to study spin-spin forces in proton-proton elastic scattering you need three major hardware items:

1. A Polarized Proton Beam
2. A Polarized Proton Target
3. A Spectrometer to clearly identify elastic events.

In Fig. 2 one can see the layout of our experiment. The downstream double-arm spectrometer detects events when the polarized proton beam elastically scatters from the polarized proton target. We measure the number of elastic events in four different initial spin states, which gives us four pure-initial-spin cross-sections

$$d\sigma/dt(\uparrow\uparrow) , \quad d\sigma/dt(\downarrow\downarrow) , \quad d\sigma/dt(\uparrow\downarrow) \text{ and } d\sigma/dt(\downarrow\uparrow) . \quad (3)$$

We have no knowledge of the final state spins and thus each pure-

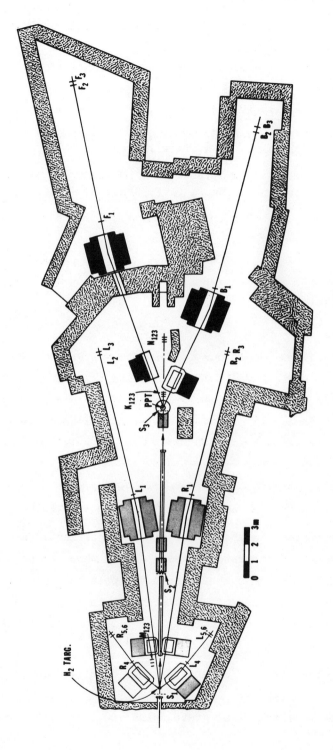

Figure 2. Layout of the experiment. The polarized beam passes through the liquid H_2 target and its polarization is measured by the L-R scattering asymmetry in the polarimeter. The beam then scatters in the polarized proton target (PPT) and the elastic events are detected by the F and B scintillation counters. The M, N, and K counters monitor the beam intensity, while the S_1, S_2, and S_3 chambers monitor the beam position.

VERY RECENT SPIN RESULTS AT LARGE $P_\perp^2$

initial-spin cross-section is a sum of two pure-4-spin cross sections. For example:

$$d\sigma/dt(\uparrow\uparrow) = d\sigma/dt(\uparrow\uparrow \to \uparrow\uparrow) + d\sigma/dt(\uparrow\uparrow \to \downarrow\downarrow) \quad . \tag{4}$$

In this talk I will just discuss the spin-parallel and spin-anti-parallel cross-sections which are defined by:

$$d\sigma/dt(\text{parallel}) \equiv \tfrac{1}{2}\left[d\sigma/dt(\uparrow\uparrow) + d\sigma/dt(\downarrow\downarrow)\right] ,$$

$$d\sigma/dt(\text{anti-parallel}) \equiv \tfrac{1}{2}\left[d\sigma/dt(\uparrow\downarrow) + d\sigma/dt(\downarrow\uparrow)\right] . \tag{5}$$

Our recent data is shown in Fig. 3 where we have plotted the ratio $d\sigma/dt$ (parallel):$d\sigma/dt$ (anti-parallel) against $P_\perp^2$. [Note that parallel is often denoted ($\uparrow\uparrow$) and anti-parallel ($\uparrow\downarrow$)]. It is quite clear that something very significant is happening at large $P_\perp^2$. Last year at the time of this meeting the ratio had reached 2 at $P_\perp^2 = 4.2(\text{GeV}/c)^2$. It has now[5] reached 4 at $P_\perp^2 = 5.09(\text{GeV}/c)^2$. This ratio may keep going up or it may level off at 4. These two possibilities are indicated on the graph by the solid and dashed lines. Our present errors are too large to distinguish between these two fascinating possibilities. We are already at $90°_{cm}$ which gives the maximum $P_\perp^2$ available at each incident energy, so we can go no further at 11.75 GeV/c. However, we can go to larger $P_\perp^2$ by increasing the ZGS energy to 12.75 GeV/c; we hope to do this in the spring. We later hope to go to much larger $P_\perp^2$ by using the proposed AGS polarized proton beam; but that will not be available for at least two years.[6]

I think that one can get a better understanding of the physics behind these huge spin effects by considering Fig. 4. This is a graph of the pure-initial-spin cross-sections themselves plotted against $\rho_\perp^2$. Notice that the spin-parallel cross-section $d\sigma/dt(\uparrow\uparrow)$ has a break at about $\rho_\perp^2 = 3(\text{GeV}/c)^2$ and, after the break, has the

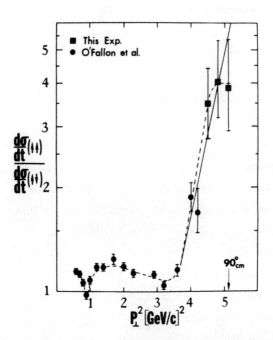

Figure 3. Plot of the ratio of the differential elastic cross sections in pure initial-spin states for p+p→p+p at P_{lab}=11.75 GeV/c. The spin-parallel cross section increases dramatically relative to the spin-antiparallel cross section at the onset of the hard-scattering component at $P_\perp^2 = 3.6$ (GeV/c)^2. The curves are only to guide the eye.

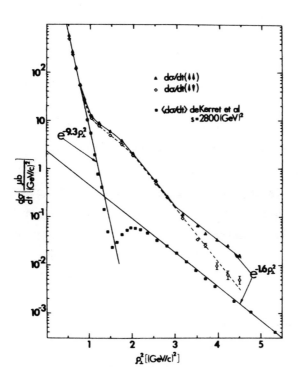

Figure 4. The 11.75 GeV/c spin-dependent proton-proton elastic cross sections are plotted against $\rho_\perp^2 = \beta^2 P_\perp^2 \sigma_{tot}(s)/38.3$, the $P_\perp^2$ scaled according to the Lorentz-contracted geometric model (Ref. 1). $(d\sigma/dt)_{\uparrow\uparrow}$ denotes scattering with the incident protons' spins parallel while $(d\sigma/dt)_{\uparrow\downarrow}$ denotes scattering with the spin antiparallel. Both spins are measured transverse to the scattering plane. Spin-averaged data from Ref. 3 are plotted for comparison, showing that the slope of the hard-scattering component is the same at 11.75 GeV/c and s=2800 GeV^2 when plotted against $\rho_\perp^2$.

same slope as the 1500 GeV ISR data which is shown for comparison. The anti-parallel cross-section, $d\sigma/dt(\uparrow\downarrow)$, appears to have no break at all. If you make the very reasonable assumption that the $e^{-1.5 p_\perp^2}$ "hard-scattering" component after the break is caused by direct scattering of the protons' constituents, then it appears that these constituents rarely scatter unless the protons' spins are parallel. This is an unexpected and possibly very important new fact about the nature of the protons' constituents.

It is quite surprising that this spin effect is so large. If you consider a spin-$\frac{1}{2}$ polarized proton to be a "polarized target" containing 3 spin-$\frac{1}{2}$ quarks, then the polarization of these quarks is only 33%. How then can you get a ratio of 4 in the spin cross-sections? It is impossible if you consider the quark-quark scatterings to be identical incoherent events. You can only get such a large effect by assuming that either the direct quark scatterings are strongly coherent or that the large-$P_\perp^2$ events are totally dominated by scatterings of one of the 3 quarks, which also contains <u>all</u> of the proton's polarization. Both these conjectures seem a little strained to me, but I guess they are possible.

In any case these spin effects are certainly unexpectedly large at high-$P_\perp^2$ where direct constituent scattering is dominant. I think these spin effects are giving us a new source of fundamental information about the nature of the proton's constituents. I also think that there may be some more surprises waiting for us.

In fact we just got quite a surprise in December when we, for the first time, measured spin-spin forces in neutron-proton elastic scattering at high energy. We used the 6 GeV/c polarized neutrons from the world's first 12 GeV/c polarized deuteron beam, which was produced by Everette Parker and the ZGS staff this fall. We scattered these polarized neutrons from our polarized proton target and measured n-p elastic scattering using a double arm spectrometer similar to the one shown in Fig. 2. The only significant change is that the F counters are replaced by a neutron counter which contains alternating layers of scintillator and brass to convert the uncharged scattered neutrons, and allow detection.

The Fermi momentum of the neutron in the deuterium smears the kinematics and makes it impossible to have as tightly constrained a spectrometer as in our normal proton experiments. Moreover the neutron detector is only about 50% efficient. The loose constraints increase the background event rate from the non-hydrogen "junk" atoms in the polarized target to about 25%. Thus the experiment is considerably more difficult that a polarized proton-proton experiment and going out to large $P_\perp^2$ was not possible. Nevertheless we carefully measured and subtracted the background by taking runs with the PPT beads replaced by hydrogen-free Teflon beads; and we calibrated the Fermi smear by simultaneously measuring p-p elastic scattering from the deuteron beam. Thus we obtained a reasonably precise measurement of spin-spin effects in neutron-proton elastic scattering at 6 GeV/c and $P_\perp^2$ = 0.8 and 1.0 (GeV/c)^2.

Our data is shown in Fig. 5 where we have plotted the spin-spin correlation parameter, A_{nn}, against $P_\perp^2$ for 6 GeV/c n-p elastic scattering; we have also shown p-p elastic data for comparison. Notice that in this small-to-medium $P_\perp^2$ region (which is all that is available at 6 GeV/c) the p-p spin-spin effects are fairly small compared to the large-$P_\perp^2$ effects, which we saw at 12 GeV/c. There is some very interesting detailed structure, but the overall size of the effects are fairly small. Notice that the spin-parallel cross-section is always larger than the anti-parallel cross-section.

The neutron-proton spin-spin effects seem to be totally different. They are quite large even at the small-$P_\perp^2$ values that we were able to reach. In fact the n-p spin-spin effects are twice as large as the p-p spin-spin effects at this $P_\perp^2$. Moreover the n-p effects have the <u>opposite</u> sign; the spin-antiparallel scattering is much more probable than spin-parallel scattering.

These results are totally unexpected. I do no understand what they signify. Why do two protons always prefer to scatter when

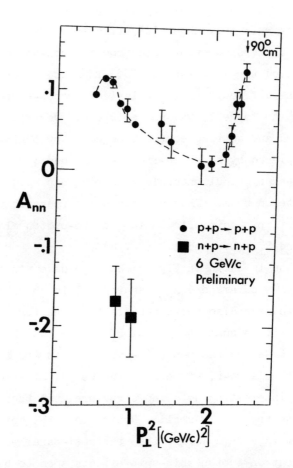

Figure 5. Plot of the spin-spin correlation parameter, A_{nn}, for pure initial-spin state nucleon-nucleon elastic at 6 GeV/c. The proton-proton and neutron-proton data are quite different.

their spins are parallel, while neutron-proton scatterings prefer antiparallel spins? Why are the n-p spin effects so large at such a small $P_\perp^2$? Spin effects in high energy physics seem to give us one surprise after another. Perhaps this is because our present understanding of high energy physics has some major flaw. Perhaps these spin experiments, which I have come to regard with affection, are trying to be helpful and point out this flaw; but we are all too fixed in our ways to see it.

REFERENCES

1. A. D. Krisch, Phys. Rev. Letters 11, 217 (1963); Phys. Rev. 135B (1964); Phys. Rev. Letters 19, 1149 (1967); Phys. Letters 44B, 71 (1973).
2. P. H. Hansen and A. D. Krisch, Phys. Rev. D15, 3287 (1977).
3. H. DeKerret et. al., Phys. Letters 62B, 363 (1976); 68B, 374 (1977). J. L. Hartmann et. al., Phys. Rev. Letters 39, 975 (1977).
4. A. D. Krisch, in New Frontiers in High Energy Physics, eds. B. Kursunoglu et. al., 17 (1978) Plenum.
5. D. G. Crabb et. al., Phys. Rev. Letters 41, 1257 (1978).
6. Preliminary Design Study for Acceleration of Polarized Protons in the Brookhaven AGS, B. Cork et. al., (unpublished report).

HADROPRODUCTION OF MASSIVE LEPTON PAIRS AND QCD*

Edmond L. Berger

Stanford University, Stanford, CA 94305 and

Argonne National Laboratory, Argonne, IL 60439**

ABSTRACT

A survey is presented of some current issues of interest in attempts to describe the production of massive lepton pairs in hadronic collisions at high energies. I concentrate on the interpretation of data in terms of the parton model and on predictions derived from quantum-chromodynamics (QCD), their reliability and their confrontation with experiment. Among topics treated are the connection with deep-inelastic lepton scattering, universality of structure functions, and the behavior of cross-sections as a function of transverse momentum.

I. INTRODUCTION

Massive lepton pair production in hadronic collisions, $h_1 + h_2 \to \ell\bar{\ell}X$, commonly known as the Drell-Yan process,[1-3] measures the ability of interacting hadrons to reconfigure their momentum into local production of a virtual photon γ^* with four-momentum Q^μ. In

---
*Work supported by the Department of Energy under contract number EY-76-C-03-0515.
**Permanent address.

the parton model, the production is imagined to occur through the annihilation of a quark from one hadron with an antiquark from the other, in the subprocess $\bar{q}q \to \gamma^* \to \ell\bar{\ell}$; $\ell = \mu, e, \ldots$ The Drell-Yan process can thus be exploited to determine the quark and antiquark structure functions of hadrons in the timelike region, $Q^2 > 0$, and to probe other important aspects of the dynamics of hadronic constituents at short distance.

Not long ago it was an open question whether a high-mass lepton pair continuum would be observed experimentally with the general qualitative features expected by the Drell-Yan model. Excellent data extending to muon-pair masses of ~15 GeV in pN collisions,[4-8] and to ~11 GeV in π^-N reactions[9-11] have removed most doubts. Although the quantitative consistency of the classical model with data as a function of all kinematic variables (target atomic number A; pair mass M; scaled longitudinal momentum $x_F = Q_L/Q_L^{max}$; pair transverse momentum $\vec{Q}_T$; and decay angles θ^*, ϕ^* in the pair restframe) has not been demonstrated in precise detail, the general picture is clear. The focus has changed to other questions.

From the perspective of quantum chromodynamics (QCD), issues of present concern include (a) whether the Drell-Yan prescription is the full answer, and (b) if so, whether constituent structure functions determined for $Q^2 > 0$ via the Drell-Yan mechanism should be, and/or are experimentally observed to be, identical to those extracted from analyses of deep-inelastic lepton scattering $\ell N \to \bar{\ell} X$ in the region $Q^2 < 0$. Here $\ell = \mu, e,$ or ν. This general topic and scaling violations are treated in Sections II and III. Comparisons with data are presented in Section IV. The transverse momentum $\vec{Q}_T$ behavior of the lepton pair production cross-section $d\sigma/d^2\vec{Q}_T \, dQ^2 \, dx_F$ is another subject of considerable interest. In QCD the dynamics at high $\vec{Q}_T$ are governed by elementary hard scattering processes analogous to those studied in large $\vec{P}_T$ purely hadronic reactions,[12] such as $h_1 h_2 \to \pi X$. In $h_1 h_2 \to \gamma^* X$, the whole jet (γ^*) is observed experimentally and, therefore, difficulties associated with trigger bias and jet decay are eliminated. This technical advantage of $h_1 h_2 \to \gamma^* X$ has not yet been exploited fully, in part because present data extend only to $|\vec{Q}_T| \lesssim 4$ GeV/c. In the small $\vec{Q}_T$ region, the $\vec{Q}_T$ dependence is influenced by soft-gluon emission processes, by non-perturbative and by bound state effects. While complicating the comparisons of data with calculations of first-order perturbative QCD processes, the physics of these soft-gluon, bound state and non-perturbative effects is of interest in its own right. I discuss transverse momentum distributions and their moments in Section V.

A general conclusion regarding QCD which emerges from comparisons with data is one of rough consistency. There are no glaring failures in regions of phase space where calculations should apply. However, the theory is not being "tested". Important calculations have yet to be made of the size and kinematic variation of non-leading, higher order corrections ($\propto 1/\log Q^2$, $1/Q^2$, ...) before one can be satisfied with the general agreement of theory with the magnitude and with the x_F and M variations of $d^2\sigma/dM\,dx_F$. For Q_T spectra, these objections are compounded by further questions concerning the applicability of perturbative QCD in the small $\vec{Q}_T$ region ($\vec{Q}_T^2 \ll Q^2$).

In this review, I attempt to survey the present situation in the phenomenology of massive lepton pair production, enumerating some unresolved issues and points of contention.

II. THE CLASSICAL DRELL-YAN MODEL AND ITS PROBLEMS

As a point of departure, I begin with the classical Drell-Yan approach.[1] The physical process is sketched in Fig. 1. We imagine that there is a known probability $q_i(x_a, \vec{k}_{Ta})$ for finding a quark of flavor i among the constituents of incident hadron a; this quark carries the fraction x_a ($x_a > 0$) of the longitudinal momentum of hadron a, as well as transverse momentum $\vec{k}_{Ta}$. Likewise, $\bar{q}_j(x_a, \vec{k}_{Tb})$ is the probability for finding an antiquark of flavor j. Later, these functions will be generalized to include explicit Q^2 dependence. In the classical parton model they do not depend on Q^2. In the par-

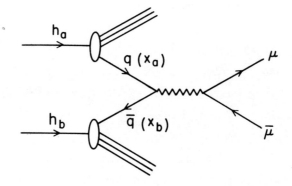

Figure 1. Basic Drell-Yan quark-antiquark annihilation mechanism for lepton pair production in hadronic collisions, illustrated here for $h_a h_b \to \mu\bar{\mu}X$; q and $\bar{q}$ denote respectively a quark and an antiquark constituent.

ton model, the distributions in $\vec{k}_{Ta}$ and $\vec{k}_{Tb}$ are also understood to be sharply limited to small energy independent values, with $\langle k_T \rangle \simeq$ 300 MeV. The classical Drell-Yan cross-section for $h_a h_b \to \ell^+\ell^- X$ is then

$$\frac{Q^2 \, d\sigma^{ab}}{2 \, dx_L \, d^2\vec{Q}_T \, d\cos\theta^*} = \frac{\pi\alpha^2}{2} \int d^2\vec{k}_{Ta} \, dx_a \, d^2\vec{k}_{Tb} \, dx_b \tag{1}$$

$$\frac{1}{3} \sum_i e_i^2 [q_i(x_a, \vec{k}_{Ta}) \, \bar{q}_i(x_b, \vec{k}_{Tb}) + \bar{q}_i(x_a, \vec{k}_{Ta}) \, q_i(x_b, \vec{k}_{Tb})]$$

$$(1+\cos^2\theta^*) \, \delta^{(2)}(\vec{Q}_T - \vec{k}_{Ta} - \vec{k}_{Tb}) \, \delta(x_L - x_a - x_b) \, \delta(Q^2 - x_a x_b s) \; .$$

In the $\ell^+\ell^-$ rest-frame, θ^* is the polar angle of a lepton ℓ with respect to the axis defined by the collinear $q\bar{q}$ system; e_i is the fractional quark charge. Displayed explicitly in the denominator of Eq. (1) is a factor 3 representing the three colors of quarks; the distributions q and $\bar{q}$ are summed over the color index.

The expression on the right hand side of Eq. (1) is the leading contribution in the $Q^2 \to \infty$ limit. Terms have been dropped which are proportional to Q^{-2} relative to the result shown. In some cases, these Q^{-2} terms may be shown to have a different dependence on, e.g., x_F and $\cos\theta^*$. For example, at large x_F in $\pi p \to \mu\bar{\mu}X$, $d\sigma/d\cos\theta^*$ is predicted[13] to vary as $\sin^2\theta^*$, instead of the oft-quoted $(1 + \cos^2\theta^*)$. I restrict my attention in this paper to values of the lepton pair mass $M > 4$ GeV, in order to exclude the J/ψ resonance region, and to ensure that conditions are satisfied for valid application of the impulse approximation. The Υ region $9.0 < M < 10.5$ GeV should also be excluded.

There are at least two phenomena which require our going beyond the basic model. These are the observed relatively large values[4-9] of $\langle Q_T \rangle$ in $hh \to \mu\bar{\mu}X$, and the scaling violations in $q(x, Q^2)$ (i.e., explicit Q^2 dependence at fixed x) observed in deep-inelastic μ and ν scattering.[14-17] Within the context of QCD both phenomena are interpreted as manifestations of gluonic radiative processes.

1. Q_T effects

The basic amplitude A_μ represented by Fig. 1 is not manifestly gauge invariant ($A_\mu \cdot Q^\mu \neq 0$) if the quark and/or antiquark are off-shell.[1,18] For example, if we consider quark a to be off-shell, the amplitude will contain a factor $\not{q}_a \gamma_\mu u(q_b)$. Therefore, $A_\mu \cdot Q^\mu \propto q_a^2$ where q_a^μ is the four-vector momentum of quark a. Using energy-momentum conservation for $h_a \to q_a + X_a$, we may easily show that

$$q_a^2 = \frac{-k_{Ta}^2}{(1-x_a)} + \ldots \qquad (2)$$

The mass terms (...) ignored in Eq. (2) are inconsequential for the present discussion. If k_{Ta}^2 is very small ($<k_{Ta}> \simeq 300$ MeV), as is required in the classical parton model, and if x_a is not near 1, the q_a^2 is negligible with respect to Q^2, and the classical model ought to be applicable. However, it is observed experimentally[4] that $<Q_T^2> \simeq$ 1.8 GeV^2 for $5 < M_{\mu\mu} < 10$ GeV at $p_{lab} = 400$ GeV/c. Moreover, a comparison of FNAL and ISR data, as shown in Fig. 2, suggest that $<Q_T^2>$ increases with p_{lab}, perhaps as fast as $<Q_T^2> \propto s$. If we attempt to explain these effects entirely with intrinsic quark transverse momenta, we deduce that $<k_{Ta}^2> \simeq \frac{1}{2} <Q_T^2> \simeq 1$ GeV^2 at 400 GeV^2. Thus, $1 \lesssim |q_a^2| \lesssim 10$ GeV^2 for $0 \lesssim x_a \lesssim 0.9$. This is an unacceptably large violation of gauge invariance. While non-asymptotic kinematic effects should play some role, the increase of $<Q_T^2>$ with s at fixed Q^2/s also cannot be explained purely with intrinsic quark and antiquark transverse momenta. Because $<k_{Ta}^2>$ depends only on x_a, $<Q_T^2>$ should in turn be independent of s at fixed Q^2/s.

2. Scaling violations

Although there is some controversy regarding the detailed interpretation of the results,[19,20] high energy data on deep-inelastic mu and neutrino scattering[14-17] cannot be described with scaling struc-

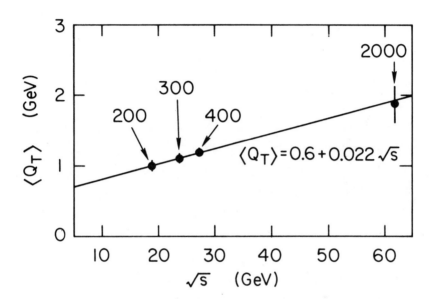

Figure 2. The energy dependence of $\langle Q_T \rangle$ for $pN \to \ell\bar{\ell}X$. Shown are Fermilab data at p_{lab} = 200, 300, and 400 GeV/c (Ref. 4) and an ISR point at $\sqrt{s}$ = 62 GeV (Refs. 6 and 8). All data are at rapidity $y \approx 0$. For the FNAL points $0.2 < M/\sqrt{s} < 0.5$, whereas for the ISR point $M/\sqrt{s} \approx 0.1$. The ISR point is an average of the values of 1.67±0.21 GeV for $6.0 < M < 8.75$ GeV quoted by the CCOR group (Ref. 6), and 2.1±0.6 GeV for $6 < M < 7$ GeV quoted by the ABCSY group (Ref. 7). The straight line fit shown is only one of many phenomenological forms which would interpolate between the FNAL and ISR data.

ture functions, $q(x)$ and $\bar{q}(x)$. It is "natural" to interpret the data instead in terms of non-scaling distributions $q(x,Q^2)$. In deep-inelastic scattering these functions are measured in the time-like $Q^2 < 0$ region. Strong theoretical guidance is necessary in order to relate these distributions obtained in the $Q^2 < 0$ deep-inelastic region to those which are appropriate in the time-like $Q^2 > 0$ region for hh → $\mu\bar{\mu}X$ in Eq. (1). Without such guidance, Eq. (1) has little predictive power, since it is unclear what meaning to attach to the functions $q(x)$ and $\bar{q}(x)$ if they are unrelated to those for deep inelastic scattering. QCD supplies a prescription for connecting the $Q^2 < 0$ and $Q^2 > 0$ regions.

In Section III, I describe QCD modifications to the classical Drell-Yan picture.

III. BEYOND THE CLASSICAL MODEL

The incident hadrons may be imagined to be beams of quarks, anti-quarks, and gluons. At a somewhat lower level of resolution, e.g., smaller Q^2, the beams also contain "bound" constituents, such as virtual mesons ($q\bar{q}$) and diquarks (qq). For the moment, I assume that the only initial constituents are the quarks, antiquarks, and gluons, and that these constituents are free, unbound. Later it will be essential to reexamine this assumption and to ask to what extent the neglect of bound state effects and substructure ($q\bar{q}$,qq,...) affects our conclusions.

There are many ways in which a massive γ^* may be produced from the interaction of the initial constituents. In QCD, or in any field theory of interacting constituents, we may draw an entire series of diagrams, ordered according to the strong coupling constant α_s. Some of these are illustrated in Fig. 3, beginning with the basic zero order diagram $q\bar{q} \rightarrow \gamma^*$. In first order in α_s, there is an _initial_ gluon set ("Compton scattering") Figs. 3(b) and (c), for which the subprocess $qG \rightarrow \gamma^*q$ is driven by the gluon distribution in one of the initial hadrons, as well as the _final_ gluon set ("two body anni-

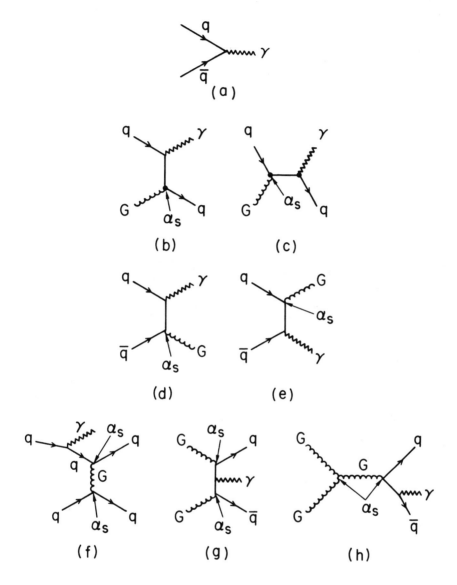

Figure 3. Series of diagrams illustrating the interactions of free quark, antiquark, and gluon constituents: (a) the basic zero order Drell-Yan process $q\bar{q} \to \gamma^*$; (b) and (c) the first order $O(\alpha_s)$ Compton processes $qG \to \gamma^* q$; (d) and (e) the first order two body annihilation process $q\bar{q} \to \gamma^* G$; (f), (g) and (h), a sample of second order, $O(\alpha_s^2)$, processes. Not drawn are many other diagrams in $O(\alpha_s^2)$ related by gauge invariance requirements to those shown.

hilation") Figs. 3(d) and (e) which a gluon is radiated into the final state $q\bar{q} \to G\gamma^*$. The initial constituents q, $\bar{q}$, or G in Fig. 3 are all intended to be on-shell, although for formal reasons one often assigns each a small off-shell value of $p^2 < 0$. Additional gluons may be radiated from the initial or final constituent lines in Fig. 3, leading to graphs of higher order in α_s.

The series of graphs in Fig. 3 suggests that the cross-section for $h_a h_a \to \gamma^* X$ receives contributions from four types of terms, expressed symbolically as

$$\sigma_{ab \to \gamma^* X} = \int q_a(x_a) \bar{q}_b(x_b) \sigma_{q\bar{q} \to \gamma^* X_1}$$

$$+ \alpha_s \int [q_a(x_a) G_b(x_b) + \bar{q}_a(x_a) G_b(x_b)] \sigma_{qG \to \gamma^* X_2} \quad (3)$$

$$+ \alpha_s^2 \int q_a(x_a) q_b(x_b) \sigma_{qq \to \gamma^* X_3} \quad (3)$$

$$+ \alpha_s^2 \int G_a(x_a) G_b(x_b) \sigma_{GG \to \gamma^* X_4}$$

$$+ (a \leftrightarrow b)$$

Here the distributions $q(x)$, $\bar{q}(x)$, and $G(x)$ describe the quark, antiquark, and gluon densities in the initial states. In Eq. (3), the symbols X_i represent states of various multiplicities of quarks antiquarks, and gluons.

Graphs which are of high order in α_s cannot be dismissed. The contribution to the integrated cross-section $d\sigma/dQ^2$ may be shown to be proportional to $(\alpha_s \log Q^2)^n$. Even if $\alpha_s \propto 1/\log Q^2$, all terms in the series are of comparable magnitude. In nucleon-nucleon collisions, it may be imagined that the $O(\alpha_s^2)$ term with $qq \to qq\gamma^*$, Fig. 3(f), would be more important than the simple annihilation term $q\bar{q} \to \gamma^*$ which feeds on the small sea in the nucleon. Fortunately, techniques have been devised to "sum" the series of terms in α_s, at at least in the leading logarithmic approximation to each order.[21-23] The procedure parallels that employed in deep inelastic processes,[20] although with less rigor. The general conclusion is that in QCD, the Drell-Yan cross-section, Eq. (1), <u>integrated over $\vec{Q}_T$</u>, should still be valid, except for the replacement of the scaling $q(x)$ and $\bar{q}(x)$ structure functions with scale-noninvariant forms $q^{DY}(x,Q^2)$ and $\bar{q}^{DY}(x,Q^2)$. This replacement takes into account the leading logarithmic contributions of all terms in the series in α_s. Moreover, analyses show that the functions $q^{DY}(x,Q^2)$ and $\bar{q}^{DY}(x,Q^2)$ should be identical to the functions $q^{DIS}(x,Q^2)$ and $\bar{q}^{DIS}(x,Q^2)$ measured in deep-inelastic electroproduction and in neutrino scattering.

The statements made above are true only to leading order in $\log Q^2$. Even when Q^2-dependent structure functions are used, one expects QCD corrections to the Drell-Yan prediction which are proportional to $1/\log Q^2$. In order α_s, these arise from the "constant" terms which remain after the dominant $\log Q^2$ divergent piece is removed and absorbed into the renormalized structure functions. It is important to calculate explicitly the expected size and kinematic variation (with Q^2, s, x_F, y) of these correction terms, since $1/\log Q^2$ is not small for values of Q^2 now accessible experimentally. As is well known, QCD fits to scale-violating effects in DIS depend sensitively on the controversial scale parameter Λ; differences in Λ are $1/\log Q^2$ effects. A few groups have undertaken an estimate of the $1/\log Q^2$ corrections in hh $\to$ μμX. Effects as large as 100% are found[24] for values of s and Q^2 typical of present experiments. The corrections are sizeable for both $\bar{p}N$ and pN processes, and are therefore not a peculiarity of the small sea distribution in the nucleon. The large size of the $1/\log Q^2$ corrections calls perturbation theory into question, for, if $1/\log Q^2$ terms are of order 100%, what of the neglected $(1/\log Q^2)^2$ terms? In Section IV.2, I show that present data are also consistent with large deviations from the leading order Drell-Yan expression.

The leading $(\log Q^2)^n$ terms in hh $\to$ μ$\bar{\mu}$X arise after an integration over $\vec{Q}_T$, and, thus, the analysis described here is relevant only for the cross-section integrated over $\vec{Q}_T$, i.e., for $d\sigma/dQ^2\, dx_F$. I will return to a discussion of $\vec{Q}_T$ effects in QCD in Section V.

In the paragraphs above, I discussed hadronic collisions as if they are really collisions of free, unbound, on-shell, colored elementary constituents. However, free quarks and gluons do not exist, and there may well be important physical effects in hh → $\mu\bar{\mu}$X associated with the fact that constituents are bound in color-singlet hadronic wavefunctions. Binding implies, amoung other things, that the constituents may be far off-shell in some kinematic regions.

Although the full bound state problem is one which is inherently nonperturbative, it has been argued[25] that QCD perturbation theory can be employed for calculations of the "hard-part" of bound-state effects, including, for example, the behavior of structure functions as x → 1. This leads to the possibility of going beyond the simple probabilistic, factorizing form of Eq. (1), and of looking for physical correlations between effects traditionally separated into the "structure function" and "elementary constituent cross-section" factors in Eqs. (1) and (3). While introducing new complexities, the bound state effects also open a broader range of possibilities for testing QCD.

One example of the above ideas is the recent calculation[13] of the pion structure function, based on the diagrams shown in Fig. 4. It was shown that there are two contributions to the structure function, and to the Drell-Yan cross-section for $\pi N \to \mu\bar{\mu}X$, one associated with transversely polarized γ^* and a second, dominant at large x_F, associated with longitudinally polarized γ^*. At large x, where the antiquark in the pion carries most of the momentum of the incident meson, the dominance of the longitudinal cross-section reflects the antiquark's "memory" of its origins in an integer-spin boson.

A different aspect of the bound-state problem is emphasized by Blankenbecler and collaborators.[18,26,27] They argue that at small Q^2 and Q_T^2, the important hadronic constituents are those associated

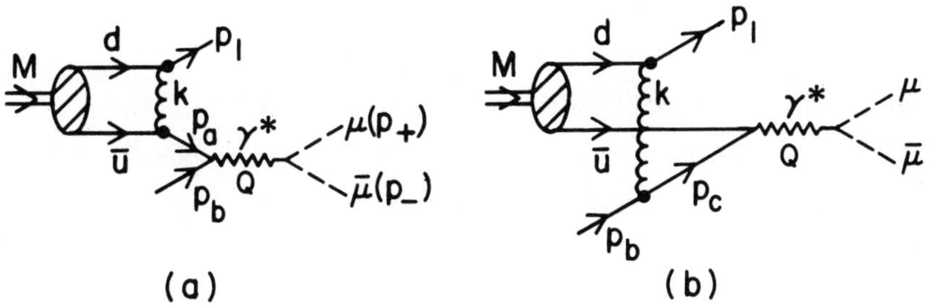

Figure 4. Diagrams for $Mq \to q\gamma^*$, $\gamma^* \to \mu\bar{\mu}$. Here M denotes a meson. Solid lines represent quarks. Symbols p_1, p_a, p_b and p_c denote four-momenta of quarks, and k is the four-momentum of the gluon. Both (a) and (b) are required by gauge invariance, although in a physical (axial) gauge, the scaling contributions as $Q^2 \to \infty$ can be identified solely with (a).

with the large distance structure of the hadron, namely, the light mesons. Thus, as Q^2 or Q_T^2 grows, corresponding to probing to smaller distances, the dominant constituent scattering processes should change gradually from meson-meson processes, to meson-quark and diquark-antiquark processes, and, finally, at the highest Q^2, to quark-gluon and quark-antiquark processes. At modest values of Q^2 and/or Q_T^2, one should explicitly include contributions to the cross-section $d\sigma/dQ^2 dQ_T^2$ associated with, e.g., the meson-quark subprocess.[18]

Finally, I should mention the possibility that non-trivial, non perturbative effects may invalidate the simple factorizing form of Eq. (1), and of all the higher order QCD corrections to it. These effects are illustrated by, but not limited to, the initial state and final-state interactions[28] sketched in Fig. 5. Instanton effects[29] have been suggested as a possibly large non-factorizing source of deviations from the Drell-Yan formula, but have not yet been shown to be numerically large enough to warrant concern.

I have done little more here than to advocate the desirability of going beyond elementary field calculations to include bound state effects properly. It is a problem of more than prosaic interest in many respects, for example, in attempts to describe the Q_T distribution (cf., Section V). In deep inelastic scattering as well as in massive lepton pair production, physical effects associated with the bound state origins of the initial constituents are no less interesting, nor less important in the presently explored interval of Q^2, than the much explored logarithmic effects due to gluonic radiative corrections.[20]

IV. COMPARISONS WITH EXPERIMENT

In the previous Section I surveyed some of the theoretical ideas of interest in connection with the cross section for $h_a h_b \to \mu\bar{\mu}X$. Here I shall <u>adopt the simplest view</u>, viz., that the cross-section integrated over Q_T is predicted exactly by the Drell-Yan formula, Eq. (1), with only one modification, which is the use of Q^2 depen-

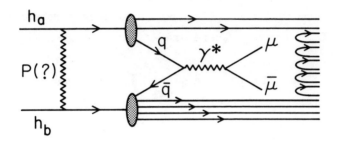

Figure 5. A diagram illustrating an initial state interaction (Pomeron exchange (?)), as well as the final-state evolution of the "spectator" quarks, from hadrons h_a and h_b, into observed color-singlet hadrons.

dent structure functions $q(x,|Q^2|)$ extracted from data on deep-inelastic lepton scattering (DIS). My aim is to compare the resulting "predictions" with data on hh → $\mu\bar{\mu}X$ to see to what extent this simple view is consistent with experiment. I focus on the absolute yield and on the s, M, and x_F variations of $d\sigma/dM\,dx_F$. There are other predictions, of course, but these have been reviewed adequately elsewhere.[2,3]

1. Valence-valence processes

(a) $\bar{p}p$. In many respects, the $\bar{p}p$ (or $\bar{p}N$) process is ideal. In the approach I have adopted the cross-section $\sigma \propto \bar{q}_{\bar{p}}(x_a,Q^2)$ $q_p(x_b,Q^2) = q_p(x_a,Q^2)\,q_p(x_b,Q^2)$ is specified entirely in terms of quark structure functions which can be extracted reasonably reliably from DIS. However, there are as yet no high-energy data on $\bar{p}N \to \mu\bar{\mu}X$.

(b) πN. The process $\pi^- N \to \mu\bar{\mu}X$ has recently been investigated in detail at high energies.[9-11] According to the model, the cross section is a sum of two pieces, $\sigma = q_\pi \bar{q}_N + \bar{q}_\pi q_N$. Since the antiquark content $\bar{q}_N$ of the nucleon is relatively small, the cross-section is represented to first approximation by the second term: $\bar{q}_\pi(x_\pi)\,q_N(x_N)$. More precisely,

$$\sigma \propto \bar{q}_\pi(x_\pi,Q^2)[\tfrac{4}{9} x_N u_N(x_N,Q^2) + \tfrac{1}{9} x_N \bar{d}_N(x_N,Q^2)] \tag{4}$$

If we adopt the values of $u_N(x_N,Q^2)$ and $\bar{d}_N(x_N,Q^2)$ determined in DIS experiments, Eq. (4) can be used to determine $\bar{q}_\pi(x,Q^2)$ from the data. However, there is even more content to Eq. (4). Recognizing that the logarithmic Q^2 variation predicted by QCD is relatively mild for $Q^2 \gtrsim 10$ GeV^2 (i.e., above the J/ψ peaks), we may begin by ignoring it in Eq. (4). Then, the two kinematic variables x_π and x_N in Eq. (4) may be reexpressed in terms of the two observables $x_F = x_\pi - x_N$ and $Q^2 = sx_\pi x_N$. The manner in which the observed non-zero value of Q_T^2 is handled leads to some uncertainty in the resolution of these equations for x_π and x_N in terms of Q^2 and x_F, but the results I shall describe are not affected in any meaningful way.

A check was made by the Chicago-Princeton group[9] that the dependence of their data on x_π and x_N agrees with the factorization property in Eq. (4); i.e., $\sigma \propto f_1(x_\pi) f_2(x_N)$. The agreement is a verification that one dominant production mechanism is at work, and that its effects are at least consistent with the factorization property of Eqs. (1) and (4). Next, the functional dependences of $\bar{q}_\pi(x_\pi)$ on x_π and of $q_N(x_N)$ on x_N were extracted from the data. A meaningful question is whether the x_N dependence of $q_N(x_N)$ from $\pi N \to \mu\bar{\mu}X$ is consistent with that seen in deep inelastic reactions. This has also been answered in the affirmative by the Chicago-Princeton collaboration, in the limited range $0.05 < x_N < 0.3$ accessible in their experiment, and for $<Q^2> \approx 25$ GeV^2. Only the shape can be checked, of course, not the absolute normalization of $q_N(x_N)$. Finally, adopting the absolute normalization of $q_N(x_N)$ from DIS data, the Chicago-Princeton group[9] used their data to extract an absolutely normalized function $x_\pi \bar{q}_\pi(x_\pi)$ for $0.3 < x_\pi < 0.95$. Their results are shown in Fig. 6; they have been fitted to the form

$$x\bar{q}_\pi(x) = A(1-x)^p \qquad (5)$$

with $p = 1.01 \pm 0.05$ and $A = 0.52 \pm 0.03$, for $<Q^2> \approx 25$ GeV^2.

The results of the Chicago-Princeton group are in good accord with the parton model, with color, but QCD cannot be said to play much of a role. The shape (x_N dependence) of $q_N(x_N)$ seems to be the same in DIS and in $\pi p \to \mu\bar{\mu}X$ for $0.05 < x_N < 0.3$ and $<Q^2> \approx 25$ GeV^2. Second, the integral $\int x\bar{q}_\pi(x)\,dx \approx 0.20$ shows that about 40% of the pion's momentum is carried by quarks and antiquarks, as seems "reasonable". If there were no color factor in Eq. (1), this fraction would be reduced to $40/3 \approx 14\%$, which seems clearly wrong. Thus, the color factor is supported. Finally, the x_π dependence

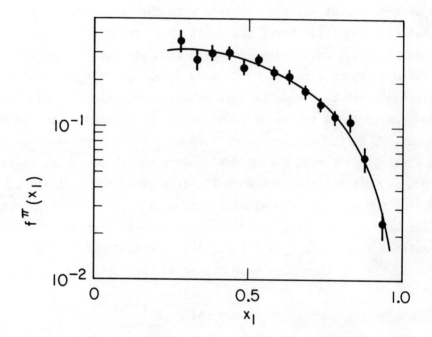

Figure 6. The spin averaged structure function $f^{\pi^-}(x) = x\bar{u}_\pi(x)$ obtained by the Chicago-Princeton collaboration, Ref. 9. The solid line is a fit to the form $a\sqrt{x}(1-x)^b$ with $a = 0.90 \pm 0.06$ and $b = 1.27 \pm 0.06$.

of $xq_\pi(x_\pi)$ is very close to the parton model/constituent counting rules prediction of $(1-x)^1$, derived, however, for <u>spinless</u> quarks.[30]

This last point deserves some discussion. The fact that the quarks carry spin 1/2 seems well established by the fact that the angular distribution $d\sigma/d\cos\theta^*$ is observed by the Chicago-Princeton group to be approximately of the form $(1+\alpha\cos^2\theta^*)$ with $\alpha \simeq 1$. Spinless quarks (or a mechanism[27] such as $\pi\pi \to \gamma^* \to \mu\bar\mu$) would require $d\sigma/d\cos\theta^* \propto \sin^2\theta^*$ because the γ^* is then necessarily longitudinally polarized. However, spin 1/2 quarks also lead to an <u>even</u> power behavior[31] of $xq_\pi(x)$ as $x \to 1$: $(1-x)^2$ or $(1-x)^0$, not $(1-x)^1$. It seems, therefore, that there is an inconsistency between the observations of $d\sigma/d\cos\theta^* \propto (1+\cos^2\theta^*)$ and $xq_\pi(x) \propto (1-x)^1$. This paradox may be resolved by a more careful study of the x_F, Q^2, and $\cos\theta^*$ dependences of the data, as described below. The $(1-x)^1$ result may be regarded as a "spin-averaged" answer.

Within a specific QCD representation of the behavior of the (spin 1/2) quarks in a pion, it was shown recently[13] that the structure function should have the form

$$xq_\pi(x) \propto (1-x)^2 + \frac{2}{9}\frac{c}{Q^2}(1-x)^0 \quad . \tag{6}$$

Here c is a constant of order 1 GeV^2. For $<Q^2> \simeq 25$ GeV^2, as in the data, the sum of terms in Eq. (6) may easily mimic the observed $(1-x)^1$ behavior. It was further shown that the scaling piece, $(1-x)^2$, in Eq. (6) is associated with transversely polarized γ^*'s and thus with a $(1+\cos^2\theta^*)$ angular distribution, whereas the non-scaling portion $(1-x)^0/Q^2$ is tied to longitudinally polarized γ^*'s and therefore to a $\sin^2\theta^*$ distribution. In the data, the angular distribution is weighted heavily by events with small values of x_F (and hence of x_π), and therefore the $(1+\cos^2\theta^*)$ term dominates. If events are selected with $x_F \gtrsim 0.7$, the effects of the non-scaling $(1-x)^0 Q^{-2} \sin^2\theta^*$ term should be enhanced. It would be instructive to verify whether the correlated x_F, Q^2, $\cos^2\theta^*$ dependences of the data can be described successfully with forms suggested in Ref. 13. Verification would confirm that bound state effects and structure functions at large x can be understood in some detail in a QCD framework. Note that Eq. (6) also implies a strong violation of scaling in πN processes (i.e., $M^4 d\sigma/dM^2 \neq f(M/\sqrt{s})$) at moderate value of s.

2. <u>Valence-ocean processes</u>

The processes most accessible to experimental investigation are pp → $\mu\bar{\mu}$X and pN → $\mu\bar{\mu}$X for which the basic interaction is the annihilation of a (valence) quark from one nucleon with an antiquark from the sea in the other nucleon:

$$\sigma \propto q_p(x_p) \bar{q}_N(x_N) + \bar{q}_p(x_p) q_N(x_N) . \qquad (7)$$

Because high-energy data, with values of $Q^2 = M^2_{\mu\bar{\mu}}$ substantially above the J/ψ resonance region, are available at several different energies, a check can be made of the classical scaling relation embodied in Eq. (1): $M^4 d\sigma/dM^2 = f(M/\sqrt{s})$. Attempts can also be made to identify the deviations from exact scaling which are demanded by QCD gluonic radiation graphs. In addition to these issues of scaling and its violation, it is interesting to compare the ocean $\bar{q}_N(x)$ distribution determined from data on pN → $\mu\bar{\mu}$X (via the Drell-Yan Eq. (1)) with that determined from deep-inelastic neutrino processes.

(a) <u>Scaling</u>. If perfect scaling holds, as implied by Eq. (1) with structure functions independent of Q^2, then the quantity $s d^2\sigma/d\sqrt{\tau}\,dy$ should be a function only of the ratio $\tau = M^2/s$ and of the lepton pair rapidity y. Using their data from pN processes at 200, 300, and 400 GeV/c, the Columbia-Fermilab-Stony Brook collaboration[4] tested this prediction at y ≈ 0.2, the only value of y at which the experimental acceptance permits a comparison of data at different energies. The results are shown in Fig. 7. Scaling is observed to hold to within 20% at y ≈ 0.2 for 0.2 < M/√s < 0.5.

This observation of scaling at 20% level is not inconsistent with the scaling violations inherent in QCD.[32] In the theory, and

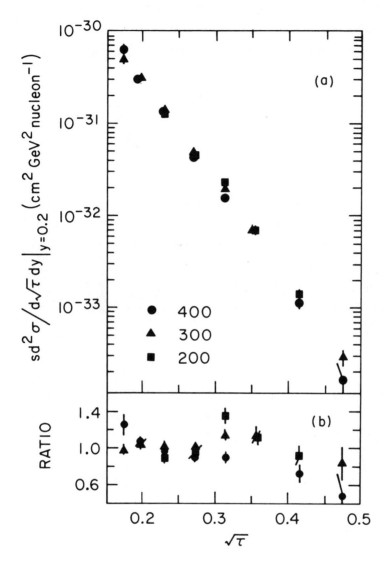

Figure 7. A test of the classical scaling prediction with data on $pN \to \mu\bar{\mu}X$ at 200, 300, and 400 GeV/c, by the Columbia-Fermilab-Stony Brook collaboration (Ref. 4). The data are centered at rapidity y=0.2 at each energy. In part (a), the scaling quantity $sd^2\sigma/dyd\sqrt{\tau}$ is shown; $\tau = Q^2/s$. A simple function $F(\tau)$ is fitted to these data points. The results in (b) are obtained by dividing those in (a) by $F(\tau)$; the scatter of points illustrates the deviations from perfect scaling.

in DIS data, the pattern of scaling violations is such that distribution functions have a "cross-over" point near x = 0.2. For x < 0. structure functions increase in magnitude as Q^2 grows, whereas for x > 0.2 they decrease. The rate of change is only logarithmic in Q^2. In $pN \to \mu\bar{\mu}X$, $x_F = 0$ implies $x_q = x_{\bar{q}} = M/\sqrt{s}$. Therefore, for $y \simeq 0$ and $M/\sqrt{s} > 0.2$, QCD provides essentially no deviation from exact scaling. For $y \simeq 0$ and $M/\sqrt{s} > 0.2$, a slow (logarithmic) decrease with s of $sd^2\sigma/d\sqrt{\tau}dy$ should be observed. For $y \simeq 0$ and $M/\sqrt{s} < 0.2$ a slow growth is predicted. These qualitative expectations are as consistent with the data in Fig. 7 as is the statement of perfect scaling. The lever arm in s in the Fermilab energy range is too small to yield any appreciable deviations from scaling unless x_F and $M/\sqrt{s}$ are large.

In an attempt to increase the lever arm, it is desirable to compare data from Fermilab with results from the CERN-ISR. However, this endeavor is unsatisfactory because the ranges of $M/\sqrt{s}$ do not overlap. One comparison is shown in Fig. 8: ISR results are clustered in the region $M/\sqrt{s} < 0.2$, whereas the Fermilab data populate $M/\sqrt{s} > 0.2$. The Fermilab results may not be extended to lower $M/\sqrt{s}$ because of the J/ψ resonance effects. To extend the ISR results to higher values of $M/\sqrt{s}$ would require data at large values of M, where rates are punishingly low.

(b) <u>Universality of the ocean?</u> If we adopt quark distribution functions determined from DIS experiments, the Fermilab data on $pN \to \mu\bar{\mu}X$ can be used to extract an absolutely normalized sea quark spectrum $\bar{q}_N(x,Q^2)$ for $16 \lesssim Q^2 \lesssim 200$ GeV^2 and $0.2 < x < 0.5$. It is interesting to compare this sea with that determined from deep-inelastic neutrino scattering. As discussed in Section III, if Q^{-2} and $1/\log Q^2$ corrections to the Drell-Yan equation are unimportant, QCD requires these two ocean spectra to be identical at the same values of x and Q^2.

In Fig. 9, I compare ocean spectra extracted from the Columbia-Fermilab-Stony-Brook data[4] on $pN \to \mu\bar{\mu}X$, and from the CERN-

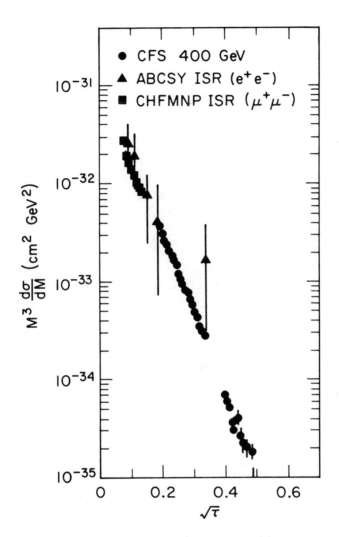

Figure 8. Comparison of FNAL data (CFS, Ref. 4) on $pN \to \mu\bar{\mu}X$ with CERN-ISR data on $pp \to \mu\bar{\mu}X$ (CHFMNP, Ref. 7) and $pp \to e^+e^-X$ (ABCSY, Ref. 8) in terms of the scaling quantity $M^3 d\sigma/dM$. Note that these data are integrated over $x_F > 0$; the figure is adapted from Ref. 7. The Υ region in the FNAL data ($\sqrt{\tau} \simeq 0.35$) is omitted.

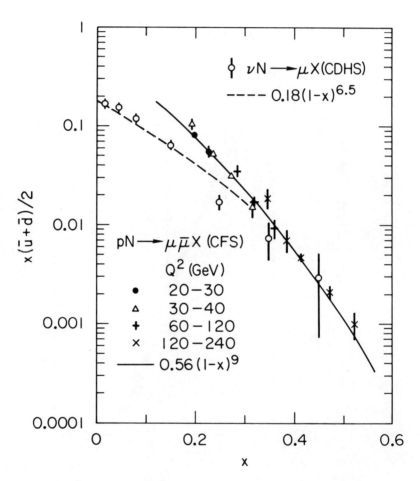

Figure 9. A comparison of the sea distribution obtained from $pN \to \mu\bar{\mu}X$, via the classical Drell-Yan formula, with that determined in deep-inelastic neutrino scattering. The neutrino points are extracted from Ref. 14.

Dortmund-Heidelberg-Saclay data[14] on $\nu N \to \mu X$ and $\bar{\nu} N \to \mu X$. Some comments about the treatment of the data are in order before conclusions may be drawn. In the pN experiment, the contributions of the strange and charm quark seas are presumably entirely negligible; they have been neglected in the fit to the data. For the results shown, it was assumed in the fit that $\bar{u}(x,Q^2) = \bar{d}(x,Q^2)$. However, if one assumes $\bar{u} \neq \bar{d}$ (e.g., $\bar{u} \propto (1-x)^3 \bar{d}$), the results for the average quantity shown in Fig. 9 do not change in any significant way. Note that values of x and Q^2 in the pN results are correlated: at $x_F = 0$, where the data are concentrated, $Q^2 = sx^2$. To break this coupling, and thus to extend the results to lower x, one needs data over a range of values of x_F. The pN points may be fitted to the simple form

$$\frac{1}{2} x(\bar{u} + \bar{d}) = A(1-x)^B , \qquad (8)$$

with $A = 0.56 \pm 0.04$, and $B = 9 \pm 1$. In this fit, the values of A and B are taken to be independent of Q^2, whereas in a QCD approach one would expect both to change logarithmically with Q^2. The meager data do not yet warrant more sophistication, but it should not be forgotten that A and B are appropriate for a wide band in Q^2: $20 < Q^2 < 200$.

Even more discussion of the νN results is necessary. The points shown were extracted from Fig. 18 of Ref. 14, and that paper should be consulted for details of the data analysis. Presented in Fig. 18 of Ref. 14 are values of the ratio

$$\frac{x\bar{q}(x) + x\bar{s}(x)}{\int (xq(x) + x\bar{q}(x)) \, dx} , \qquad (9)$$

where $\bar{q}(x) = \bar{u}(x) + \bar{d}(x) + \bar{s}(x)$. In the text of Ref. 14 it is stated that $\int F_2 \, dx = \int (x \, q(x) + x\bar{q}(x)) \, dx = 0.44 \pm 0.02$ for $30 < E_\nu < 90$ GeV, and $\int F_2 \, dx = 0.45 \pm 0.03$ for $90 < E_\nu < 200$ GeV. The portion of the strange sea may be estimated from dimuon production in neutrino interactions:[33]

$$\frac{\int x\bar{s}(x)\, dx}{\int (xq(x) + x\bar{q}(x))\, dx} = 0.025 \pm 0.01 \quad . \tag{10}$$

Moreover, in Ref. 14 it is determined that

$$\frac{\int (x\bar{q} + x\bar{s})\, dx}{\int (xq + x\bar{q})\, dx} = 0.16 \pm 0.01 \quad . \tag{11}$$

Combining Eqs. (10) and (11), I find that

$$[\bar{u}(x) + \bar{d}(x)] \simeq 0.688[\bar{q}(x) + \bar{s}(x)] \quad . \tag{12}$$

I used this last expression in translating points from Fig. 18 of Ref. 14 to my Fig. 9. In Ref. 14, a fit is made of the form

$$x[\bar{q}(x) + \bar{s}(x)] \propto c(1-x)^d \quad , \tag{13}$$

with $d = 6.5 \pm 0.5$. I show this form in Fig. 9, with my determination of normalization coefficient $c = 0.18$, obtained through the same manipulations I described above.

The neutrino data provide a representation of the sea for $0.05 \lesssim x < 0.25$ and $5 \lesssim Q^2 \lesssim 20$ GeV^2. This kinematic range overlaps that of the pN data only for x near 0.2 and $Q^2 \simeq 20$ GeV^2. In this region of overlap, the magnitude of the sea determined in νN processes is at most one-half that observed in pN $\to \mu\bar{\mu}X$. If the drastic assumption is made that the strange sea can be ignored completely in the νN data, the νN points in Fig. 9 would be moved up by a factor of 1.5, leaving them still at a level of at most 75% of the pN points near $x = 0.2$.

The differences in the parameters of the fits shown in Fig. 9 are not necessarily significant since the fits are done in different x and Q^2 ranges. However, near $x = 0.2$ and $Q^2 = 20$ GeV^2 it is evident that the oceans extracted from νN and pN data do not agree. The effective sea distributions are not universal.

Can one explain this discrepancy? Obviously the data may not be sufficiently precise, but a factor of two seems too much to assign to purely experimental problems. It is possible that the (valence) quark structure function used in the analysis of $pN \to \mu\bar{\mu}X$ is incorrect, biasing the final results for the sea distribution. I have not checked this point, but, again the required factor of two is unlikely. Part of the discrepancy between the ocean spectra in Fig. 9 may be explained with the type of Q^2 variation which QCD provides. As Q^2 grows, the magnitude of the sea is predicted to increase at small x; the effective power p in a fit of the form $(1-x)^p$ should also grow. These two effects are consistent with the trends indicated in Fig. 9. However, again, the magnitude of the discrepancy seen in Fig. 9 is too great. One obvious explanation is that the discrepancy should be attributed to neglect of the $1/\log Q^2$ (or $1/Q^2$) QCD correction terms discussed in Section III. Published estimates[24] suggest that the first order $1/\log Q^2$ effects may indeed be as large as 100%. If the comparison shown in Fig. 8 is substantiated by further experimental analyses, it is evident that QCD calculations must be pushed to higher order in all processes, including a quantitative treatment of all $1/\log Q^2$ (and $1/Q^2$) effects.

V. TRANSVERSE MOMENTUM SPECTRA

As mentioned in Section II, the fact that $<Q_T>$ is relatively large and seems to be an increasing function of s requires a treatment of transverse momentum spectra which goes beyond the classical parton model. The data cannot be explained satisfactorily simply by assigning quarks and antiquarks an "intrinsic" non-perturbative transverse momentum spectrum, although such a contribution is surely a part of the full picture. In a first-order perturbative QCD approach,[3,34,35] the large Q_T tail of the Q_T distribution in $hh \to \mu\mu X$ is generated by the $O(\alpha_s)$ graphs shown in Fig. 10. These QCD con-

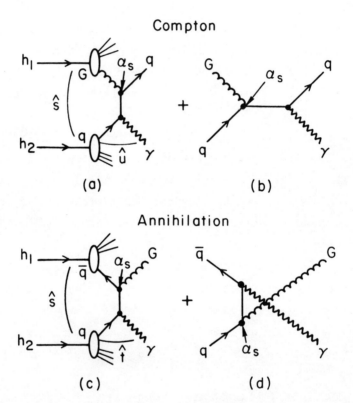

Figure 10. The first order, $O(\alpha_s)$, QCD contributions to $h_1 h_2 \to \gamma X$ at large Q_T. Shown are the "Compton" set $Gq \to \gamma q$ and the two body annihilation set $\bar{q}q \to G\gamma$.

tributions can explain naturally the growth of $<Q_T>$ with s. However, since data now extend only to $|\vec{Q}_T| \lesssim 4$ GeV/c, there is not a large region in Q_T in which we are justified in comparing the model with present data on $d\sigma/d^2\vec{Q}_T$. The region $Q_T < 1$ GeV/c is certainly outside the scope of the perturbative QCD approach, and even the intermediate region $Q_T \lesssim 4$ GeV/c seems to be below the range of applicability of the simple $O(\alpha_s)$ approach. I will describe first what has been done with the $O(\alpha_s)$ approach and then discuss modifications,[36-38] as well as other approaches.[18,39,40] In the "successful" fits to data on the transverse momentum distribution, success is achieved largely through the introduction of small Q_T effects which have little or nothing to do with QCD.

1. Comparison with $O(\alpha_s)$ QCD contributions

The distribution $Ed\sigma/d^3\vec{Q}$ which I obtained from the $O(\alpha_s)$ QCD graphs is shown in Fig. 11 as a function of Q_T for $pN \to \mu\bar{\mu}X$ at 400 GeV/c, $7 < M_{\mu\mu} < 8$ GeV, and $y = 0$. Shown for comparison are data from the Columbia-Fermilab-Stony Brook collaboration. Details of the calculation may be found in Ref. 3. Simple QCD cross-sections obtained from the graphs in Fig. 10 are convoluted with quark, antiquark and gluon densities in the initial hadrons. The initial and final quarks and gluons are taken to be massless and on-shell. The initial quark structure functions are obtained from data on DIS. The initial antiquark distribution is deduced from my fit[3] to the cross-section $d\sigma/dMdy$ for $pN \to \mu\bar{\mu}X$, integrated over $\vec{Q}_T$. For the initial gluon density, I use

$$xG(x) = \frac{1}{2}(p+1)(1-x)^p \tag{14}$$

with $p = 6$. I set the strong coupling "constant" $\alpha_s = 0.3$. I ignore Q^2 dependence in α_s and in the structure functions. To first order in α_s this neglect is immaterial; corrections are of higher order in α_s.

Evident in Fig. 11 is the divergence ($\propto Q_T^{-2}$) associated with the

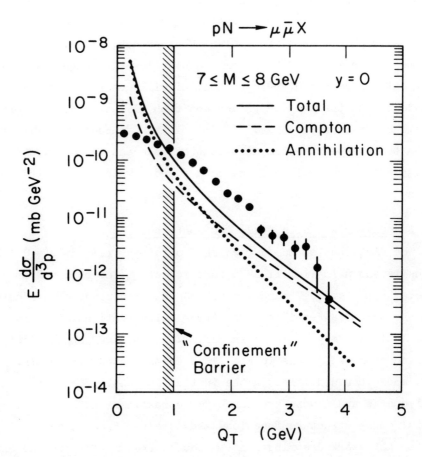

Figure 11. Data from Ref. 4 are shown on the Q_T distribution of $pN \to \mu\bar{\mu}X$ at 400 GeV/c and y=0 in the dimuon mass interval $7 \leq M \leq 8$ GeV. Shown also are calculations of the $O(\alpha_s)$ QCD expectations for this distribution, from Ref. 3. The solid curve marked "total" is obtained from an incoherent addition of the Compton and annihilation contributions. Indicated by cross-hatching is the critical $Q_T \approx 1$ GeV below which the perturbation calculation is inapplicable. The comparison of theory and experiment is similar for other values of M (not shown).

"mass singularity" in the $O(\alpha_s)$ amplitudes when the exchanged quark goes on-shell. The large tail of the distribution is controlled by the large x behavior of the quark, antiquark, and gluon densities in the initial hadrons. The "Compton" $(qG \to \gamma^* q)$ contribution in Fig. 11 falls off less rapidly with Q_T than the annihilation $(q\bar{q} \to \gamma^* G)$ contribution because in nucleons the initial gluons are harder (have larger <x>) than the initial antiquarks. In $\bar{p}N$ processes, roles are reversed, and the large Q_T behavior is controlled by the two body annihilation process $q\bar{q} \to \gamma^* G$. Therefore, in pN reactions, the large $Q_T \gamma^*$ is balanced in Q_T by a quark jet, whereas in $\bar{p}N$ (and $\pi^- N$) reactions, the large $Q_T \gamma^*$ is balanced by a gluon jet. This suggests that in order to isolate gluon jets it would be useful to trigger on a large $Q_T \gamma^*$ produced in $\bar{p}N$ or $\pi^- N$ reactions.

In Fig. 11, I have sketched a wall at $Q_T = 1$ GeV/c below which the theory is clearly inapplicable. Below $Q_T = 1$ GeV/c, perturbation theory breaks down, and various non-perturbative, bound state, soft gluon, coherence, color shielding, and other effects surely dominate the behavior of the Q_T spectrum. However, the comparison of theory and data in Fig. 11 suggests that the $O(\alpha_s)$ QCD result is inadequate even in the intermediate region $4 > Q_T > 2$ GeV/c. Two aspects of the discrepancy may be noticed: the theory falls below the data by a factor of 2 in absolute normalization, and the concavities do not seem to match well. The implementation of various schemes for removing the Q_T^{-2} divergence (such as off-shell kinematics or quark masses) can only worsen the discrepancy in absolute rate. The theoretical yield can be raised only if one adopts unreasonably large values of α_s, or structure functions whose normalization is too large.[41]

2. Beyond $O(\alpha_s)$

Attempts have been made to go beyond the $O(\alpha_s)$ calculation. It is desirable to try to sum contributions to all orders in α_s, at least in the leading logarithm approximation. When only one momentum variable grows in unbounded fashion, as at large Q_T with fixed small Q^2, or at large Q^2 in DIS with an integral done over Q_T, a summation over leading logarithms to all orders in perturbation theory modifies the classical parton model by the simple replacement of structure functions with scaling violating forms, such as $q(x, \log Q_T^2)$. However, when both Q_T and Q^2 are large, as we require now, the presence of the two large mass scales complicates the analysis. Nevertheless, as a result of summing over leading logarithms, Dokshitzer, Dyakonov, and Troyan[23,36] have succeeded in showing that the lowest order cross-section is modified by a function depending on $\log(Q^2/Q_T^2)$, for $Q_T^2 \ll Q^2$. The higher order gluonic radiation terms result in an effective "form-factor" in Q_T^2. In addition to the limitation $Q_T^2 \ll Q^2$, the DDT answer is also valid only for

$$\frac{\alpha_s(Q_T^2)}{\pi} \log\left(\frac{Q^2}{Q_T^2}\right) \ll 1 \quad , \tag{15}$$

i.e., $Q_T^2 \gg \Lambda Q$, where Λ is the usual scale parameter in QCD. Kajantie and Lindfors[37] compared results obtained in the simple $O(\alpha_s)$ approach with those from the DDT method. Two important qualitative conclusions emerge from their analysis.[37] The DDT result provides the type of concavity of the distribution $d\sigma/dQ_T^2$ which is seen in the data (i.e., as $Q_T \to 0$, the distribution begins to flatten). A second improvement is in the absolute rate, with the DDT curve above the simple $O(\alpha_s)$ result by about a factor of 2, as the data require. No detailed comparisons of the DDT results with data have been published as yet. However, the improvements noted by Kajantie and Lindfords indicate that the DDT method is promising.

An alternative approach for including the effects of soft gluon emission was proposed recently by Parisi and Petronzio.[38] Their formal results differ from those of DDT, but they also find that incorporation of the soft gluon effects improves the agreement of the Q_T spectrum with data.

3. Higher-twist effects

A different approach to physics in the intermediate Q_T region is based on the constituent interchange model (CIM), or, equivalently "higher-twist" QCD contributions.[39] In the CIM approach, it is stressed that there is important bound or quasi-bound substructure in hadrons, in addition to the "free" quarks, gluons, and antiquarks. Examples are mesonic ($q\bar{q}$) and diquark (qq) constituents of hadrons. Since the virtual mesonic ($q\bar{q}$) substates are colorless, they are presumably relatively long-lived. The scattering of such substates can be shown to yield cross-sections which may overwhelm those from the simple elementary constituent scattering terms in some regions of phase space. In inclusive large p_T hadronic scattering, e.g., pp → πX, the CIM contributions provide a cross-section falling like p_T^{-8}, resembling the data more in the region $2 < p_T < 8$ GeV/c than the simple p_T^{-4} contributions.[12] In Fig. 12(a) I illustrate the CIM contributions (M+q→γ*q) to massive γ* production. The analogous terms for large p_T hadron production are shown in Fig. 12(b).

Detailed studies of the CIM contributions to massive lepton pair production have been made by Blankenbecler and Duong-van,[18] and by others.[40] The absolute rate of the CIM contribution to pp → γ*X at fixed Q_T is fixed by parameters determined in CIM fits to pp → πX at large p_T. In Fig. 13 I present a comparison of the CIM contribution with the simple $O(\alpha_s)$ QCD results. For $Q_T < 5$ GeV/c, the CIM contribution is larger than the $O(\alpha_s)$ QCD result. As Q_T grows at fixed Q^2, the CIM cross-section falls as Q_T^{-6} whereas the $O(\alpha_s)$ terms behave as Q_T^{-4}.

The results in Fig. 13 suggest that a proper description of the data on pp → γ*X in the intermediate Q_T region requires the CIM terms. Away from $Q_{\parallel} = 0$, the incoherent addition of the CIM and $O(\alpha_s)$ QCD contributions seems to be justified because the flow of color is not identical in the two cases. After the meson M is removed, the remaining hadron is still colorless, whereas after a gluon is removed, the hadronic residue is colored. However, this

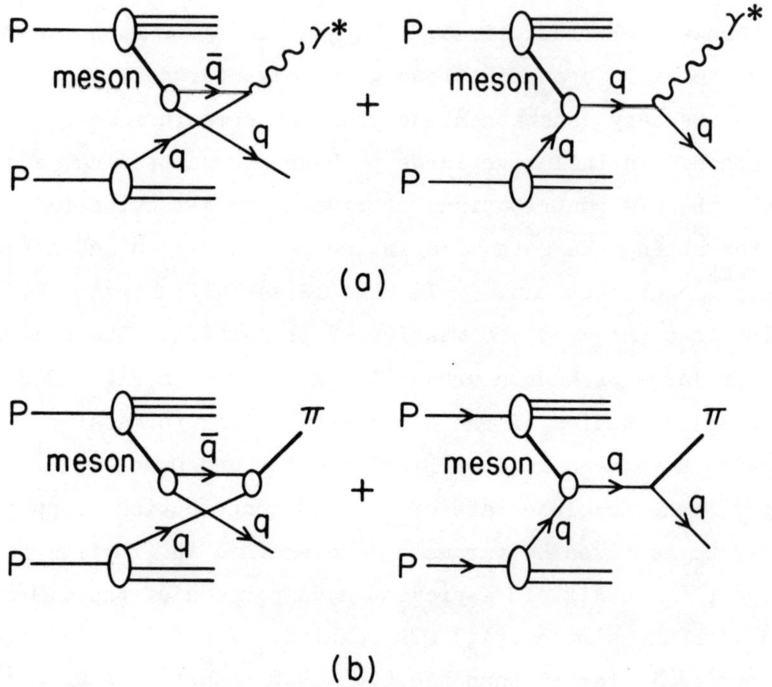

Figure 12. (a) The CIM diagrams for the process $Mq \to \gamma^* q$;
(b) CIM diagrams for $Mq \to \pi q$.

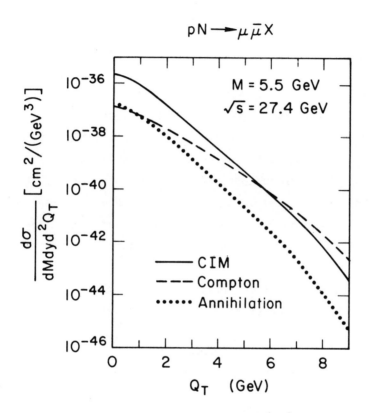

Figure 13. A comparison of the CIM and QCD $O(\alpha_s)$ contributions to the muon pair cross-section $d\sigma/dM\,dy\,dQ_T^2$ for $pN \to \mu\bar{\mu}X$ at $\sqrt{s} = 27.4$ GeV, $y=0$, and $M=5.5$ GeV. This figure is taken from Ref. 40. The mass-singularity divergences at $Q_T=0$ have been removed through the use of off-shell kinematics for initial and final constituents.

issue is somewhat clouded since the manner by which color neutrality is restored after constituent scattering is almost never addressed. Another aspect of a (possible) double counting problem has to do with overall normalization, even if incoherence is assumed. The normalization of the initial constituent densities has to be done in a consistent manner.

4. Small Q_T

Regardless of whether one follows the simple $O(\alpha_s)$ QCD approach or the CIM approach, the region $Q_T < 1$ GeV/c requires special treatment. Fits to the data extending into this region have been proposed and, as fits go, they are successful. However, the fits in this region, whether successful or not, have little to do with the basic dynamics of the various approaches.

If the initial and final constituents are massless, the mass-singularity divergence of the cross-section at $Q_T = 0$ is present in both the CIM and QCD approaches. Simple kinematic arguments indicate, however, that the constituents are not massless and not on-shell. For example, if we apply energy and momentum conservation constraints to the vertex $h_a \to qX$, we may derive the expression

$$q^2 = -\left[\frac{\vec{k}_T^2 + xm_X^2}{(1-x)}\right] + xm_a^2 . \quad (16)$$

This is the square of the four vector momentum of constituent q. The constituent q (whether quark, gluon, meson, ...) carries longitudinal momentum fraction x and transverse momentum $\vec{k}_T$. I have assumed that the initial hadron as well as the final hadronic system X are on-shell. While this seems to be a sensible assumption, it is by no means obvious that X is on-shell, and even if so, what mass m_X to assign to it, since X is not color neutral in the diagrams which most people draw.

If the initial constituents are assigned the off-shell space-like four momentum required by Eq. (16), with m_X^2 treated as a positive free parameter, then the mass-singularity is removed from the physical region (unless x and $\vec{k}_T^2$ are both zero), and $d\sigma/d\vec{Q}_T^2$ no longer diverges at $\vec{Q}_T = 0$. Implementing these kinematic requirements in our $O(\alpha_s)$ computation, and integrating over the "intrinsic" $\vec{k}_T$ spectra, we would find that the theoretical curves in Fig. 11 no longer explode as $Q_T \to 0$. An excellent fit to the shape of the experimental spectra (although by no means to the absolute yield) can be achieved by proper choices of the parameters m_X^2 and $\langle \vec{k}_T^2 \rangle$. Likewise, an excellent fit can be achieved with the CIM approach.[40] Some extra care is required in dealing with off-shell initial constituents, inasmuch as the most naive procedure can lead to gauge dependence of the final answers. Proper inclusion of the off-shell effects is required in studies of large p_T processes, as well as in massive lepton pair production, especially when the hard-scattering cross-sections are convoluted with rather large intrinsic transverse momentum contributions.

Although rightfully emphasizing proper kinematic constraints, the successful fits in the small Q_T region do not address the basic question of whether the large $\vec{Q}_T$ theories are correct. For $Q_T <$ 1 GeV/c, the comparison with data is being carried out in a region whether the models are inapplicable, and where the shape of the theoretical curves is controlled by parameters, such as m_X^2 and $\langle k_T^2 \rangle$, introduced by hand. In a sense the fits are dangerous in that they mislead one into thinking that more of the real small Q_T problem is solved and/or that the models are more universally applicable than is the case. The validity of the $O(\alpha_s)$ and of the CIM contributions should be judged on the behavior of the data at "large" Q_T.

Similar comments may be made regarding an entirely different method for handling data at small Q_T, proposed by Altarelli, Parisi, and Petronzio,[34] and followed by others.[41] They subtract away the infinity at $Q_T = 0$ in the simple $O(\alpha_s)$ approach and then smear this difference with an intrinsic k_T spectrum whose parameters are adjusted to fit the general shape of the experimental spectrum $d\sigma/d\vec{Q}_T^2$. A fit to the absolute normalization is enforced by requiring that the integral of the final result yield the observed integral.

5. Energy dependence

I have concentrated here so far on the description of the Q_T dependence of $d\sigma/dQ_T^2$ at fixed energy. However, as emphasized before,[3] the real test of QCD is in the energy dependence at fixed Q_T, or at fixed $Q_T/\sqrt{s}$. Indeed, insofar as Q_T dependence itself is concerned, a Gaussian fit to the data in Fig. 11 would do quite well at fixed s.

The statement of energy dependence in the $O(\alpha_s)$ QCD approach is provided by the expression

$$\frac{s^3 d\sigma}{dQ^2 dx_F dQ_T^2}\bigg|_{\text{large } Q_T} = f(Q^2/s, x_F, x_T) \quad , \tag{17}$$

where $x_T = 2Q_T/\sqrt{s}$. It may be contrasted with the classical scaling relation for the integrated cross-section,

$$\frac{s^2 d\sigma}{dQ^2 dx_F} = g(Q^2/s, x_F) \quad . \tag{18}$$

At small Q_T, where intrinsic k_T effects control the physics, the scaling relation presumably is a simple generalization of Eq. (18); viz.,

$$\frac{s^2 d\sigma}{dQ^2 dx_F dQ_T^2}\bigg|_{\text{small } Q_T} = g(Q^2/s, x_F, Q_T^2) \quad . \tag{19}$$

All of the expectations, Eqs. (17), (18), and (19), are modified by modest logarithmic dependence on Q^2 (and Q_T^2) when the full scale-noninvariant dependences of structure functions and of α_s on Q^2 are included in the calculation. Specifically, Eq. (17) is replaced by

$$\left.\frac{s^3 d\sigma}{dQ^2 dx_F dQ_T^2}\right|_{\text{large } Q_T} = \tilde{f}(Q^2/s, x_F, x_T; \log Q^2, \log Q_T^2) \, .$$

The contrast between Eqs. (17) and (19) is striking. No tests have been made as yet of these crucial expectations. In going from p_{lab} = 400 GeV/c at FNAL ($\sqrt{s} \simeq 27$ GeV) to $\sqrt{s} \simeq 53$ GeV at the ISR, one must double both Q_T and Q in order to keep x_T and Q^2/s fixed. Although this is admittedly a difficult task, given the luminosity limitations at the ISR, it would be valuable to verify even the expected trends over a more limited interval in Q_T.

6. Moments: mass dependence at $x_F = 0$

The use of moments of the Q_T distribution may facilitate comparisons between theory and data, in that the problematic low Q_T region can be deemphasized. For example, we may consider the quantity

$$\langle Q_T^n \, d\sigma/dQ_T^2 \rangle = \int dQ_T^2 \, \frac{Q_T^n \, d\sigma}{dQ_T^2} \, . \tag{20}$$

For simplicity, I suppress temporarily the Q^2 and x_F dependences of $d\sigma/dQ_T^2$. Since $d\sigma/dQ_T^2$ diverges as Q_T^{-2} in the simple $O(\alpha_s)$ QCD approach, the integral in Eq. (20) converges for all $n \geq 2$. Although the integrand has no infrared or mass singularity divergence for $n \geq 2$, this does not imply that the answer is <u>insensitive</u> to the small Q_T region. Only for n much larger than 2 is the small Q_T region sufficiently deemphasized in the final result. Unfortunately, as n grows, so do the errors on the experimental results. As a compromise, I limit myself to n = 2. Dividing Eq. (20) by the full cross-section integrated over Q_T (obtained from data, or from Eq.

(1) with Q^2 dependent structure functions), one may define

$$\langle Q_T^2 \rangle_{QCD} = \frac{1}{\sigma} \int Q_T^2 \, dQ_T^2 \left\{ \frac{d\sigma^{compt}}{dQ_T^2} + \frac{d\sigma^{ann}}{dQ_T^2} \right\} . \qquad (21)$$

This is the $O(\alpha_s)$ contribution to $\langle Q_T^2 \rangle$ if I use the first order perturbative QCD expressions for $\sigma^{compt}(qG \to q\gamma^*)$ and for $\sigma^{ann}(q\bar{q} \to \gamma^*G)$ on the right hand side. Obviously, $\langle Q_T^2 \rangle_{QCD}$ cannot be expected to reproduce $\langle Q_T^2 \rangle_{exp}$ for the same reasons that the $O(\alpha_s)$ distribution does not reproduce $d\sigma/dQ_T^2$. Thus we define

$$\langle Q_T^2 \rangle_{exp} = \langle Q_T^2 \rangle_{QCD} + \langle Q_T^2 \rangle_{correction} \qquad (22)$$

where the correction term, added in quadrature, stands not only for such obvious physical effects as the neglected intrinsic transverse momentum contribution, due to confinement effects, but also covers a multitude of other errors of omission and comission.

A comparison of $\langle Q_T^2 \rangle$ with data is presented in Fig. 14. Shown is the Q^2 dependence at y=0 in pN collisions at fixed energy, p_{lab}= 400 GeV/c. The two-body annihilation contribution has a fixed magnitude and shape, determined by the QCD cross-section and by the known quark and antiquark structure functions (Q^2 variation of these densities is ignored). The Q^2 dependence of the Compton term is somewhat less certain since the gluonic density is not as well known. To obtain the results shown, I use Eq. (14), with power p=6. Choosing instead p=5 (or 7), I would find a curve for the Compton contribution which rises more (or less) rapidly with Q^2 for M>4 GeV. The overall normalization of the Compton contribution may also be increased if I depart from the assumption embodied in Eq. (14) and assign more than 50% of the hadron's momentum to the gluonic component. However, these uncertainties are minor. In Fig. 14, the net QCD contribution[3] to $\langle Q_T^2 \rangle$ is seen to rise with Q^2 until M≈6 GeV,

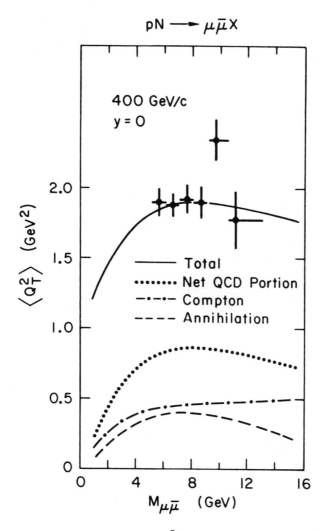

Figure 14. Shown are values of $\langle Q_T^2 \rangle$. The short dashed line illustrates the value obtained from the annihilation process $q\bar{q} \to G\mu\bar{\mu}$, and the dot-dashed line represents the contribution of the Compton process $qG \to q\mu\bar{\mu}$. The dotted line is the net QCD contribution to $\langle Q_T^2 \rangle$ to first order in α_s; it is obtained by addition of the Compton and annihilation portions. The solid line is obtained from the "net QCD" curve by addition of the constant "confinement" contribution 1.04 GeV^2. The data are from Ref. 4. All curves are calculated for the process $pN \to \ell^+\ell^- X$ at $p_{lab} = 400$ GeV/c and y=0.

and then the flatten off near a level of ≈ 0.8 GeV^2, a factor of two or more beneath the experimental result.

Before I discuss the discrepancy in the absolute value of $\langle Q_T^2 \rangle$, it is worth noting that the shape of the Q^2 dependence of $\langle Q_T^2 \rangle_{QCD}$ is in reasonable agreement with experiment. This flat behavior of $\langle Q_T^2 \rangle_{QCD}$ vs. Q^2 for $M \gtrsim 6$ GeV may be contrasted with the naive expectation that

$$\langle Q_T^2 \rangle_{QCD} \propto Q^2 \quad . \tag{23}$$

Because QCD is not a soft field theory, one should indeed expect that at fixed $M/\sqrt{s}$ (and fixed y or x_F)

$$\langle Q_T^2 \rangle_{QCD} \propto Q^2 f(Q^2/s) \quad . \tag{24}$$

The flat behavior of $\langle Q_T^2 \rangle_{QCD}$ with Q^2 for $M \gtrsim 6$ GeV shows that the scaling function $f(Q^2/s)$ has a strong dependence on Q^2 at fixed s. This strong dependence is supplied by the rapidly falling antiquark and gluonic structure functions. Stated otherwise, the flat behavior of $\langle Q_T^2 \rangle_{QCD}$ with Q^2 at 400 GeV/c provides an indirect measurement of the fact that the effective power p in Eq. (14) is greater than 4 or 5.

To reach the experimental level of $\langle Q_T^2 \rangle \approx 1.8$ GeV^2 in Fig. 14, one must add ≈ 1 GeV^2 to the $O(\alpha_s)$ QCD result. Part of this supplement is surely associated with the non-perturbative, intrinsic $\langle k_T^2 \rangle$ carried by each constituent, due to the constituent's confinement in a hadron of limited transverse size. In their global fit to $d\sigma/dQ_T^2$, Altarelli, Parisi, and Petronzio[34] also found that their "intrinsic" component has $\langle k_T^2 \rangle \approx 1$ GeV^2.

If we divide 1 GeV^2 equally between each of the two initial constituents, we find that we are asking that confinement supply a rather hefty $\langle k_T^2 \rangle \approx 0.5$ GeV^2, or $\langle k_T \rangle = 600$ MeV, instead of the more typical 300 MeV. A similarly large number is invoked in some aspect of fits[42] to data on correlations at large p_T; ρ mesons[43] as well as hadronic clusters are observed to be produced with $\langle p_T \rangle \approx 600$ MeV. Nevertheless, $\langle k_T \rangle \approx 600$ MeV is uncomfortably large for several reasons. First, it raises the technical issue of our neglect of off-shell kinematics. Returning to Eq. (16), we see that if our fit requires $\langle k_T^2 \rangle \approx 0.5$ GeV^2, we are treading close to inconsistency in ignoring the off-shell nature of the initial constituents. Second, there is a more important issue of principle to be faced. In evaluating the $O(\alpha_s)$ QCD terms, we pretend to have identified the essential hard scattering, large Q_T effects. However, in the end, we must hide more than half of the observed $\langle Q_T^2 \rangle$ in an uncalculated contribution. All of the large Q_T is not "out in the open", as it should be.

I know of no explicit calculation of $\langle Q_T^2 \rangle_{QCD}$ in the DDT approach,[23] but it seems clear that the residual unexplained portion of $\langle Q_T^2 \rangle$ will be decreased if the DDT method is implemented. This effect is correlated with the fact that the cross-section $d\sigma/dQ_T^2$ in the DDT approach flattens as Q_T decreases. In the DDT approach, one may well achieve values of the intrinsic $\langle k_T \rangle$ closer to the expected 300 MeV.

In the CIM approach, it should be mentioned that Duong-van and Blankenbecler[18] calculated values of $\langle Q_T^2 \rangle \approx 1.8$ GeV^2 and $\langle Q_T \rangle \approx 1.2$ GeV, which later data fully supported. In obtaining their results, they used full off-shell kinematics. Thus, the large Q_T CIM effects and the "intrinsic" contributions are combined directly in their answer.

7. **Moments: x_F and quantum number dependences**

In the previous paragraphs I discussed the behavior of $\langle Q_T^2 \rangle_{QCD}$ as a function of Q^2 at fixed x_F, concentrating on pN collisions at 400 GeV/c for comparisons with data. Data are available also on the x_F and s dependences of $\langle Q_T^2 \rangle$ in pN collisions,[4,5] and on the Q^2 and x_F dependences in π^-N collisions[9] at 225 GeV/c.

The $\pi^- N$ data allow us to examine a situation in which the valence-valence process is dominant. At large Q_T, the two-body annihilation reaction $q\bar{q} \to \gamma^* G$ should dominate the Compton process $qG \to \gamma^* q$. For the $\pi^- N$ calculations, I used a gluon density in the pion suggested by constituent counting rules,[30] again normalized to 50% of the pion's momentum:

$$xG_\pi(x) = 2(1-x)^3 \ . \tag{25}$$

At the time I did the calculations reported here, the quark structure function of the pion had not been determined. Thus, I tried a few different forms for $xQ_\pi(x)$, to see what differences arise in the final results. Shown in Fig. 15 are values of $<Q_T^2>_{QCD}$ which I obtained from my $O(\alpha_s)$ calculations, for three choices of the quark structure function of the pion. As before, I employ massless and on shell initial and final elementary constituents in the computations. The three choices are : (a) "FF", the pion structure function suggested in Ref. 44, for which $xq_\pi(x) \to (1-x)^0$ as $x \to 1$; (b) "Dao et al.", from Ref. 45, for which $xq_\pi(x) \to (1-x)^{2.55}$ as $x \to 1$; and (c) a purely ad-hoc form whose explicit structure is

$$xq_V(x) = 0.35 (1-x)^{0.6} \tag{26}$$

$$xq_\pi(x) = xq_V(x) + 0.1 (1-x)^5 \tag{27}$$

All three "models" in Fig. 15 yield essentially identical results for $x_F < 0$, in the target nucleon's "fragmentation region". For $x_F > 0$, differences are marked, both in normalization and x_F dependence, due to the different behavior of the three pion structure functions. Nevertheless, all three models fail to reproduce the size of $<Q_T^2>_{exp}$, which is again roughly a factor of two above the calculations at 225 GeV/c.

In order to compare results with πN data on $<Q_T>$, I again add

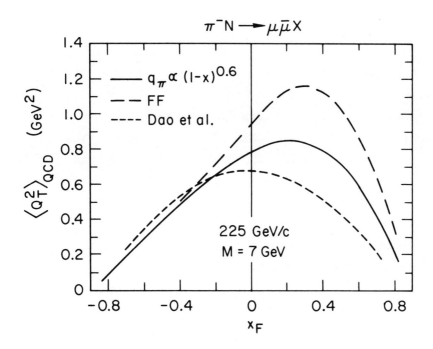

Figure 15. The $O(\alpha_s)$ QCD results for $\langle Q_T^2 \rangle$ are shown as a function of x_F for $\pi^- N \to \mu\bar{\mu}X$ at 225 GeV/c and $M_{\mu\bar{\mu}} = 7$ GeV. Values are presented for three choices of the quark structure function of the pion.

"confinement" contribution in quadrature to the QCD $O(\alpha_s)$ result, using the same value $<Q_T^2>_{conf} = 1.04$ GeV^2, determined in the fit to pN collisions at 400 GeV/c and y=0. To limit arbitrary freedom, I assume that the value 1.04 GeV^2 is independent of Q^2, s, and x_F, and of quantum number effects. To obtain a value of $<Q_T>$ from the resulting $<Q_T^2>$, one must assume a form for the distribution $d\sigma/dQ_T^2$, which I take to be the form fitted to the pN data.[4] Thus,

$$<Q_T> = 0.859 <Q_T^2>^{1/2} . \qquad (28)$$

A comparison of the calculated values of $<Q_T>$ with data on the Q^2 dependence at $x_F = 0.2$ is shown in Fig. 16, and with data on the x_F dependence in Fig. 17. The models with relatively flat quark structure functions are capable of reproducing both the magnitude and general shape of $<Q_T>$ as a function of both x_F and Q^2. Judging from the fact that the results obtained with the $(1-x)^{0.6}$ form pass systematically through the lower portion of the error flags in Figs. 16 and 17, I imagine that calculations based on the "true" spin-averaged structure function $\propto (1-x)^1$, will do somewhat worse. However here again, a proper calculation should be done including a full treatment of spin effects.[13] The agreement of these "predictions" with the data is satisfactory.

It is interesting to observe that

(i) $<Q_T>_{225 \text{ GeV/c}}^{\pi N} \approx <Q_T>_{400 \text{ GeV/c}}^{pN}$; (Data)

(ii) $<Q_T^2>_{QCD}^{\pi N, 225 \text{ GeV/c}} \approx <Q_T^2>_{QCD}^{pN, 400 \text{ GeV/c}}$; (Theory) .

This suggests that the discrepancy between $<Q_T^2>_{QCD}$ and experiment, from whatever origin, is roughly universal. Valence-valence processes are not endowed with a much larger or a much smaller "intrinsic" portion of $<Q_T^2>$ than the valence-ocean pN reaction.

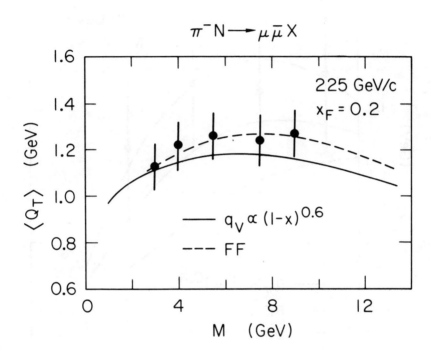

Figure 16. For $\pi^- N \to \mu\bar{\mu} X$ at $p_{lab} = 225$ GeV/c, calculated values of $\langle Q_T \rangle$ are compared with data as a function of muon pair mass, at $x_F = 0.2$; results are shown for two choices of the quark structure function of the pion.

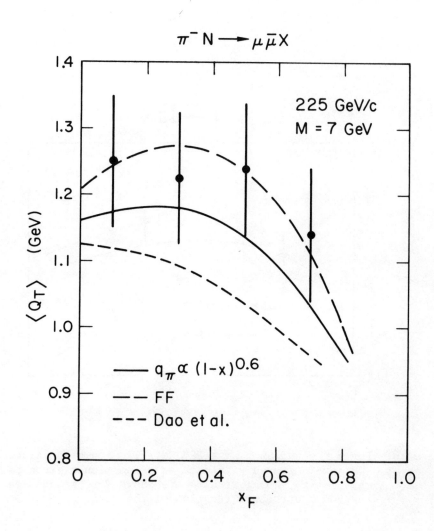

Figure 17. For $\pi^- N \to \mu\bar{\mu}X$ at p_{lab} = 225 GeV/c, QCD expectations for $<Q_T>$ are compared with data as a function of x_F at $M_{\mu\bar{\mu}}$ = 7 GeV; theoretical results are shown for three choices of the quark structure function of the pion.

The comparisons with the data in Figs. 16 and 17 suggest that improved agreement would be obtained if the "intrinsic" $\langle k_T^2 \rangle$ contribution is modestly larger in pions that in nucleons, and if the intrinsic $\langle k_T^2 \rangle$ grows with x. More explicitly, if we imagine a quark structure function $q(x,k_T^2)$, it should be true that $\langle k_T^2(x) \rangle_\pi$ grows slowly with x, and that $\langle k_T^2(x) \rangle_\pi > \langle k_T^2(x) \rangle_N$. Although little on the x_F dependence of $\langle Q_T^2 \rangle$ is available from pN collisions, the results[4,5] show a rather flat behavior of $\langle Q_T^2 \rangle$ vs. x_F, in disagreement with the results of my $O(\alpha_s)$ calculation in Fig. 18. This comparison suggests again that the intrinsic $\langle k_T^2(x) \rangle_N$ should grow with x.

In an interesting paper based largely on simple kinematic arguments, Ellwanger shows that $\langle k_T^2(x) \rangle$ is determined by the x dependence of the structure functions.[46] In particular, he derives

$$\langle k_T^2(x) \rangle = \frac{x(1-x)}{xq(x)} \int_x^1 \left(\frac{m_X^2}{(1-x')^2} - m_a^2 \right) q(x') \, dx' . \quad (29)$$

Here m_X is the mass of the (on-shell) system X which results from the hadronic dissociation $h_a \to qX$, and m_a is the mass of the incident hadron h_a. This equation would yield an absolute determination of $\langle k_T^2(x) \rangle$ except for uncertainties associated with m_X^2 (which itself may be a function of x). Taking m_X^2 to be constant, Ellwanger obtains curves of $\langle k_T^2(x) \rangle$ which grow with x. Choosing $xq(x) \propto (1-x)^3$, appropriate for nucleons, and $m_X = 1$ GeV, he finds that $\langle k_T^2(x) \rangle_{\text{valence}}$ changes from about 0.25 GeV^2 at x=0.2 to 0.45 GeV^2 at x=0.8. Ellwanger shows that as $x \to 1$,

$$\langle k_T^2(x) \rangle \to m_X^2/(n-1) , \quad (30)$$

where n is the power in an expansion of $xq(x) \propto (1-x)^n$, valid near x=1. Therefore, although $\langle k_T^2(x) \rangle_{\text{sea}}$ also grows with x, it is smaller at a given x than $\langle k_T^2(x) \rangle_{\text{valence}}$. Whereas the desired growth of the intrinsic contribution with x is obtained, the size of $\langle k_T^2(x) \rangle_{\text{valence}}$

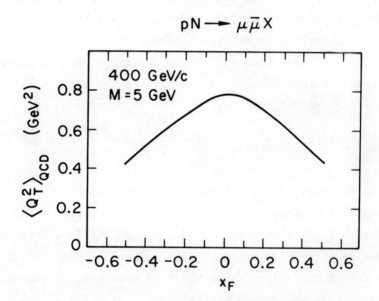

Figure 18. A prediction for the dependence of $\langle Q_T^2 \rangle_{QCD}$ on the longitudinal momentum fraction x_F of the lepton pair in pN collisions at 400 GeV, at the lepton pair mass M = 5 GeV

$+\langle k_T^2(x)\rangle_{sea}$ cannot be made as large as the required 1 GeV^2 unless $m_X^2 \approx 3$ GeV^2, which is perhaps too large. This is another argument against attributing all of the discrepancy between $\langle Q_T^2\rangle_{exp}$ and $\langle Q_T^2\rangle_{QCD}$ to intrinsic transverse momentum effects. Note, finally, that Eq. (30) implies that

$$\langle k_T^2(x)\rangle_\pi > \langle k_T^2(x)\rangle_N \quad , \tag{31}$$

unless the value of m_X^2 is much smaller in π dissociation that in N dissociation.

8. Energy dependence of $\langle Q_T^2\rangle$

On purely dimensional grounds we may argue that $\langle Q_T^2\rangle_{QCD} \propto sg(Q^2/s,x_F)$ for large Q^2. Assuming that the non-QCD "intrinsic" portion of $\langle Q_T^2\rangle$ is independent of s at fixed Q^2/s and x_F, as seems reasonable, we deduce that

$$\langle Q_T^2\rangle = a(Q^2/s,x_F) + sb(Q^2/s,x_F) \quad . \tag{32}$$

Allowing for scaling violations, the function $b(Q^2/s) \to b(Q^2/s)/\log Q^2$. If there is some cross-talk between the QCD and "intrinsic" contributions, a term $\propto \sqrt{s}$ may also be present on the right hand side of Eq. (32).

To first order in α_s, the function $b(Q^2/s,x_F)$ in Eq. (32) may be obtained directly from a computation of the graphs in Fig. 10. Results are shown in Fig. 19; they may be used for predicting values of $\langle Q_T^2\rangle$ at various energies. For example, at $M/\sqrt{s} = 0.1$, I calculate $b = 0.73 \times 10^{-3}$ GeV^{-2}, implying that at $M/\sqrt{s} = 0.1$ and $\sqrt{s} = 62$ GeV, $\langle Q_T^2\rangle = 1.0 + 2.8 = 3.8$ GeV^2. This expectation agrees well with the measurement of $\langle Q_T^2\rangle_{exp} = 4.2 \pm 0.9$ GeV^2 reported by the CERN-Columbia-Oxford-Rockefeller ISR collaboration.[6] A comparison of the predicted energy dependence with Fermilab data[4] is shown in Fig. 20.

The energy dependence of $\langle Q_T^2\rangle$ expected from QCD arguments is certainly consistent with experiment. In the CIM approach,[18] $\langle Q_T^2\rangle$ is

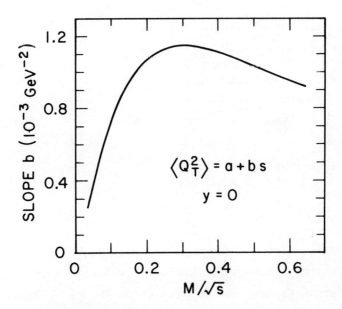

Figure 19. In the expression $\langle Q_T^2 \rangle = a + bs$, the slope b computed from the first order QCD graphs is shown as a function of $M/\sqrt{s}$. These slopes can be used to obtain predictions for $\langle Q_T^2 \rangle$ at various lab energies and lepton pair masses.

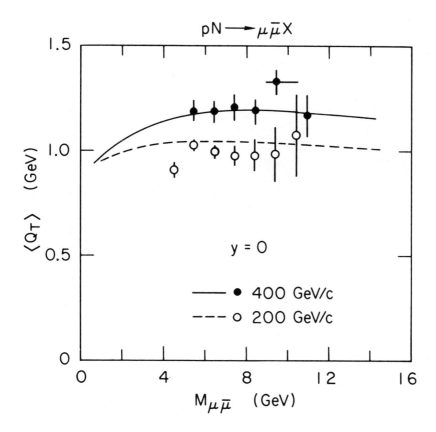

Figure 20. Shown are the $O(\alpha_s)$ QCD expectations for the mean transverse momentum $\langle Q_T \rangle$ of lepton pairs produced in $pN \to \mu\bar{\mu}X$ at y=0 and p_{lab} = 200 and 400 GeV/c, compared with CFS data (Ref. 4). Details of the theoretical calculation may be found in Ref. 3.

expected to rise slightly in the Fermilab range, and to tend to a constant as $s \to \infty$. The ISR results rule out CIM as the full answer, but, as remarked earlier, present data on $d\sigma/dQ_T^2$ in the intermediate Q_T range appear to require the CIM terms.

As discussed in Section V.5, convincing tests of QCD must be carried out at "large Q_T", where reasonably reliable predictions are available for the energy, x_F, and Q_T dependence of $d\sigma/dQ_T^2$, including absolute normalization, and systematics associated with the quantum numbers of the incident beams.[3] Moments of the Q_T distribution tend to be unreliable guides since a great deal of the integral comes from the low Q_T region where use of the theory is inappropriate.

VI. CONCLUSIONS AND OUTLOOK

The ever improving quality of high-energy data[4-11] on massive lepton pair production now permits reasonably detailed tests of several theoretical ideas. More progress may be expected within the coming year or so as new results emerge from experiments at Fermilab and at CERN. In pN collisions, exploration of a broad range in $x_F = x_p - x_N$ will complement the excellent data now confined largely to $x_F = 0$ and make possible a thorough investigation of the (valence quark structure function of the nucleon in the timelike ($Q^2 > 0$) region, especially for large x_p ($x_p > 0.5$). In both pN and πN reactions, we can look forward to accurate measurements of the full angular distribution $W(\theta^*,\phi^*)$ in the lepton pair rest frame, including variations with x_F and Q_T. These distributions should yield some insight into bound state mechanisms,[13] and, at large enough $\vec{Q}_T$, test features of perturbative QCD.[47] Visionaries may seek signals in $W(\theta^*,\phi^*)$ associated with weak-electromagnetic interference (γ^*,Z^0). Extending the range of measurements of $d\sigma/d^2\vec{Q}_T$ to higher $\vec{Q}_T$ poses experimental challenges, but seems necessary for clean tests of the ideas reviewed in Section V. Better data on the Q^2 and s dependence of cross-sections may permit the identification of scaling violation expected theoretically in $d\sigma/dQ^2 dx_F$ and in $d\sigma/dQ^2 dx_F d^2\vec{Q}_T$.

Little information exists now on the properties of the hadronic spray which accompanies a massive lepton pair. At large Q_T, the virtual photon should be balanced by a quark jet in pN collisions and by a gluon jet in $\pi^- N$ and $\bar{p}N$ reactions. It would be useful to design experiments to investigate these companion systems. In addition, two roughly longitudinal jets of hadrons should be present in each event, one the debris of the projectile, and the other of the target. The charge and momentum distributions in these longitudinal jets are expected to yield further insight into hadronic structure.[48]

The proven viability of the Drell-Yan process as a probe of hadronic structure should encourage future high-energy studies with antiproton beams and with polarized nucleon beams and targets.

On the theoretical side, the necessity for QCD, or for some similar theoretical scheme, seems evident for at least two reasons: scaling violations, and the observed large values of $<Q_T>$. However, in obth deep inelastic scattering and massive lepton pair production, much has to be done yet to identify and separate the logarithmic scaling violations associated with asymptotic freedom from the perhaps more mundane scaling violations of an inverse power nature ($\propto Q^{-2}$), expected on general physical grounds.[49] Detailed work is desirable, leading to quantitative estimates of the magnitude of the Q^{-2} terms and of their functional dependence on various kinematic variables such as x_F and $\vec{Q}_T$. In the realm of logarithmic, $\log Q^2$ QCD effects, there is already evidence in both deep inelastic processes[20] and in massive lepton pair production that one must go beyond the leading logarithm approximation in computing observables for comparisons with data. The next to leading, non-factorizing $1/\log Q^2$ correction terms in lepton pair production may be as large as the leading terms (cf., Section IV.2) in the range of Q^2 now being explored ($20 \leq Q^2 \leq 100$ GeV^2). If so, a new method must be devised to reorder the perturbation series in $\log Q^2$.

Techniques are also being developed[36,38] for handling the domain in which both Q^2 and Q_T^2 are large, but more work along these lines is needed, as well as on the problem of continuing the results to small Q_T^2. Finally, it would be highly desirable to go beyond the stage of elementary constituent scattering cross sections and of probabilistic formulas such as Eqs. (1) and (3), and to deal more directly with the fact that the incident and final hadronic states are bound systems.

ACKNOWLEDGEMENTS

In preparing this report, I have benefitted greatly from discussions with many colleagues at SLAC.

REFERENCES

1. S. D. Drell and T. M. Yan, Phys. Rev. Lett. 25, 316 (1970), 25, 902 (1970), and Ann. Phys. (N.Y.) 66, 578 (1971). For a recent general discussion, see C. S. Lam and Wu-Ki Tung, Phys. Rev. D18, 2447 (1978).
2. For a recent review of the data, see L. Lederman, Proceedings of the 19th International Conference on High Energy Physics, Tokyo, 1978. Edited by S. Homma, M. Kawaguchi and H. Miyazawa.
3. E. L. Berger, Proceedings of the 1978 Vanderbilt Conference, AIP Conference Proceedings Series, No. 45, Edited by R. S. Panvini and S. E. Csorna; and in Hadron Physics at High Energies, 1978 Marseille (France) Workshop, Edited by C. Bourrely, J. Dash and J. Soffer.
4. Columbia-Fermilab-Stony Brook Collaboration (FNAL), J. K. Yoh et al., Phys. Rev. Lett. 41, 684 (1978) and 41, 1083 (1978), plus references cited therein, as well as Ref. 2.
5. Seattle-Northeastern-Michigan-Tufts Collaboration (FNAL), S. Childress et al., Report No. NUB-2370, submitted to the Tokyo Conference, 1978.

6. CERN-Columbia-Oxford-Rockefeller Collaboration (ISR), A. L. S. Angelis et al., Report No. COO-2232A-69, September 1978, submitted to the Tokyo Conference, 1978.
7. CERN-Harvard-Frascati-MIT-Naples-Pisa Collaboration (ISR), F. Vannucci, contribution to the Karlsruhe Summer Institute, 1978.
8. Athens-Brookhaven-CERN-Syracuse-Yale Collaboration (ISR), C. Kourkoumelis et al., Brookhaven Report BNL-25075 (1978) and references cited therein, plus I. Mannelli, Tokyo Conference report.
9. Chicago-Illinois-Princeton Collaboration (FNAL), C. B. Newman et al., Phys. Rev. Lett. $\underline{42}$, 951 (1979); G. E. Hogan et al., Phys. Rev. Lett. $\underline{42}$, 948 (1979); and K. J. Anderson et al., Phys. Rev. Lett. $\underline{42}$, 944 (1979).
10. Birmingham-CERN-Munich-Neuchatel-Ecole Polytechnique-Rutherford Collaboration (CERN-SPS), M. J. Corden et al., Phys. Lett. $\underline{76B}$, 226 (1978). See also J. Alspector et al., Phys. Lett. $\underline{81B}$, 397 (1979).
11. Saclay-Imperial College-Southampton-Indiana Collaboration (CERN-SPS), M. A. Abolins et al., Saclay report DPhPE 78-05 (August 1978), paper submitted to the Tokyo Conference, 1978.
12. For views of dynamics at large p_T, see M. Jacob and P. Landshoff, Phys. Reports $\underline{48}$, 285 (1978); S. J. Brodsky, SLAC-PUB-2298, Lectures presented at the La Jolla Summer Workshop on QCD, 1978, and SLAC-PUB-2217, invited talk presented at the Symposium on Jets in High Energy Collisions, Copenhagen, 1978; R. D. Field, Jr. Tokyo, 1978 (op. cit.); and S. D. Ellis and R. Stroynowski, Rev. Mod. Phys. $\underline{49}$, 753 (1977).
13. E. L. Berger and S. J. Brodsky, Phys. Rev. Lett. $\underline{42}$, 940 (1979).
14. CERN-Dortmund-Heidelberg-Saclay Collaboration (CERN-SPS), J. G. H. deGroot et al., CERN Report November 1978, submitted to Zeitschrift fur Physik.
15. Aachen-Bonn-CERN-London-Oxford-Saclay Collaboration, P. C. Bosetti et al., Nucl. Phys. $\underline{B149}$, 13 (1979) and $\underline{B142}$, 1 (1978).

16. Harvard-Chicago-Illinois-Oxford Collaboration, W. A. Loomis et al. Fermilab Report 78/94-Exp. (1978), and H. L. Anderson et al., Phys. Rev. Lett. 40, 1061 (1978).
17. M. D. Mestayer, SLAC Report No. 214 (1978) and references therein.
18. M. Duong-van, K. V. Vasavada and R. Blankenbecler, Phys. Rev. D1 1389 (1977); M. Duong-van and R. Blankenbecler, Phys. Rev. D17, 1826 (1978).
19. See, for example, L. Abbott, SLAC-PUB-2296 (1979), to be published in the Proceedings of the 1979 Coral Gables Conference, Orbis Scientiae, 1979; I. Schmidt and R. Blankenbecler, Phys. Rev. D16 1318 (1977); and H. Harari, SLAC-PUB-2254 (1979), submitted to Nucl. Phys.
20. For an excellent review of the uses of QCD in deep inelastic scattering, including higher order corrections, see A. J. Buras, Fermilab report 79/17-THY (1979). See also G. C. Fox, Caltech Report 68-658 (1978), Invited review presented at the Purdue Conference on Neutrino Physics, Lafayette, Indiana.
21. H. D. Politzer, Phys. Lett. 70B, 430 (1977), and Nucl. Phys. B129, 301 (1977); C. T. Sachrajda, Phys. Lett. 73B, 185 (1977); H. Georgi, Phys. Rev. D17, 3010 (1978); G. Altarelli, G. Parisi and R. Petronzio, Phys. Lett. 76B, 351 (1978); J. Frenkel, M. J Shailer and J. C. Taylor, Nucl. Phys. B148, 228 (1979).
22. R. K. Ellis et al., Phys. Lett. 78B, 281 (1978); C. H. Llewellyn Smith, Schladming Lectures, Oxford Report 47/78 (1978); D. Amati R. Petronzio and G. Veneziano, Nucl. Phys. B140, 54 (1978), and B146, 29 (1978); S. Gupta and A. H. Mueller, Columbia Report CU TP-139 (1979); S. B. Libby and G. Sterman, Phys. Rev. D18, 3252 (1978); R. C. Hwa and J. Wosiek, Rutherford Report RL-78-82 (19
23. Y. L. Dokshitser, D. I. D'Yakonov and S. I. Troyan, "Inelastic Processes in Quantum Chromodynamics," from the XIIIth Winter School of Leningrad B. P. Konstantinov Institute of Nuclear Physics (1978), SLAC Translation No. 183.

24. G. Altarelli, R. K. Ellis and G. Martinelli, MIT Report CTP No. 776 (1979); J. Kubar-Andre and F. E. Paige, Phys. Rev. D19, 221 (1979); K. Harada, T. Kaneko and N. Sakai, CERN Report TH-2619 (1979); J. Abad and B. Humpert, Phys. Lett. 80B, 286 (1979), and Wisconsin Report COO-881-57 (1978); J. Kripfganz and A. P. Contogouris, McGill Report 79-0049 (1979).
25. See, e.g., S. J. Brodsky and G. P. Lepage, SLAC-PUB-2294 (1979) and references therein.
26. C. T. Sachrajda and R. Blankenbecler, Phys. Rev. D12, 3624 (1975).
27. T. Goldman, M. Duong-van and R. Blankenbecler, SLAC-PUB-2283 (1979).
28. C. E. DeTar, S. D. Ellis and P. V. Landshoff, Nucl. Phys. B87, 176 (1975).
29. J. Ellis, M. K. Gaillard and W. J. Zakrezewski, Phys. Lett. 81B, 224 (1979); R. D. Carlitz and D. B. Creamer, University of Pittsburgh, Report No. PITT-208 (1978).
30. S. J. Brodsky and G. Farrar, Phys. Rev. Lett. 31, 1153 (1973); V. Matveev et al., Lett. Nuovo Cimento 7, 719 (1973); F. Ravndal, Phys. Lett. 47B, 67 (1973); A. Vainshtein and V. Zacharov, Phys. Lett. 72B, 368 (1978); R. Blankenbecler and S. J. Brodsky, Phys. Rev. D10, 2973 (1974).
31. S. D. Drell, D. J. Levy and T. M. Yan, Phys. Rev. D1, 1617 (1970); G. Farrar and D. Jackson, Phys. Rev. Lett. 35, 1416 (1975).
32. For estimates of the effects of scaling violatons on $d/dQ^2 dy$, see R. D. Field, Ref. 12.
33. M. Holder et al., Phys. Lett. 69B, 377 (1977).
34. G. Altarelli, G. Parisi and R. Petronzio, Phys. Lett. 76B, 351 (1978); Phys. Lett. 76B, 356 (1978); R. Petronzio, CERN Report TH-2495 (1978).
35. H. Fritzsch and P. Minkowski, Phys. Lett. 73B, 80 (1978); C. Michael and T. Weiler, contribution to the XIIIth Rencontre de Moriond, Les Arcs, France (1978); K. Kajantie and R. Raitio, Nucl. Phys. B139, 72 (1978); F. Halzen and D. Scott, Phys. Rev.

D19, 216 (1979), D18, 3378 (1978), and Phys. Rev. Lett. 40, 1117 (1978); K. Kinoshita and Y. Kinoshita, Prog. Theor. Phys. 61, 526 (1979); M. Gluck and E. Reya, Nucl. Phys. B145, 24 (1978); H. Georgi, Phys. Rev. Lett. 42, 294 (1978).

36. Yu. L. Dokshitzer, D. I. Dyakonov and S. I. Troyan, Phys. Lett. 78B, 290 (1978).
37. K. Kajantie and J. Lindfors, Nucl. Phys. B146, 465 (1978).
38. G. Parisi and R. Petronzio, CERN Report Ref. TH-2627 (1979).
39. R. Blankenbecler, S. J. Brodsky and J. F. Gunion, Phys. Rev. D6, 2652 (1972).
40. C. Debeau and D. Silverman, Phys. Rev. D18, 2435 (1978), and SLAC PUB-2187 (1978).
41. See some of the papers listed in Ref. 35.
42. R. P. Feynman, R. D. Field and G. C. Fox, Phys. Rev. D18, 3320 (1978), and Nucl. Phys. B128, 1 (1977).
43. British-French-Scandinavian Collaboration, M. G. Albrow et al., CERN Report, December 1978, submitted to Nucl. Phys.
44. R. D. Field and R. P. Feynman, Phys. Rev. D15, 2590 (1977).
45. F. T. Dao et al., Phys. Rev. Lett. 39, 1388 (1977).
46. U. Ellwanger, Heidelberg Report HD-THEP-78-19 (1978).
47. See, e.g., R. Thews, Arizona Report, January 1979; K. Kajantie, J. Lindfors and R. Raitio, Phys. Lett. 74B, 384 (1978); E. L. Berger, J. Donohue and S. Wolfram, Phys. Rev. D17, 858 (1978).
48. See, e.g., T. A. DeGrand, SLAC Report: SLAC-PUB-2182 (August 1978) and T. A. DeGrand and H. I. Miettinen, Phys. Rev. Lett. 40, 612 (1978).
49. See, e.g., I. Schmidt and R. Blankenbecler, Ref. 19; and E. L. Berger and S. J. Brodsky, Ref. 13.

PROGRAM

ORBIS SCIENTIAE 1979

MONDAY, JANUARY 15, 1979

Opening Address and Welcome

SESSION I: THE FUTURE OF THE THEORY OF GENERAL RELATIVITY IN THE EVOLUTION OF PHYSICS AND/OR THE FUTURE OF PHYSICS IN THE EVOLUTION OF THE THEORY OF GENERAL RELATIVITY

Moderator: Behram Kursunoglu, University of Miami

Dissertators: P.A.M. Dirac, Florida State University
"DEVELOPMENTS OF EINSTEIN'S THEORY OF GRAVITATION" (45 min.)

Behram Kursunoglu, University of Miami
"A NON-TECHNICAL HISTORY OF THE GENERALIZED THEORY OF GRAVITATION, DEDICATED TO THE ALBERT EINSTEIN CENTENNIAL" (45 min.)

Douglas M. Eardley, Yale University

SESSION II: CURRENT DIRECTIONS IN GENERAL RELATIVITY

Moderators: Stanley Deser, Brandeis University
Dieter Brill, University of Maryland

Dissertators: R. Arnowitt, Northeastern University
"SUPERSYMMETRY FORMULATED IN SUPERSPACE" (45 min.)

Bryce DeWitt, University of Texas-Austin
"QUANTUM GRAVITY"

Leonard Parker, University of Wisconsin
"APPLIED QUANTUM GRAVITY: APPLICATIONS OF THE SEMICLASSICAL THEORY" (45 min.)

Stanley Deser, Brandeis University

SESSION III: SPIN IN HIGH ENERGY PHYSICS

Moderator: Alan D. Krisch, University of Michigan

Dissertators: Donald Franklin Moyer, Northwestern University
"REVOLUTION IN SCIENCE: THE 1919 ECLIPSE TEST OF GENERAL RELATIVITY"(30 min.)

Paul Souder, Yale University

Jabus B. Roberts, Rice University

Alan D. Krisch, University of Michigan

SESSION IV: DIRECT EXPERIMENTAL EVIDENCE FOR CONSTITUENTS IN THE NUCLEON FROM SCATTERING EXPERIMENTS

Moderator: Alan D. Krisch, University of Michigan

Dissertators: Rodney L. Cool, Rockefeller University

Frank J. Sciulli, California Institute of Technology
"EVIDENCE FOR QUARKS FROM NEUTRINO-NUCLEON SCATTERIN" (50 min.)

Karl Berkelman, Cornell University
"DIRECT EXPERIMENTAL EVIDENCE FOR CONSTITUENTS IN THE NUCLEON FROM ELECTROMAGNETIC SCATTERING EXPERIMENTS" (50 min.)

SESSION V: GRAND UNIFIED THEORIES

Moderator: Sheldon Glashow, Harvard University

Dissertators: Abraham Pais, Rockefeller University
"PHYSICS AFTER τ AND T"

Dimitri Nanopoulos, Harvard University
"PROTONS ARE NOT FOREVER"

Itzhak Bars, Yale University
"GAUGE HIERARCHIES IN UNIFIED THEORIES"

Paul H. Frampton, Ohio State University
"ANOMALIES, UNITARITY AND RENORMALIZATION"

ORBIS SCIENTIAE 1979, PROGRAM

SESSION VI: GRAND UNIFIED THEORIES (continued)

Moderator: Sydney Meshkov, National Bureau of Standards

Dissertators: Nicholas P. Samios, Brookhaven National Laboratory
                  "CHARM PARTICLE PRODUCTION BY NEUTRINOS"

                  David B. Cline, University of Wisconsin
                  "TECHNIQUES TO SEARCH FOR PROTON INSTABILITY
                  TO 10^{34} YEARS"

                  S.P. Rosen, Purdue University
                  "CHARGED AND NEUTRAL-CURRENT INTERFERENCE: THE
                  NEXT HURDLE FOR WEINBERG-SALAM"

                  Kenneth Johnson, Massachusetts Institute of
                    Technology
                  "THE QUARK MODEL PION AND THE PCAC PION" (45 min.)

                  Francis Low, Massachusetts Institute of Technology
                  "QUARK MODEL EIGENSTATES AND LOW ENERGY SCATTERING"
                  (45 min.)

                  Richard E. Norton, University of California,
                    Los Angeles
                  "ON THE EQUATIONS OF STATE IN MANY BODY THEORY"
                  (45 min.)

SESSION VII:

Moderator: Heinz R. Pagels, Rockefeller University

Dissertators: Marshall Baker, University of Washington-Seattle
                  "DYSON EQUATIONS, WARD IDENTITIES, AND THE
                  INFRARED BEHAVIOR OF YANG MILLS THEORIES"
                  (45 min.)

                  Robert Carlitz, University of Pittsburgh
                  "INSTANTONS AND CHIRAL SYMMETRY" (45 min.)

                  Laurence Yaffe, Princeton University
                  "QCD AND HADRONIC STRUCTURE" (45 min.)

                  Alfred Mueller, Columbia University
                  "HIGH ENERGY PREDICTIONS FROM PERTURBATIVE Q.D.C."
                  (45 min.)

SESSION VIII:

Moderator: Stephen D. Ellis, University of Washington-Seattle

Dissertators: Stephen D. Ellis, University of Washington-Seattle
"PERTURBATIVE QCD: AN OVERVIEW"

L.F. Abbott, Stanford Linear Accelerator Center
"TOPICS IN THE QCD PHENOMENOLOGY OF DEEP-INELASTIC SCATTERING"

Jeffrey F. Owens, Florida State University
"QUANTUM CHROMODYNAMICS AND LARGE MOMENTUM TRANSFER PROCESSES"

Lowell S. Brown, University of Washington-Seattle
"TESTING QUANTUM CHROMODYNAMICS IN ELECTRON-POSITRON ANNIHILATION AT HIGH ENERGIES"

Edmond L. Berger, Stanford Linear Accelerator Center

PARTICIPANTS

Larry Abbott
Stanford Linear Accelerator Center

Carlos Aragone
Universidad Simon Bolivar

Richard Arnowitt
Northeastern University

Marshall Baker
University of Washington

Michael Barnett
Stanford Linear Accelerator Center

Itzhak Bars
Yale University

Carl Bender
Washington University

Edmond L. Berger
Stanford Linear Accelerator Center

Karl Berkelman
Cornell University

Stephen Blaha
Williams College

Richard Brandt
New York University

Dieter Brill
University of Maryland

Lowell S. Brown
University of Washington

Arthur A. Broyles
University of Florida

Robert Carlitz
University of Pittsburgh

Peter Carruthers
Los Alamos Scientific Laboratory

Lay Nam Chang
Virginia Polytechnic Institute
 & State University

George Chapline
Lawrence Livermore Laboratory

Gordon R. Charlton
Department of Energy

Mou-shan Chen
Center for Theoretical Studies

Norman H. Christ
Columbia University

David B. Cline
University of Wisconsin

Rodney L. Cool
Rockefeller University

Fred Cooper
Los Alamos Scientific Laboratory

Leon N. Cooper
Brown University

Dennis B. Creamer
Fermi Laboratory

Stanley R. Deans
University of South Florida

Stanley Deser
Brandeis University

Bryce Dewitt
University of Texas

P.A.M. Dirac
Florida State University

Douglas M. Eardley
Yale University

Stephen D. Ellis
University of Washington

Hiroshi Enatsu
Ritsumeikan University

Paul H. Frampton
Ohio State University

Donald A. Geffen
University of Minnesota

Sheldon Glashow
Harvard University

Maurice Goldhaber
Brookhaven National Laboratory

O.W. Greenberg
University of Maryland

Terry Haigiwara
Brandeis University

Leopold Halpern
Florida State University

Barry J. Harrington
University of New Hampshire

Richard Haymaker
Louisiana State University

Jarmo Hietarinta
University of Maryland

Bambi Hu
University of Houston

Rudolph C. Hwa
University of Oregon

Kenneth Johnson
Massachusetts Institute of
  Technology

Michio Kaku
City College of New York

Ronald Kantowski
University of Oklahoma

Robert L. Kelley
University of Miami

Frank Krausz
Center for Theoretical Studies

Alan Krisch
University of Michigan

Behram Kursunoglu
Center for Theoretical Studies

Willis E. Lamb, Jr.
University of Arizona

Don Lichtenberg
Indiana University

F.E. Low
Massachusetts Institute of
  Technology

K.T. Mahanthappa
University of Colorado

William J. Marciano
Rockefeller University

Sydney Meshkov
National Bureau of Standards

A.J. Meyer, II
The Chase Manhattan Bank

Stephan Mintz
Florida International University

Don Moyer
Northwestern University

Alfred Mueller
Columbia University

D.V. Nanopoulos
Harvard University

Pran Nath
Northeastern University

Richard E. Norton
University of California at
  Los Angeles

Patrick J. O'Donnell
University of Toronto

Reinhard Oehme
University of Chicago

PARTICIPANTS

John R. O'Fallon
Argonne Universities Association

Martin Olsson
University of Wisconsin

Oliver E. Overseth
University of Michigan

Joseph F. Owens III
Florida State University

Heinz R. Pagels
Rockefeller University

Abraham Pais
Rockefeller University

William F. Palmer
Ohio State University

Leonard Parker
University of Wisconsin

Zohreh Parsa
New Jersey Institute of Technology

Arnold Perlmutter
Center for Theoretical Studies

Stephen S. Pinsky
Ohio State University

Enrico Poggio
Massachusetts Institute of
  Technology

L.G. Ratner
Argonne National Laboratory

A.L. Read
Fermi Laboratory

Jabus Roberts
Rice University

Fritz Rohrlich
Syracuse University

S.P. Rosen
Purdue University

Douglas A. Ross
California Institute of Technology

Nicholas P. Samios
Brookhaven National Laboratory

Robert Savit
University of Michigan

Jonathan F. Schonfeld
University of Minnesota

Frank Sciulli
California Institute of Technology

L.M. Simmons, Jr.
Los Alamos Scientific Laboratory

Paul Souder
Yale University

E.C.G. Sudarshan
University of Texas

Katsumi Tanaka
Ohio State University

Vigdor L. Teplitz
U.S. Army

Kent M. Terwilliger
University of Michigan

Wu-yang Tsai
University of Miami

Hung-Sheng Tsao
Rockefeller University

Walter W. Wada
Ohio State University

Kameshwar C. Wali
Syracuse University

Ling-Lie Wang
Brookhaven National Laboratory

Hywell White
Brookhaven National Laboratory

Laurence G. Yaffe
Princeton University

INDEX

Abbott, L. F.	327-346
Anomalies, Unitarity and Renormalization	133-138
Applequist - carazzone theorem	125
Avignone and Greenwood spectrum	180
Bag model (calculations)	210
Bag model, MIT	192ff
Baker, M.	247-266
Berger, E. L.	455-514
Black holes	105
Brown, Lowell S.	373-393
Carlitz, Robert D.	267-283
Cerenkov radiation/light	159ff
Charged and Neutral-Current Interference: The Next Hurdle for Weinberg-Salam	175-191
Charm Particle Production by Neutrinos	139-155
Chiral Symmetry	267-274
Cline, David B.	157-173
Cross-Section, Total Neutrino-Nucleon	7ff
Deep Inelastic Scattering	40ff
Dyson Equations	247-266
Elastic Scattering	31
Ellis, Stephen D.	313-325
Equations of State in Many Body Theory, On the	221-245
Fermi system/transform	221ff
Fourier transform	32ff, 225ff
Frampton, Paul H.	133-137
Gauge Hierarchies in Unified Theories	115ff
Glashow-Weinberg mixing angle	179
Glashow-Weinberg theorem	86
High Energy Predictions From Perturbation QCD	303-311
High-Q^2 Results	328ff
Instantons and Chiral Symmetry	267-284
Jaffee (prediction, flavor nonet, etc.)	210ff
Jets	347ff, 306ff, 320ff
Johnson, K.	191-205
Krisch, A. D.	441-453

Kronecker delta	226ff
Low, F. E.	207-219
Minkowski spacetime	304
MIT bag model	192ff, 285ff
Mueller, A.	303-311
Nachtmann moments	332ff
Nambu-Goldstone mode	276
Nanopoulos, D. V.	91-114
Narrow resonance approximation (NRA)	241ff
Neutrino-Nucleon Differential Cross-Section	12ff
Neutrino-Nucleon Scattering at Low Energies	5ff
Neutrino Probe, The	3ff
Neutron and Proton Targets, Comparison of	9ff
Norton, R. E.	221-245
Owens, J. F.	347-371
PCAC Octet	191ff
Pais, A.	79-90
Partons	12ff, 349ff
Parton model, quark	66, 313ff, 332ff
Perturbative QCD: An Overview	313ff
Physics After τ and T	79ff
Pion scattering	332
Pion Decay Constant, The	201ff
Poisson probability distribution	300ff
Problem of R, The	25ff
Protons Are Not Forever	91ff
QCD^3	96ff
QCD and Hadronic Structure	285-302
QCD Phenomenology of Deep-Inelastic Scattering, Topics in the	327-346
Quantum Chromodynamics and Large Momentum Transfer Processes	347-372
Quantum Chromodynamics (QCD)	73ff, 92ff, 192, 285-302
Quantum Chromodynamics (QCD) Evidence for q^2-Dependence	22ff
Quantum Chromodynamics, Testing in Electron-Positron Annihilation at High Energies	373-394
Quantum gravity field theory	270ff
Quark Model, Consequences of the Simple	20ff
Quark Model Eigenstates and Low Energy Scattering	207-220
Quark parton model	66
Quark Model Pion and the PCAC Pion, The	191-206
Quarks from Neutrino-Nucleon Scattering, Evidence for	1-30
Quarks - Point-like Structure	1ff
Rosen, S. P.	175-190

INDEX

Samios, N. P.	139-155
Scattering, Deep In-Elastic	40ff
Scattering, Elastic (electron), etc.	31ff
Scattering Experiments, ...	31ff
Scattering, Neutrino-Nucleon at Low Energies	5ff
Sciulli, F.	1-30
Slavnov-Taylor identities	247
Sommerfield-Watson factors	227
Souder, P. A., and V. W. Hughes	395-440
SU(5) Model, Phenomenology of the	146ff
T-amplitudes/graphs	234ff
Techniques to Search for Proton Instability to 10^{34} Years	157-174
T-Graphs, Evaluation of	235ff
Unitarity	133ff
Ward Identities	247-266
Weinberg angle	110
Weinberg-Salam model	178ff
Wigner mode	276
Yaffee, Laurence	285-301
Yang-Mills Theories, Infrared Behavior of	247-266